AF839965

# Construction Conflict Management and Resolution

# Construction Conflict Management and Resolution

*Edited by*

PETER FENN and ROD GAMESON
*University of Manchester*
*Institute of Science and Technology*
*(UMIST)*

---

Proceedings of the First International
Construction Management Conference,
The University of Manchester
Institute of Science and Technology (UMIST),
25–27 September 1992

---

Routledge
Taylor & Francis Group
LONDON AND NEW YORK

First published 1992 by E & F N Spon

This edition published 2013 by Routledge
2 Park Square, Milton Park, Abingdon, Oxon OX14 4RN
52 Vanderbilt Avenue, New York, NY 10017, USA

First issued in paperback 2020

*Routledge is an imprint of the Taylor & Francis Group, an informa business*

First edition 1992

© 1992 Peter Fenn and Rod Gameson

All rights reserved. No part of this book may be reprinted or reproduced or utilised in any form or by any electronic, mechanical, or other means, now known or hereafter invented, including photocopying and recording, or in any information storage or retrieval system, without permission in writing from the publishers.

Notice:
Product or corporate names may be trademarks or registered trademarks, and are used only for identification and explanation without intent to infringe.

A catalogue record for this book is available from the British Library

Library of Congress Cataloging-in-Publication data available

ISBN 13: 978-0-367-57994-4 (pbk)
ISBN 13: 978-0-419-18140-8 (hbk)

## Acknowledgements

The organizers gratefully acknowledge the special support of the following organizations and firms in the sponsoring of the Conference.

International Council for Building Research, Studies and Documentation (CIB)

The Centre for Dispute Resolution (CEDR)

Dibb Lupton Broomhead, Solicitors

High Point plc

Masons, Solicitors and Privy Council Agents

J.E. Price Associates,
Construction Contracts Consultants

FASS (The Federation of Associations of Specialists and Sub-Contractors)

# Contents

# Contributors

## KEYNOTE SPEAKERS

**His Honour Judge J. Newey QC**
Senior Official Referee, London, UK

**R. Baden Hellard**
Polycon Group of Consultants, London, UK

**C. Cooper**
American Arbitration Association/The Asia/Pacific Center for the Resolution of Internations Business Disputes, San Francisco, USA

**A. Houghton**
Des Voeux Chambers, Hong Kong

**Dr A. Lavers**
Fishburn Boxer Reader in Law, Department of Estate Management, Oxford Polytechnic, UK

**Dr K. Mackie**
Chief Executive of the Centre for Dispute Resolution, London, UK

**D. Miles**
Partner, Glovers Solicitors, London, UK

**Professor A. Rahim**
Department of Management, Western Kentucky University, USA

**M. Smith**
High-Point, Birmingham, UK

## AUTHORS

**E. Antohie**
Department of Management, Polytechnic Institute of Iasi, Romania

**Dr Sadi Assaf**
College of Environmental Design, King Fahd University of Petroleum & Minerals, Dhahran, Saudi Arabia.

**Dr Abdulmohsen Al-Hammad**
College of Environmental Design, King Fahd University of Petroleum & Minerals, Dhahran, Saudi Arabia.

**D. Baker**
Faculty of Environmental Studies, University of Waterloo, Ontario, Canada

**B. Bentley**
Dibb Lupton Broomhead, Sheffield, UK

**D. Bishop**
David Bishop Associates, Doncaster, UK

**G. Bowles**
Department of Civil Engineering, Surveying & Building, Dundee Institute of Technology, Dundee, UK

**D. Carter**
School of Architecture & Building Engineering, University of Liverpool, Liverpool, UK

**D. Cheetham**
School of Architecture & Building Engineering, University of Liverpool, Liverpool, UK

**Professor S. Clegg**
Department of Management, University of St Andrews, St. Andrews, UK

**B. Colledge**
School of the Environment, Leeds Polytechnic, Leeds, UK

**C. Cosma**
Department of Management, Organisation and Building Economics, Polytechnic Institute of Iasi, Romania

**Dr C. Cree**
Coopers & Lybrand Deloitte, London, UK

**D. Dalton**
Department of Civil Engineering, University of Leeds, Leeds, UK

**R. Davies**
Federation of Associations of Specialists & Sub-contractors (FASS), Stevenage, UK

**I. Eilenberg**
Department of Building & Construction Economics, Royal Melbourne Institute of Technology, Australia.

**Dr R. Fellows**
School of Architecture and Building Engineering, University of Bath, Bath, UK

**J. Franks**
School of Building Services and Construction Management, South Bank University, London, UK

**A. Gale**
Department of Building Engineering, University of Manchester Institute of Science and Technology, Manchester, UK

**P. Gardiner**
School of Engineering and Computer Science, University of Durham, UK

**S. Green**
Department of Construction Management & Engineering, University of Reading, Reading, UK

**H. Gow**
Department of Civil Engineering, Surveying and Building, Dundee Institute of Technology, Dundee, UK

**M. Hancock**
School of Architecture & Building Engineering, University of Bath, UK

**Professor D. Henderson**
College of Law, University of Kentucky, .USA

**D. Hollands**
Arbitrator & Mediator, Auckland, New Zealand

**Professor M. Horner**
Department of Civil Engineering, University of Dundee, Dundee, UK

**D. Jones**
Department of Civil & Structural Engineering, University of Manchester Institute of Science and Technology, Manchester, UK

**P. Kennedy**
Department of Building & Surveying, Glasgow Polytechnic, Glasgow, UK

**Professor D. Langford**
Barr Chair of Construction, University of Strathclyde, UK

**Sir C. Leeds Bt**
Universite de Nancy II, France

**J. Lewis**
School of Architecture & Building Engineering, University of Liverpool, Liverpool UK.

**R. Lupasteanu**
Department of Management, Polytechnic Institute of Iasi, Romania

**P. McGowan**
Department of Civil Engineering, University of Dundee, Dundee, UK

**A. McLellan**
Faculty of Environmental Studies, University of Waterloo, Ontario, Canada

**P. Nicholson**
School of Architecture, University of Nottingham, UK

**M. Powell**
Department of The Built Environment, Anglia Polytechnic, Essex and Cambridge, UK

**R. Quick**
Morris Fletcher & Cross, Brisbane, Australia

**S. Revay**
Revay and Associates Limited, Montreal, Canada

**Professor J. Scrivener**
Department of Architecture and Building, University of Melbourne, Australia

**Dr G. Siedel**
School of Business Administration, The University of Michigan, Ann Arbor, Michigan, USA

**Dr N. Shehadeh**
Department of Civil Engineering, Leeds University, Leeds, UK

**J. Simmons**
Department of Mechanical Engineering, Heriot-Watt University, Edinburgh, UK

**J. Sommerville**
Glasgow College of Building and Printing, Glasgow, UK

**Professor T. Stipanowich**
College of Law, University of Kentucky, USA

**Professor P. Thompson**
Department of Civil & Structural Engineering, University of Manchester Institute of Science and Technology, Manchester, UK

**P. Turner-Wright**
Leadbitter, Oxford, UK

**C. Wall**
Commercial, Mediation & Arbitration Services Limited, Hong Kong

**V. Watts**
Department of Architecture and Building, University of Melbourne, Australia

**Dr R. Zakieh**
Department of Civil Engineering, University of Dundee, Dundee, UK

**R. Zikmann**
Zikmann & Associates, Sydney, Australia

# Referees

**Michael O'Shea**
Masons Solicitors, Manchester

**Dr Will Hughes**
University of Reading, Reading, UK

**Richard Collins**
New South Wales, Australia

**Colin Wall**
CMA, Hong Kong

**Professor Aldo Norsa**
Instituto Universitario di Architettura di Venezia, Italy

**Claude Mathurin**
Engineer, Place de la Madeleine, Paris

**Vera Van Houtte**
Advocate, Netherlands

**Sebastian Toombs**
Royal Incorporation of Architects in Scotland, Edinburgh, UK

**Professor D Bishop CBE**
St Albans, UK

**Professor Gerard Blachere**
France

**Professor Colin Davidson**
University of Montreal, Montreal, Canada

**Hans Sundstrom**
AB Bostadsgaranti, Sweden

**Antonio C Canda**
Consejo General de Colegios, Madrid, Spain

**Fleming Lethan**
Ministry of Housing & Building, Denmark

**Professor Kunio Kawagoe**
Japan Building Equipment, Tokyo, Japan

# Preface

The aim of this book is to examine and investigate techniques involved with the management and resolution of conflict arising in construction projects, both in the United Kingdom and around the world. Papers have been received from ten countries. The book has been produced with a number of objectives in mind. Firstly, it is hoped that readers will be provided with a greater understanding of the field having been exposed to the views of experienced practitioners. Secondly, it presents an opportunity for academics to disseminate their research findings. Finally it acts a source of reference to be consulted in connection with professional practice, research and teaching.

Papers have been grouped into topic areas which reflect the key areas of construction conflict. Part One begins with a rapporteur section, by Anthony Lavers, which reviews the papers and provides a summary of the key points emerging from them. This is followed by a keynote paper by Judge John Newey QC. Part Two considers numerous aspects of how conflict can be managed through the many phases of a project.

Part Three deals with established adjudicative procedures for dealing with conflict, and Part Four looks at emerging methods for resolving disputes, collectively termed Alternative Dispute Resolution (ADR).

Finally, Part Five concentrates on educational issues and considers methods to be utilised to prepare professionals to deal with conflict more effectively.

Like any construction project, the production of this book has been a team effort. Therefore we would like to thank the following people for their contributions. Firstly, the authors who contributed papers. Secondly, Anthony Lavers for producing the rapporteur section. Thirdly, our panel of referees for scrutinising the papers, and finally, Moira Kynnersley and Lisa Kerfoot for their invaluable assistance in the editing, collating and production of the final version of this book.

The book will be launched at an International Conference on Construction Conflict held at UMIST, Manchester, in September 1992, where a number of the papers will be presented. This will provide a forum where professionals and academics can debate this important area.

Peter Fenn and Rod Gameson

# Part One

## Introduction

This section introduces the theme of construction conflict setting the scene for the remainder of the book.

'Construction conflict: management and resolution - analysis and solutions' (Lavers) is a rapporteurs report on the overall content of the book, which attempts to draw together themes and conclusions from the numerous approaches proffered by the authors.

'The Construction Industry' (Newey), the conference opening address, discusses the nature of construction conflict and disputes.

# CONSTRUCTION CONFLICT: MANAGEMENT AND RESOLUTION ANALYSIS AND SOLUTIONS

ANTHONY LAVERS
School of Estate Management, Oxford Polytechnic, U.K.

Abstract
This paper attempts to draw together the principal themes of the UMIST First International Conference on Construction Conflict: Management and Resolution and to identify the most important perspectives of the papers presented. The subjects covered include the phenomena of conflict and their management, experiences of traditional resolution mechanisms ie litigation and arbitration. Alternative Dispute Resolution (ADR) and education/attitude changes.
Keywords:
Conflict, Construction Disputes, Conflict Management, Conflict Resolution, Construction Litigation, Arbitration, ADR, Alternative Dispute Resolution, Construction Education.

## 1 Introduction

The analysis and solutions of the title of this paper are not mine. They are those advanced by the authors of the papers at the UMIST First International Construction Management Conference on Construction Conflict: Management and Resolution. My role in this paper and at the Conference is to try to draw together themes and conclusions from the individual approaches of the respective authors. The themes of this paper are loosely based upon those used for classification of the papers for presentation (where appropriate) and for publication, namely Construction Conflict, Claims, Litigation and Arbitration, Alternative Dispute Resolution (ADR) and Education and Attitude Change.

## 2 Background

The background against which this conference is held is of a process and an industry in which conflict has risen significantly over the last three decades. Commentators who have attempted to quantify the extent of conflict are agreed that this has increased, although there are sub-trends which are worthy of note. Peter Fenn[1], in a paper referred to by Fellows[2] saw a 500% increase in the initiation of litigation in the twenty years to 1986, although cases actually coming to the courts remained approximately constant, which may of course be merely descriptive of their capacity and of the fact that they are working at or near it. Judge Newey's figures[3] show a 100% increase in litigation in the period 1973-80 with increases of approximately 15% per annum in the period 1980-89. Judge Newey, in this paper ascribes the increase in litigation and arbitration at least in part to "changes in Common and Statute Law which have made it easier to bring claims". Certainly this would be consistent with his figures for 1990 and 1991 which show no increase and a decrease in litigation respectively, which could be seen partly as the result of the brake being applied to negligence claims in tort as a result of D and F Estates v Church of England Commissioners[4], Department of Environment v Thomas Bates[5] and Murphy v Brentwood District Council[6]. This sub-trend, although welcomed in some quarters, should not be over-estimated: absence of tortious remedies has created the collateral warranty explosion and viewed from another angle can simply leave injured parties uncompensated. In any event, a reduction in litigation observable from the Official Referees Court does not mean a commensurate reduction in conflict and dispute, nor was Judge Newey suggesting that it does. His reference to companies in liquidation making litigation unprofitable to pursue is surely right and does not indicate a lessening of conflict.

So, the background is one of conflict and of dispute, some of which ends in formal claims, in litigation or arbitration, which is the subject of Section 4 of this paper. Whether this is inevitable or avoidable, positive or negative is discussed in several papers referred to in Section 3 below.

There is dissatisfaction with the existing traditional mechanisms for resolving disputes. That fact underlies many of the papers in Sections 4 and 5 and is strongly articulated by Davies[7] taking the view of specialist contractors and sub-contractors in bemoaning "the arrival in strength of the legal profession" in construction disputes in the early 1980s. It should be noted here that there is a dispute in taxonomy as to what constitute 'traditional' and what 'alternative' methods. Whereas the majority of authors assimilate arbitration to litigation to distinguish from ADR, Eilenberg[8] says that ADR "is regarded as including arbitration". If excessive legalism is perceived as central to the problem[9], the degree of involvement may be used as an indicator to assist in the classification. Thus the informal tribunals mentioned by Eilenberg in the State of Victoria as excluding legal representation could properly be regarded as part of ADR, whereas similar mechanisms with professional advocates and expert witnesses could be regarded as quasi-litigation; part of the traditional method of resolving disputes to which an alternative is sought. Normally, then arbitration should be assimilated with litigation (and it is, throughout Sections 3,4 and 5 of this paper), unless features are built into it, such as the exclusion of legal representation, which prevent it acquiring the characteristics mentioned above.

Because of the dissatisfaction with traditional methods of conflict resolution in the construction industry, a wide range of options has been and continues to be explored. The attention of the authors of the papers referred to in Section 6 of this paper is concentrated on the earliest chronological stage of the continuum of a construction project, namely the education and training of the personnel who will be involved, especially the professional consultants. Hancock[10] states that "problems and conflicts within the construction industry are a result of misunderstanding and a lack of perception founded in our education of construction industry professionals". The basic belief of these authors is that the inculcation of different attitudes can help avoid conflict. The second stage, chronologically, also depends upon avoidance of conflict, more mechanistically through selection and tuning of procurement vehicles; the contractual and other relationships between the parties in a construction project. Colledge[11] in particular sees the commercial and contractual relationships between the parties as fundamental to a reduction of conflict and several papers propose techniques for avoidance of disputes through better or more systematic preparation and communication. A simple example of the latter is supplied by Judge Newey: "if an untried technology is to be used, the Employer should be warned and his consent obtained"[12]. This suggestion is, of course, redolent of Judge Newey's decision in Victoria University of Manchester v Hugh Wilson[13]: based on the simple truth that there is less room for subsequent disagreement if the designer 'takes the client with him/her'.

The next stage assumes that disputes do arise notwithstanding efforts to avoid them, but seeks to reduce any harmful effects. Dispute management is advocated as a means of recognising conflict and dealing with it efficiently. Rahim[14] states that "Organisational conflict must not necessarily be reduced, suppressed or eliminated, but managed to enhance individual, group and organisational effectiveness". This view can be characterised as 'pragmatic' ie accepting the inevitability of disputes and concentrating on their management rather than their complete eradication (although not excluding minimisation).

While it would be artificial and in some situations simply wrong to distinguish between management and resolution of disputes, a difference of emphasis can be observed in some papers between handling a dispute as it arises and its eventual outcome. Baden Hellard[15] advocates the appointment of a contract management adjudicator and the idea of an interim reference point to foresee, identify and manage points of disagreement is explored in more detail by other authors, perhaps most interestingly by Wall[16] (under the heading of ADR) who records the implementation of such a system in Hong Kong. There may here be a difference of taxonomy between those who see this referee as managing disputes as they arise and those who see the purpose as resolution of disputes which have arisen, albeit in an early form. It is not likely that both would be used in the same project and they can properly be regarded as different versions of similar concepts.

Ultimately, on any meaningful analysis of the industry at present, disputes will arise which cannot be nipped in the bud. Valuable work has been done on improving conflict resolution by the authors of these papers. There is research on existing systems of litigation and arbitration such as Quick's paper[17] on arbitration costs and the Watts and Scrivener[18] paper reviewing construction litigation in the Supreme Courts of New South Wales and Victoria and the Court of Appeal of Australia. There are, of course, advocates of ADR systems, led by Cooper[19] and Mackie[20], and there are papers from five countries in this section which offer instructive comparisons in approach. There are proponents too of conflict resolution methods which appear novel but which are in reality older even than litigation and arbitration. Houghton[21] in his view of the Far East speaks of "the Chinese perspective of compromise", and other authors, including Nicholson[22] have seen the advantages in looking at less antagonistic practices from older, often oriental, civilisations. The reasons for this preference for the avoidance of open conflict may be cultural. Koh Kim Chuan[23] observed this well-known but little understood phenomenon in 1981 when he described "our Chinese mentality" which "abhors any attendance in the Court of Law" and explained this in terms of 'face': "'maintaining one's face' or 'giving one's opponent face' have much to do with the tendency not to bring disputes into the open".

The background to this Conference then can be summarised as consisting of three propositions. First, there is a perceived growth, subject to some sub-trends, in conflict in the construction industry. It is salutary and perhaps even sad to note Houghton's[24] statement "There is no doubt ... that throughout South East Asia, with states such as Hong Kong and Singapore in the forefront, there is a tendency to follow the current western thinking and to have disputes resolved by third party intervention" although there are also signs of modern alternatives. So conflict is a feature of the international industry.

Second, there is dissatisfaction with existing legal and legalistic methods of conflict resolution, chiefly litigation and, often, arbitration. While their perceived deficiencies vary between jurisdictions, they are generally seen as frequently very costly, time-consuming, inconvenient and tending to intensify and exacerbate existing conflict, to the detriment of working relationships.

Third, research is being conducted by scholars from many nations and a wide range of disciplines and professional backgrounds to find better ways of dealing with the phenomena of conflict: education, contractual and systemic avoidance, management and improved as well as alternative forms of dispute resolution.

## 3 The phenomena of construction conflict and conflict management

This sub-heading is used advisedly, because it cannot be said that there is a single phenomenon of conflict. A dispute between a contractor and a client, or the client's architect, over a loss and expense claim is qualitatively different from a tenant seeking to sue an engineer in tort following a major structural failure. Aspects of conflict will vary with different types and sizes of project, different procurement systems, different legal regimes and different personnel.

Nevertheless, this by no means renders the study of conflict useless, although it makes generalisation difficult. On the contrary, the authors of the UMIST papers are in broad agreement that understanding of conflict is fundamental to an ability to identify disputes at a stage when it is still possible to avoid or manage their development. Zikmann[25] emphasises the importance of understanding conflict and adopts an analysis which embraces interest conflicts, structural conflicts, value conflicts, relationship conflicts and data conflicts. On the question of whether some form of conflict in construction is inevitable, the preponderance of opinion is that it is. Smith[26] is clear on this point: "Construction conflicts are ... endemic in the industry. The reasons for them flow from the way the industry functions". Baden Hellard[27] agrees that "conflict is a particular feature of construction" and cites the existence of some 94 different standard contract forms in the U.K. industry as a major contribution to this inevitability. Clegg[28] develops this further. He answers Fenn's question[29]: "Why do a substantial percentage of construction contracts end in serious dispute?" with the conclusion "Because it is rational for them to do so". Clegg refers to the tendency of contracts to generate dispute because of the externality of interpretation; contracts cannot "specify their own indexicality" by providing how they will be read of used. Langford, Kennedy and Sommerville[30] agree that "conflict between contracting companies may be inevitable".

This majority view, that the nature of the construction process makes conflict inevitable in some form, to some extent, can be characterised as 'pragmatic', as contrasted with the 'long-term strategic'. The former says 'conflict exists and will continue to do so. We will avoid and reduce it where possible, but the central question is, how do we deal with it?". The 'long term strategists' including several of the authors in the Education section, do not find the inevitability of conflict a positive or fruitful subject and concentrate on tackling root and branch the attitudes and practices in the industry and its professions which generate disputes.

Given that the existence of a degree of conflict is necessary, is this essentially negative? Certainly, the authors generally concentrate on the damaging effects of conflict. All of the papers on avoidance of disputes are predicated upon negative consequences following from them. Turner-Wright[31] sees "diminishing project performance levels induced by non-interaction, frustration and non-aligned perceptions of each other's and the project's goals" and Colledge[32] comments on the poor recent record of the industry "with respect to its performance and achievement of time and cost objectives". The NEDO Reports[33] have also seen conflict as a damaging factor in the construction process and these are cited by Colledge as well as by Smith and others. But Smith sees the tension of the contractual relationship as not only inevitable but to some extent functional. He distinguishes functional from dysfunctional conflict, which is consistent with Rahim's view[34] that it is management not suppression or even reduction of conflict which is crucial. His espousal of Partnering is not inconsistent with this position; creative tension between partners is an acknowledged phenomenon. Gardiner and Simmons[35] also classify functional and dysfunctional conflict and speak of the possibility of a project manager being able to "harness the functional outcome of conflict, resulting in project change for the better; and limiting the damage done by dysfunctional conflict".

The measures proposed for dealing with dysfunctional conflict can be conveniently although not restrictively, divided into avoidance/ minimisation and management. To regard the two as mutually exclusive would be to misunderstand those papers such as Baden Hellard's[36] and Revay's[37] which see dispute avoidance as part of an integrated conflict management strategy. As Revay puts it "conflict management does not start when the dispute first raises its ugly head". Nevertheless, it may be useful to differentiate between work which is centred upon avoidance/minimisation and that which relates to handling disputes if the avoidance techniques fail or break down. Certainly choice of an appropriate procurement method is part of the avoidance/minimisation range. Colledge[38], as has been mentioned, proposes an economic model which enables the adoption of a transaction specific approach to forming contractual relationships. Certain types of procurement method can be said to avoid certain types of conflict. Nicholson[39] recommends Build Operate Transfer in appropriate cases, where the contractor operates the building or facility constructed for an agreed period in order to generate the revenue to pay the contract sum, before transferring it to the client for a nominal amount. Cosma[40] sees the marketing of kit-form houses for self-build as a way of avoiding time and payment disputes which an individual owner and small-scale contractor might find difficult to resolve cheaply and efficiently. This is not potentially applicable only to Rumania, where Cosma's work was done; she notes that in the United States 20% of single family residences are built by home-owners, a sector where total or partial transfer of work from contractor to client may well be beneficial in conflict avoidance. But it is not only the type of procurement method selected which may be relevant to conflict avoidance. The substance and, indeed, the spirit of the contract may also be of great importance. Fellows[41] concludes that "hard, bad, unfair bargains" work against the interests of the construction industry and those who work in it. He ascribes this to the law of Karma, but the legal system may produce exactly the same effect. Seeking to exact too heavy an imposition from the other side may not operate as intended. In Rosehaugh Stanhope v Redpath Dorman Long[42] and Beaufort House Development v Zimmcor International[43] the UK Court of Appeal declined to give effect to a purported provision in a construction management arrangement which would have given the clients' construction managers absolute discretion to determine what loss had been occasioned by delay and to claim immediately for it. The Court of Appeal felt that such a provision was so potentially onerous upon the contractor that if should not be enforced, by reason of the contra proferentem rule[44]. This does not mean that construction management needs be rejected outright, although Revay[45] would have it so, since he sees it as a source for conflict. It is meant to emphasise that a more balanced agreement is a preferable option as a means of avoiding conflict. Had the clients in these cases not attempted to impose such onerous provisions, the contract would have arguably been enforceable in full before the courts.

The majority of Conflict papers, however, concentrate on management of conflict rather than avoidance itself, either expressly by advocating the adoption of management systems or techniques or implicitly by recommending approaches which could be utilised in a management strategy. Basic requisites of management of conflict according to Rahim[46] are diagnosis and intervention. Diagnosis may well be facilitated by the use of classifications of conflict such as those used by Zikmann[47] and Gardiner and Simmons[48]. Also of value in diagnosis of conflict may be data on the most likely sources. Revay[49] gives a list of most frequent causes for claims which are mainly client deficiencies or consultant deficiencies. This is Canadian research. Watts and Scrivener[50] have several categories of sources or dispute in their Australian data which suggest contractor deficiencies (or alleged deficiencies), as well, indeed their data shows 15% of all disputes arising from alleged contractor/sub-contractor deficiencies resulting in attempted determination. Rahim's other requisite for intervention, is supported by Zikmann,[51] who distinguishes between aggressive and creative responses as types of active response to conflict and recommends that "The emphasis is on identifying creative and workable solutions which can satisfy the needs and dispel the fears of the parties involved". Both Cree[52] and Lewis, Cheetham and Carter[53] see conflict management as susceptible to a project management approach, indeed the former paper assimilates disputes with projects. The reasoning proceeds thus: disputes can be regarded as projects, projects need management, disputes need project management. Cree proposes a decision tree to optimise choices in the project management of a dispute. The Lewis, Cheetham and Carter paper refers to Risk Management as one of the capabilities required of the Client's Project Manager and consistently with Rahim's recommendation, gives three phases of Risk Management: risk identification, risk analysis and risk response. Fellows adds a fourth, viz risk allocation. Lewis, Cheetham and Carter's paper adduces a project as an illustration of this approach, namely an Anglican vicarage in a New Town in the North West of England. The hall-mark of their approach is a qualitative not quantitative analysis.

Attitude is widely regarded as crucial in intervention. On the positive side there is an insistence by some authors upon an integrated and integrative approach. Langford, Kennedy and Sommerville[54] set out to explain "Informed project management" with a capacity for anticipating zones of conflict with team members "bound together by mutually set, internalised goals, rather than by contractual arrangements alone". Fellows[55] continuing his argument against the competitive ethos of the construction process seeks a recognition "that all involved are in business with operational imperatives which have some degree(s) of commonality". Turner-Wright[56] too speaks of "the need for a higher degree of integration within a construction site management team".

On the negative side concern is expressed by some authors at the attitudes of some personnel involved in the management process. Lapusteanu and Anthohle's paper[57] contains some apparently pessimistic observations on the continuity of problematic management attitudes between communist and post-communist Rumania where the personnel are often the same as before. Leeds[58] expresses reservation about French style in managing conflicts, which he characterises as inflexibility in negotiation, the adoption of unyielding positions and abrupt termination of discussions. These characteristics arise from French dislike of compromise, which suggests to a French negotiator a 'lose-lose' result, with neither side satisfied. Compromise can come to mean a "dishonest opportunistic or shady deal" (une compromission) or a flawed result (un compromis boiteaux). It must here be said that Leeds regards the record of the French industry on conflict avoidance rather than management as much stronger, and he too sees a more optimistic development in the last 25 years of a preference for 'concertation', being integration to minimise the effects of conflict. Perhaps the gravest reservations about attitudes of personnel concerned are expressed by Davies[59], who sees the central role of the legal professions as inconsistent with efficient conflict management and whose contribution might have found favour with Jack Cade's supporter, Dick, the butcher of Ashford[60].

The positive proposals are carried forward into specific techniques. Baker and McLellan[61], dealing with an especially emotive and polycentric form of dispute concerning mineral extraction in Canada, see dispute management as the creation of windows of opportunity for the mutual exchange of incentives and concentration upon the satisfaction of the aspirations and objectives of the other party(ies) as a means of obtaining a settlement. Of the most specific techniques or tools put forward for dispute management, mention has already been made of the decision tree of Cree[62], the method of analysis of Lewis, Cheetham and Carter[63] and the model indicating variables in main-contractor sub-contractor relationships of Langford, Kennedy and Sommerville[64]. To these should be added Green's[65] proposal of a formal decision-making model for use during the briefing/outline design stages. Green suggests how the model might work through a case study of a new laboratory; the primary objective being the establishment of a shared understanding of design objectives, rather than what the paper describes as unrealistic objectives of optimisation or maximisation of value.

## 4 Claims, litigation and arbitration

The overriding question posed by the inclusion of this section in the UMIST Conference is that posed by Smith[66] in his expectation that authors "will further question whether the existing dispute resolution systems can live up to the expectations raised" (by perceptions of enforcement of rights and obligations). In short, can the existing mechanisms, chiefly litigation and arbitration be made to work, or at least to work better, in resolving claims and other disputes? Three of the papers in this section offer specific and detailed proposals for improving the operation of existing conflict-resolution arrangements. Of these, Colledge's[67] economic model of commercial relationships can be regarded as an attempt to make construction contracts work better as anticipations of and provision against conflict, by adopting a transaction-specific approach. McGowan el al[68] have produced a paper which identifies the need to evaluate systematically the effects in terms of time and cost of variations and other changes, desired or enforced. Their solution is an application of the concept of 'resource significance', based on the premise that the capacity to separate material costs from resource/fixed costs is fundamental to objective evaluation and thus the possibility of settlement. This paper advocates a particular contractual regime, namely the New Engineering Contract, (presumably, in appropriate cases) as creating the right environment to permit such a process.

Bentley[69] advocates the wider use of a known technique, namely adjudication, not as an alternative to current procurement and dispute resolution methods but as a valuable addition to existing provision. He notes with approval the presence of adjudication clauses in a growing number of major standard form contracts and sees certain features of the technique as highly beneficial. Most notably, the ability to obtain interim decisions, during the continuation, of the project, within a short time scale, which are binding until subsequent litigation or arbitration, may reduce the, potential damage caused by conflicts. Bentley concludes that, while ultimate success will depend upon the attitudes of the protagonists (and the Courts), "there seems little doubt that adjudication has merit as a dispute resolution procedure, offering benefits not otherwise available in the traditional procedures". An affinity may be remarked between some forms of adjudication and the Dispute Resolution Adviser discussed by Wall[70], which is referred to in Section 5 below.

The other papers in this Section can be regarded as descriptions and analyses of existing arrangements for dispute resolution and their characteristics, especially characteristic deficiencies. Assaf and Al-Hammad[71] give an account of the contractual provisions for dispute resolution in Saudi Arabia's 1988 Standard Public Works Contract. Points of interest here are the provisions for calculation of liquidated damages, which in some respects seem to be regarded as closer, and acceptably so, to a penalty, with an upper limit on the total payable of 10% of the value of the contract. There is provision for reference of disputes as to interpretation of contract which cannot be resolved mutually, to the Board of Grievances (Diwan Al-Mathelem) for final judgement. Watts and Scrivener[72], a doctoral research student/supervisor team have already produced important data on sources of dispute, including sub-groups referring to causes, and on 'triggers' which bring the dispute to litigation. The second stage of the ongoing research is to focus on documentation weaknesses and failure in administration techniques. Documentation weaknesses are identified by Revay[73] as area requiring improvement to achieve better dispute avoidance, and this second stage may also yield significant results. Quick[74] has produced a detailed analysis including extensive case law, of the application of the so-called 'English Rule' (ie the costs follow the judgment) and the 'American Rule' (ie parties bear their own costs, win or lose) in UK and Australian arbitration proceedings. Quick sees attempts to displace the remarkably durable 'English Rule' by repeated experiments in the UK and Australia with the 'American Rule' as a form of 'Costs ADR', although he notes recent modifications in the US operation of the 'American Rule' which ironically bring it closer to the 'English Rule'.

## 5 Alternative Dispute Resolution (ADR)

The authors whose papers are included in this section explore and in varying degrees advocate a range of dispute resolution models and techniques. They are all properly classified as ADR. The doubt of the Master of the Rolls "whether there is any such thing as ADR" quoted by Miles[75] is respectfully rejected. The models and techniques discussed are alternatives to litigation and, despite Eilenberg's reservation, mentioned in Section 2 above, to arbitration. Mackie[76] offers 'Appropriate' instead of 'Alternative' Dispute Resolution, which has some attractions, but may minimise the force of the distinction with litigation/arbitration.

The best known forms of ADR may be taken to be mediation and conciliation. Both are well documented, especially as a result of comparatively extensive use in the United States and consequently none of the papers offers straightforward description of these models. Stipanowich and Henderson,[77] do however seek to rebut the principal anecdotal criticisms of mediation (and of mini-trials) and in doing so re-assert some of the strengths of these ADR models. The principal criticisms which they tackle are that there is a damaging admission of weakness in seeking to explore alternative models and that mediation (and mini-trial) reveal trial strategies and information. The research for the Forum on the Construction Industry and Litigation Section of the American Bar Association carried out by the University of Kentucky College of Law rejected both criticisms decisively. Mediation was seen as appropriate where the parties wish to maintain an ongoing relationship, where privacy and confidentiality were important, where a quick resolution was needed and where an economical process was needed by both parties. Mini-trial was seen as offering similar advantages but was relatively less favoured. These models were regarded as inappropriate where the dispute involved a novel question of law, where the credibility of witnesses was at stake or the good faith of the other side was seriously in doubt.

Nicholson[78], while approving the harmony engendered by the "Japanese cultural heritage of non-argument" adopts de Bono's view that to "replace the dialectic argument system of conflict resolution with a new idiom" the "intervention of a third party is essential". Mackie[79] and Cooper[80] proffer similar views as to the essential personal qualities of the mediator. Cooper regards the mediator as the 'agent of reality' who forces the disputants to review the situation since "No dispute can settle until one or both parties begin to question their belief in their own position" Mackie, too, wants the mediator to take a "high-profile, active part in negotiations".

ADR is seen as especially valuable for the smaller scale disputes where cost of traditional options for resolution may be prohibitive. Quick[81] found no evidence for the assertion that ADR could operate for around 3% of the cost of arbitration, but Miles' paper[82] expresses strongly the concern felt as to how disputes for sums below £50,000 or even £100,000 can be economically conducted by traditional means. The managing director of a leading UK developer is known to have said (perhaps unwisely) that his firm would be unlikely to pursue litigation all the way to the High Court for under £250,000. It is at these modest levels that the work of Eilenberg[83] on low-cost, small-value Residual Dispute Settlement in the State of Victoria will be of interest. The exclusion of legal representation and a costs structure designed to discourage the use of expert witnesses keeps the cost to the parties to less than £100 per day each.

Siedel[84] comes the closest of any paper to an exposition of the workings and merits of a whole ADR system with the review of mini-trial. There is a possible contradiction between Siedel's assertion that mini-trial is "considered by many experts to be the most successful of the new methods of alternative dispute resolution" and the findings of Stipanowich and Henderson[85] that mediation was generally preferred to mini-trial. Perhaps mediation in the US is not to be regarded as a new method of ADR whereas mini-trial is. Probably more significant in Siedel's paper is the account of the use of the ADR Pledge. 500 of the top US and US based multi-national corporation have now signed pledges which bind them, in good faith rather than law, to explore negotiation or ADR before pursuing litigation, with any party which has made a similar statement. Siedel includes precedents of the formula of the wording for these ADR pledges, which may be a powerful influence upon the behaviour of corporations with 'clean' images to maintain.

In a similar way, Hollands[86] reports on the inclusion of 'Amicable Settlement Clauses' in the 1987 editions of the FIDIC Civil Engineering and Electrical and Mechanical Engineering Contracts which oblige the parties to come to the negotiating table to attempt the tasks of "identifying problems, establishing facts, clarifying issues, developing settlement options and reaching agreement". This may sound optimistic, but Hollands readily concedes that it may not achieve these aims where relationships are seriously soured or where one party has no intention of settling. The value of an obligation to seek amicable settlement is that it is a 'window of opportunity' for the parties to limit losses, contain damage and preserve working relationships, while retaining control over the process.

The most difficult paper to classify is Wall's[87], because the Dispute Resolution Adviser (DRA) has some of the tasks of conflict management and even avoidance. Nevertheless, his account of the DRA system is a unique contribution to the ADR discussion at the UMIST conference because it is a technique which has just been implemented for the first time in Hong Kong; although the US Army Corps of Engineers has had Dispute Review Boards, which are conceptually similar, for some time, and Project Arbitration has some analogous features. The DRA system, selected from a range of traditional and ADR options, is being used currently in Hong Kong, following the Adviser's joint appointment in December 1991 by the Hong Kong Government's Architectural Services Department and the contractor carrying out refurbishment on the Queen Mary Hospital there. Outside of the US, where such techniques are better known, there should be considerable interest in this experiment and its outcome. It may not be an overstatement to say that the project has the power significantly to advance the cause of ADR, or presumably, to retard it, if it is not seen as successful. The details of operation of the system in Wall's paper repay careful study.

## 6 Education and attitude change

It is not only the authors of the papers in this section who regard attitudinal change of the personnel engaged in the construction process as essential if dispute avoidance, management and resolution are to be improved. There is a measure of agreement about the need for better approaches and systems of work. It is significant that Smith[88], Nicholson[89] and Wall[90] all arrive at the Partnering philosophy as having the benefit, in the words of Smith, of "a joint commitment to common goals in a long term relationship with mutual expectations of trust and co-operation replacing arms length contractual relationships". Leeds[91] remarks upon the French movement to Concertation, a concept with similar attributes, and other authors are clearly thinking along comparable lines. But it is surely the case that no such proposals will in fact be widely adopted or even accepted while traditional attitudes prevail. Thus it is that the focus for the medium-to long-term future switches to Education. Mackie[92], in evangelistic vein, calls for "a powerful campaign to achieve a change of mind-set". Miles[93] is disturbed by "a general lack of awareness of what ADR is and what it seeks to achieve". His paper includes reference to practices which seem to demand education in other directions: "Contracts signed long after the workmen enter the site". Capper[94] gave an amusing but instructive anecdotal example of a 'topping-out' ceremony which he had attended where the construction team, seeking congratulations for finishing on time and within budget were asked what contractual system had achieved this result. The reply was that this had not yet been decided. Davies[95] too complains that there are deficiencies in the management and contractual skills achieved in training, especially of architects.

The suggested classification of the Education authors as strategists rather than pragmatists is not intended in any way to belittle their contributions. On the contrary, very few of the proposals of the pragmatists are likely to be adopted or used properly without the change in 'mind-set' which only Education (which includes training and continuing professional development) can supply.

All of the Education authors have recommendations for amelioration of conflictual behaviour through education. There is a degree of consistency between three of them, which finds echoes amongst some non-Education authors, notably Turner-Wright about the need for an integrated approach to construction education. Franks[96] presents the most detailed treatment of how such integration has been attempted notably at South Bank, and what benefits may be expected, which he summarises as a reduction of confrontational attitudes and improved collaboration, especially between professionals. Hancock[97] calls, more generally, for "an improved balance between the technological and human requirements of society" and for "A return to a less specialised form of education and a clear understanding of the difference between education and training". Bishop[98], giving the Quantity Surveyor's perspective, also deplores the divisive tendency of construction education which he sees as a major cause of the 'them and us' mentality which underlies many conflicts. Franks[99] can be regarded as speaking for these 3 authors and many others when he calls for "a genuine commitment to common education" failing which "it is difficult to see an end to the conflict culture which has bedeviled the construction industry for far too long".

Finally, Powell[100] and Gale[101] address ethical issues of construction education. Powell argues for an ethical basis to construction education and insists that "The discussion of ethical issues must begin at the beginning of a student's career". Gale calls for a widening of the base of female representation within the construction disciplines. However, these demands are made not only on grounds of equity but as measures offering a genuine contribution to reduction of conflictual behaviour. Powell believes that the inculcation of an ethical approach "will lead to personal growth and development" which are antithetical to negative attitudes. Gale's research suggests that a greater concentration of feminine attributes in the construction process could benefit an industry which "is conflictual because it has a male culture". These attributes include (inter alia) a greater faculty of self-criticism and more democratic, less leader-oriented conduct of discussions. The importance of the Education papers consists to some extent in the detailed proposals for reform of content, but chiefly, in the fact that, in the words of Bishop[102] "the key to a more productive future is in the word ATTITUDE".

## 7 Overview

The background against which the Conference was called and held was of increased conflict in the construction industry, of dissatisfaction with traditional conflict resolution methods and of an increased willingness in the industry to explore alternative solutions.

There was considerable discussion of and attempts to classify, types of conflict. Its inevitability and functionality were considered. Proposals for the avoidance of dysfunctional conflict and the management of inevitable or unavoided conflict were advanced.

One section of the papers included was devoted to review of existing conflict resolution methods, their deficiencies and proposals for their improvement.

Given that existing conflict resolutions have inherent deficiencies which cannot be easily repaired by any of the methods, albeit beneficial, in the previous section, alternative dispute resolution (ADR) methods were discussed. Discussion centred on demand, to some extent on the range of models and techniques utilised in different countries, and on experiences of the use of those models and techniques.

Proposals for improved attempts at dispute avoidance, dispute management and dispute resolution would all require the inculcation of different, less conflictual attitudes, as well as some changes of technical substance in education and training for construction personnel. Proposals for reforms to achieve this were advanced.

The Conference benefitted from the submission of papers by authors in 10 countries from 4 continents and from most of the disciplines concerned with the construction process including architects, engineers, quantity surveyors, contractors, lawyers, project managers and academics. The gender distribution of the authors supported the view that the construction industry is disproportionately male.

The interest generated by the Conference may be regarded as conducive to the formation of a Working Commission, possibly under the auspices of the International Council for Building Research and Documentation for the co-ordination of further research in this general area on an international basis.

## 8 References


1 Fenn, P. (1991) **Managing Corporate Conflict and Resolving Disputes on Construction Projects**, Proceedings of the 7th Annual Conference of the Association of Researchers in Construction Management.

2 Fellows, R. (1992) **Karming Conflict,** Proceedings of the UMIST Conference.

3 Newey, J. (1992) **The Construction Industry**, Proceedings of the UMIST Conference.

4 (1988) 2 All E.R. 992.

5 (1990) 3 W.L.R. 457.

6 (1990) 3 W.L.R. 414.

7 Davies, R.S. (1992) **Construction Conflict - The Specialist Contractors View**, Proceedings of the UMIST Conference.

8 Eilenberg, I.M. (1992) **Whither Small Value Residential Dispute Settlement in Australia?**, Proceedings of the UMIST Conference.

9 Lavers, A.P. (1991) **Law and Legalism in Construction**, Proceedings of the 4th Yugoslav Symposium on Building.

10 Hancock, M. (1992) **Education of Construction Professionals to Improve the Built Environment.** Proceedings of the UMIST Conference.
11 Colledge, B. (1992) **Construction Contracts: Towards a New Relationship.** Proceedings of the UMIST Conference.
12 (op cit)
13 (1984) 2 Construction L.R. 43
14 Rahim, A. (1992) **Managing Conflict in Organisations.** Proceedings of the UMIST Conference.
15 Baden Hellard, R. (1992) **Construction Conflict - Management and Resolution.** Proceedings of the UMIST Conference.
16 Wall, C.J. (1992) **The Dispute Resolution Adviser in the Construction Industry**, Proceedings of the UMIST Conference.
17 Quick, R.W. (1992) **Costs in Arbitration Proceedings,** Proceedings of the UMIST Conference.
18 Watts, V. and Scrivener, J. (1992) **Review of Australian Building Disputes Settled by Litigation.** Proceedings of the UMIST Conference.
19 Cooper, C.A. (1992) **Mediation, The Experience in the United States.** Proceedings of the UMIST Conference.
20 Mackie, K.J. (1992) **Alternative Dispute Resolution and Construction Disputes.** Proceedings of the UMIST Conference.
21 Houghton, A. (1992) **Alternative Dispute Resolution - A Far East Perspective.** Proceedings of the UMIST Conference.
22 Nicholson, M.P. (1992) **Peace, Love and Harmony.** Proceedings of the UMIST Conference.
23 Koh, K.C. (1981) **Arbitration for the Construction Industry: the Singapore Scene.** Proceedings of the Joint Conference of the Singapore Institute of Arbitrators/Chartered Institute of Arbitrators.
24 (op cit)
25 Zikmann, R. (1992) **Successful Conflict Management.** Proceedings of the UMIST Conference.
26 Smith, M. (1992) **Facing Up to Conflict in Construction.** Proceedings of the UMIST Conference.
27 (op cit)
28 Clegg, S.R. (1992) **Contracts Cause Conflicts**, Proceedings of the UMIST Conference.
29 (op cit)
30 Langford, D.A., Kennedy, P. and Sommerville, J. (1992) **Contingency Management of Conflict: Analysis of Contract Interfaces.** Proceedings of the UMIST Conference.
31 Turner-Wright, P. (1992) **Construction Management Integration: An Analysis of the Degree of Integration between Construction Professionals and Project Performance.** Proceedings of the UMIST Conference.
32 (op cit)
33 NEDO (1988) **Faster Building for Commerce**, NEDO, London and
NEDO (1991) **Partnering: Contracting without Conflict,** NEDO London.

34 Supra.
35 Gardiner, P.D. and Simmons, J.E.L. (1992) **The Partnership Between Conflict, Change and Project Management Strategy,** Proceedings of the UMIST Conference.
36 (op cit)
37 Revay, S.G. (1992) **Can Construction Claims be Avoided?** Proceedings of the UMIST Conference.
38 (op cit)
39 (op cit)
40 Cosma, C. (1992) **Do-it-youself Homes - More or Less Conflict Problems?** Proceedings of the UMIST Conference.
41 (op cit)
42 (1990) 50 BLR 91
43 (1990) 50 BLR 69
44 Lavers, A.P. (1990) **The Rosehaugh Stanhope decisions.** Proceedings of the 3rd Annual Conference of the Centre for Construction Law and Management, King's College, London.
45 (op cit)
46 (op cit)
47 (op cit)
48 (op cit)
49 (op cit)
50 (op cit)
51 (op cit)
52 Cree, C.A. (1992) **Conflict Management,** Proceedings of the UMIST Conference.
53 Lewis, J. Cheetham, D.W. and Carter, D.J. (1992), **Avoiding Conflict by Risk Management - The Role of the Client's Project Manager,** Proceedings of the UMIST Conference.
54 (op cit)
55 (op cit)
56 (op cit)
57 Lapusteanu, R. and Antohle, E (1992) **Transition and Management of Uncertain Resolution,** Proceedings of the UMIST Conference.
58 Leeds, C. (1992) **The French Approach to Handling Conflicts and to Negotiating: Certain Notable Features,** Proceedings of the UMIST Conference,
59 (op cit)
60 Shakespeare, W. (1592) **Henry VI,** Part 2, Act IV Scene II.
61 Baker, D.C. and McLellan, A.G. (1992) **Substantive Techniques for Conflict Resolution: Aggregate Extraction in Southern Ontario,** Proceedings of the UMIST Conference.
62 (op cit)
63 (op cit)
64 (op cit)
65 Green, S.D. (1992) **Resolving Conflict in the Formulation of Building Design Objectives,** Proceedings of the UMIST Conference.
66 (op cit)
67 (op cit)

68 McGowan, P.H., Horner, R.M.W., Zakieh, R., Jones, D. and Thompson, P.A. (1992) **The Role of Integrated Cost and Time Models in Conflict Resolution**, Proceedings of the UMIST Conference.
69 Bentley, B. (1992) **Adjudication Procedures - A Temporary Diversion?** Proceedings of the UMIST Conference.
70 (op cit)
71 Assaf, S.A. and Al-Hammad, A.M. (1992) **Construction Contractors Liability in Saudi Arabia**. Proceedings of the UMIST Conference.
72 (op cit)
73 (op cit)
74 (op cit)
75 Miles, D. (1992) **The Problems of Using ADR in the Construction Industry**, Proceedings of the UMIST Conference.
76 (op cit)
77 Stipanowich, T.J., and Henderson, D.A. (1992)**Mediation and Mini-trial of Construction Disputes,** Proceedings of the UMIST Conference.
78 (op cit)
79 (op cit)
80 (op cit)
81 (op cit)
82 (op cit)
83 (op cit)
84 Siedel, G.J. (1992) **The Use of Mini-trials to Resolve Construction Disputes**, Proceedings of the UMIST Conference.
85 (op cit)
86 Hollands, D.E. (1992) **FIDIC Study on Amicable Settlement of Construction Disputes**, Proceedings of the UMIST Conference.
87 (op cit)
88 (op cit)
89 (op cit)
90 (op cit)
91 (op cit)
92 (op cit)
93 (op cit)
94 Capper, P. (1990) **3rd Annual Conference of the Centre for Construction Law and Management**, King's College, London.
95 (op cit)
96 Franks, J. (1992) **Construction Conflict Management - The Role of Education and Training**, Proceedings of the UMIST Conference.
97 (op cit)
98 Bishop, D. (1992) **Planning for Disputes - Educating Construction Management**, Proceedings of the UMIST Conference.
99 (op cit)
100 Powell, M. (1992) **Conflict in the Context of Education in Building Ethics,** Proceedings of the UMIST Conference.

101 Gale, A.W. (1992) **The Construction Industry's Male Culture Must Feminize if Conflict is to be Reduced: The Role of Education as Gatekeeper to a Male Construction Industry,** Proceedings of the UMIST Conference.

102 (op cit)

# THE CONSTRUCTION INDUSTRY

HIS HONOUR JUDGE JOHN NEWEY QC
Senior Official Referee, London, England

Abstract
This paper is the opening keynote conference address. The structure of the industry is discussed and the nature of construction conflict and disputes is outlined.

## 1 Construction conflict: management and resolution

The Construction Industry is judged by most criteria to be the largest in the United Kingdom. The Industry also undertakes much work overseas, particularly in the Middle East and Far East. Many professionally qualified persons are exclusively or partly concerned with the Industry, including Town Planners, Architects, Landscape Architects, Civil, Structural, Mechanical and Electrical Engineers, General Practice, Valuation, Building and Quantity Surveyors, Geologists, Accountants, Insurance Brokers and Solicitors. There are also engaged in it Building and Engineering Contractors, House Builders and Jobbing Builders, together with a host of specialists, who usually work as Sub-Contractors, in demolition, piling, steel erection, reinforced concrete, cladding, glazing, roofing, chimney lining, damp proofing, heating and air conditioning, decoration, shopfitting, bricklaying and other activities. The Industry is dependent upon suppliers for all the materials which it uses except for the ground upon which it builds and often upon hirers for cranes, cherry pickers, pumps, props and other equipment. Local Authorities, Agencies and Government Departments are concerned with regulating its activities. The Industry's clients vary between the developer of a massive office block or a Highway Authority creating a motorway to a church requiring a new vestry or a poor widow wanting her house repainted.

Work may be carried out under elaborate ad hoc contracts requiring weeks of negotiation and careful drafting, or under standard forms such as the Royal Institute of British Architects' for professional purposes or one of the main contracts or sub-contracts prepared by the Joint Contracts Tribunal on which most sections of the Industry are represented and have rights of veto, or under simple contracts in writing or reached by correspondence or orally.

Sometimes, of course, work is performed without any contract having been reached, when the doer may be able to recover payment in quasi contract.

In parallel with contracts requiring work to be done there are often "collateral contracts" warranting its performance between Designers such as Architects or Engineers of a new development and the intended first tenants of it and between Employers and nominated Sub-Contractors.

## 2 Disputes

Since the Industry is so large, there are so many individuals, companies, partnerships and Authorities engaged in it, construction work has to be carried out on open sites in conditions very different from those in a factory, failures by one or more can affect all engaged in a project and work often takes substantial periods during which economic conditions can alter, it is inevitable that disputes arise.

The London Official Referees' Courts deal with all High Court and some smaller construction cases arising in London and the South East and with many High Court cases arising elsewhere in England and Wales. Between about 1973 and 1980 there was about 100% increase in the number of cases brought to the courts and in most years after that until 1989 there was an increase of about 15%.

In 1990 there was no significant increase over 1989 and in 1991 there was a decrease. In 1973 there were three full time Official Referees who were sometimes assisted by other Judges; now there are seven full time Official Referees who are assisted by nineteen Official Referee Recorders (Queen's Counsel in private practice, who sit for not less than four weeks a year) and regularly by other Judges. I do not know of any statistics for construction cases which are commenced before part-time Provincial Official Referee or before Arbitrators, but increases are probably much the same.

One undoubted reason for more construction litigation and arbitration has been the changes in Common and Statute Law which have made it easier to bring claims. Another reason has been increased "claims consciousness". Other reasons suggested are the use of new and sometimes imperfectly understood technologies and the disappearance of the old fashioned Site Agent who exercised real control over everything which went on and was not afraid to given hints to professionals.

It would be pleasing to think that the recent fall in the volume of cases is due to better quality work, increased reasonableness, success by Adjudicators or successful resort to Alternative Dispute Resolution. The last two have probably played a part, but I think that the main reasons for decrease have been recent decisions of the House of Lords restricting claims in tort for economic loss and sadly the number of companies which have gone into liquidation.

Courts and Arbitrators backed by adequate powers of enforcement are essential to ensure general compliance with obligations and compensation for victims.

Nonetheless involvement in litigation or arbitration, especially in cases which go to trial or hearing, is at best a misfortune and at worst a catastrophe even for the successful parties. Counsel's, Solicitor's and Expert's fees are substantial.

Directors, Partners or Principals and staff are distracted from their usual work and have to devote time to instructing solicitors and attending court, perhaps day after day, which could be better devoted to earning.

Inevitably there is an interval between the commencement of proceedings and their determination. In the London Official Referees' Courts fixed dates for hearings are given on summonses for directions; cases expected to last for under eleven days which can be taken by Recorders or Visiting Judges are usually fixed for about nine months ahead and cases expected to last for over ten days which can only be taken by Official Referees, fifteen to eighteen months ahead. The position is similar before provincial Official Referees and Arbitrators. Cases cannot usually be prepared for trial in shorter periods than these, but during them even quite large companies can experience liquidity problems, while for examples a house holder and his family may have to live in a defective house for want of means to carry out remedial work.

## 3 Preventing and settling disputes

Where disputes are concerned, prevention is much better than cure! Employers should decide what they want and designers, contractors, sub-contractors and suppliers should ensure that they understand what is expected of them. Parties should then enter into proper contracts and not rely on letters of intent or other nebulous arrangements.

Standard forms of contract are often criticised, but since construction work is complicated it is inevitable that contracts governing it are also complicated. Obviously designers should design properly; junior staff should not be left to carry out major responsibilities without careful supervision; calculations should be checked and rechecked; and if an untried technology is to be used the Employer should be warned and his consent to it obtained. Contractors and others should not undertake work unless they are sure that they can perform it; they should submit realistic tenders and not hope to make work profitable by subsequent submission of claims. Agents and foremen appointed to site should be capable of providing effective leadership. Contractors who consider that part of a design is unbuildable or unsupervisable should inform the designer at once and not wait for difficulties to arise. Designers administering contracts and contractors and others carrying them out should be determined to work together and to avoid disputes.

If disputes arise, the best time for resolving them is as early as possible. That is why I think that the use of Adjudicators named in advance as now required by most JCT contracts is such an admirable idea. If negotiations on site have failed, a meeting between directors or the like may still be tried; as the late Lord MacMillan said: "Jaw jaw is better than war war". Mediation in any of its forms is probably best invoked early. If a dispute involves the construction of a contract the parties should apply to an Official Referee by Originating Summons or under Order 14A or to a legally qualified Arbitrator to give an immediate decision on it.

If, notwithstanding all efforts to the contrary, a dispute continues and goes to the Official Referees' Court, the Official Referee to whom the case is allocated will be careful not to discuss settlement with the parties, but he will endeavour by ordering disclosure of documents, exchange of experts' reports and cross service of statements of witnesses of fact to ensure that each party knows the details of the other's case and is able to form a realistic view as to the prospects of success. The Official Referee will also order a meeting of experts to endeavour to agree technical facts and to narrow issues. In the result about 85% of cases settle between summonses for directions and dates fixed for their trial. Generally Provincial Referees and Arbitrators proceed in a similar manner with similar results.

## Part Two

Construction Conflict

Construction conflict seems inevitable. Researchers and practitioners in other disciplines have developed formalized conflict management systems. Conflict management is taught in American business schools. These papers discuss a number of techniques and situations to which conflict management may usefully be applied.

'Facing up to conflict in construction' (Smith) describes the concepts of functional and dysfunctional conflict and identifies sources of expertise helpful to professionals in dealing with conflict connected with the procurement of buildings.

'Construction conflict - management and resolution' (Baden Hellard) suggests reasons for conflict in 3 principal phases of a construction project; establishing the brief, during design detailing and contract construction, and emphasises the importance of total quality management.

'Managing disputes' (Cree) puts forward ideas of good project management techniques to handle disputes relating to motivation and the direction by project managers of specialists to achieve client's objectives.

'Successful conflict management' (Zikmann) highlights the inevitability of conflict and suggests that the success of a building project is concerned with managers identifying and responding to various forms of conflict.

'Construction conflict - the specialist contractors view' (Davies) traces the development of the construction process, concentrating on contractual issues and settlement of disputes from the specialist contractors viewpoint.

'Contingency management of conflict : analysis of contract interfaces' (Langford, Kennedy and Sommerville) describes the sources of conflict found in different procurement methods, and proposes a model of variables concerning the relationships between main sub-contractors and trade organisations.

'Avoiding conflict by risk management - the role of the client's project manager' (Lewis, Cheetham and Carter) discusses the role of the clients project manager in the application of risk management, and outlines a methodology of risk management with its application being illustrated by two case studies.

‘Resolving conflict in the formulation of building design objectives’ (Green) details an example of the use of simple multi-attribute rating technique (SMART) by presenting a case study of a new laboratory illustrating the benefits of developing a formal decision model during briefing and outline design stages.

‘The relationship between conflict, change and project management strategy’ (Gardiner and Simmons) puts forward a model for modifying project management strategies, based upon the findings of research interviews conducted to identify project conflict and change.

‘Karming conflict’ (Fellows) contends that it is crucially important to prevent conflicts and disputes arising rather than concentrating on dispute resolution, and suggests that improvements could be made by considering the notions of peoplism and Karma.

‘Contracts cause conflicts’ (Clegg) argues that contracts cause rather than eliminate conflict and uses data collected from construction sites to illustrate this idea.

‘Construction management integration: an analysis of the degree of integration between construction professionals and project performance’ (Turner-Wright) analyses the effect of integration on site management teams, and evaluates an integration model relating to the concept of construction professionals working in unison.

‘The French approach to handling conflicts and to negotiating: certain notable features’ (Leeds) looks at the negotiating model of dominating-integrating and puts forward the concept of concertation to describe the mediation process in France.

‘Substantive techniques for conflict resolution: aggregate extraction in southern Ontario’ (Baker and McLellan) uses aggregate mining in Ontario to illustrate means used to reduce conflict amongst disputing parties, such as compensation strategies.

‘"Do it yourself homes" - more or less conflict problems’ (Cosma) describes economic changes in Rumania, particularly relating to state influence, with regard to contracts between clients and builders and how this has affected the settlement of disputes.

‘Transition and management of uncertain resolution’ (Lupasteanu and Antohie) discusses the influence of communist society upon the Romanian construction industry, which has led to more competition, presents decision models, and leadership and management theories.

# FACING UP TO CONFLICT IN CONSTRUCTION

MARTIN C.G. SMITH
Managing Director, UK and Offshore, High-Point, Birmingham, England.

**Abstract**
The paper identifies sources of expertise helpful to technically trained construction professionals concerned with construction procurement/contractual arrangements and the conflicts which ensue. The necessity to understand the perceptions of rights given by law and the reality in practice is commented on. Mention is made of the Chartered Institute of Arbitrators for training and qualification.

The relevance of the concepts of functional and dysfunctional conflict discussed by the sociologists Simmel and Coser is explained.

The challenge offered by the work of the manufacturing statistician/production engineers Crosby, Deming and Juran in the modern concept of Partnering, to traditional arms length contractual relationships is described.

The opportunities offered by awareness of the work of Carl Rogers, most well known of the people-centred psychotherapists, in understanding relationships including the state of mind of the individual as it affects his ability to resolve conflicts and the ways by which people can be helped is introduced.

The services offered by CEDR may reflect that expertise.

The work of John Childs, the sociologist, in "British Management Thought" is used to warn of the need to accept theories of management or contractual procedure, even if traditional, with care.

The paper concludes by expressing the sentiment that this conference is a step in the right direction of practical research inside the context in question, before conclusions are drawn.
Keywords: Conflict, Functional and Dysfunctional, Psychotherapy, Partnering, Construction Law, Management Theory, Dispute resolution.

## 1 Introduction

The theme of this conference is exciting and my company is delighted to have been one of the sponsors. It has given those of us who are practitioners the opportunity to enjoy a forum and to introduce concepts established by leaders in other fields which we have found to be illuminating. We will be further assisted in the process by many of the contributors here who are specialists in their field.

My own contribution arises from experience in a number of countries over the last ten years, as a construction professional and Director of an international consultancy concerned with construction project counselling and monitoring for commercial risk and in commercial and technical problem resolution from quite small sub-contractor issues to large disputes involving adjudication, arbitration and litigation, some of which are on the public record. During this period of living on a diet of heavy commercial administration, engineering, the interpretation of contracts and problem resolution, I have searched for expertise and its source material which has something to offer practitioners like myself. Our consultancy has a firm preference for problem resolution without the use of expensive formal procedures and we regard it as a failure when these are required. In this paper I have tried to highlight some of that source material and identify questions that continue to concern us.

Longer ago than I care to remember the university engineering department I attended, stimulated me with an experimental course it had introduced containing a subject called Industrial Anthropology. At the time, we thought it an incredible title, if not a concept. Today, I am not so sure. In this conference one could say our speakers are talking about Construction Anthropology, because what our speakers may be, or what I hope they will be talking about, is the efficiency of the construction community in regard to its management and resolution of conflict. A construction conflict is not in my opinion to be regarded solely as a one off situation concerning two parties in isolation. Construction conflicts are, after all, endemic in the industry. The reasons for them flow from the way the industry functions and the techniques of resolution adopted today, will have a fundamental, but maybe indirect effect, on how the industry evolves for tomorrow in the shape and size of firms which remain and the way they relate. It is proper that, as recognised in the conference, the industry should study the pattern and implication of its conflicts.

Accurate mutual perception of what actually happens, aided by up to date expertise which will give beneficial insight, is what we are all looking for to play our part in helping the construction community evolve.

The concepts I introduce in this paper flow from sociology, psychotherapy, manufacturing and statistical/production engineering.

An appreciation of the practical experience and development of the principles of commercial relationships and procedures for enforcement of rights established by law in various jurisdictions is fundamental if one is to identify objectives and the practical difficulties in achieving them. The training and exam syllabus offered by the Chartered Institute of Arbitrators has been found to be particularly beneficial in this regard.

Perhaps application of some of the expertise to be presented today will help the practice of the law further approach an ideal of service to the construction community.

## 2 Competition, functional conflict and dysfunctional conflict

Two or more parties have the same goal to beat a record, win a race or obtain a contract. They are in **Competition**, not having any direct dealings with each other (if there is a direct interaction, for example they get in each other's way, **conflict** arises). It is common ground in most societies that competition causes people to strive and is beneficial.

One view of a contract is that it describes a task objectively which has been competed for and "won". It is, therefore, no more than the formal arrangement for the administration of this task.

However, questions to clarifying the work content, quality and time, are bound to arise, requiring a dialogue and the familiar constituents of the commercial interface and the criteria of management teams on each side to manage it, emerge. In that dialogue I think it important to face the issue that on a particular matter, **conflict (with a small c)** exists. One party contends one thing, the other something else. It requires work by both parties to resolve. This inescapable consequence of a contract is therefore a **functional conflict.**

This is an area where **sociologists** may be able to provide us with useful insight. **Lewis Coser** in his book **"The Functions of Social Conflict"**, published in 1956, commented on **George Simmel's** work on conflict. Those authors recognised that many sociologists assumed conflict was always dysfunctional, that co-operation led to efficiency and noted "a decreasing concern with the theory of conflict and a tendency to replace analysis of conflict by the study of 'tensions', 'strains' and 'psychological malfunctions'."

I am informed that these two authors are still in vogue today together with the benchmark they established. **Conflict** is for real. When it is an inescapable part of the contracting system we have chosen, it is **functional**. It needs recognising and responding to, not pretence or the assumption either that it need not exist, or that of itself, it is a bad thing. We have then to separate carefully the **dysfunctional conflict** we don't want and see what can be done.

When any conflict is apparently ended some would contend that this is all that matters and signifies the resolution of functional conflict (as well as dysfunctional conflict). Before settlement, one party may not regard himself as having a problem at all, usually the one with the money!

**Conflict (with a capital C)**, in the pejorative sense, seems to be related to that context "if two parties are in a ditch having a fight and one is on top of the other, they are both in the ditch!" i.e. they are both suffering. Some would define this as a dysfunctional conflict.

I reject those definitions.

It is only by consideration of the construction community as a whole that one may propose which conflicts are functional and which are dysfunctional, difficult and controversial as that process might be. It might or it might not be in the communities' interest that the little guy runs out of money and stops arguing.

The practical importance of what I am saying, is that in my view, in addressing this subject, we should be, considering where and how we see the benefits of procurement arrangements (which are inescapably associated with conflict) and how these procurement arrangements should be optimised in regard to the functioning of the construction community as a whole. We should look wider than the definitions of functional and dysfunctional conflict I suggested. **I consider Functional conflict is essentially a construction community problem, when it is an inescapable consequence of our trading relationships. Dysfunctional conflict may have arisen if the actions of the parties have gone beyond what we may recognise as a functional conflict.**

My reasons for introducing this concept are twofold.

**Firstly**, it is being seriously challenged as to whether a traditional arms length contractual relationship between two companies, when each is dependent on the other, is necessarily the most efficient commercial relationship.

When considering the costs of managing the interface for both parties from the beginning, particularly if the interface is a problem one which is habitually connected with formal dispute resolution, it may be found that expenditure outweighs the benefits. If the context requires a close technical interrelationship, a hard commercial interface may be particularly inappropriate.

The concept of **Partnering** derived from Japanese/American industrial experience but also recognisable in the relationships such as between Marks and Spencer and its suppliers, or the total service the 19th century Architect or Engineer prided himself he gave his client, has to be seriously considered. There is a joint commitment to common goals in a long term relationship with mutual expectations of trust and co-operation replacing arms length contractual relationships.

Those that argue for "accountability" and the benefits of arms length contractual relationships will have to make their case carefully. There are powerful arguments for other ways of organisations working together to create a project using the Partnering philosophy. Authors such as **Deming, Crosby** and **Juran**, on whom other speakers may expand further, were American manufacturing statisticians or production engineers interested in product development in the manufacturing industry. They worked in post war Japan. They dealt head-on with the issue of functional conflict, by seeking at every stage to create organisational arrangements between parties which could lead to emphasis on shared long term goals and the promotion of trust to achieve these goals. There is recognition that this cannot be achieved without effort and commitment. A number of major players in the offshore and process plant industry see their way to lower construction costs through Partnerships, as distinct from arms length traditional contractual relationships between operator/client and contractors.

**Secondly**, our legal tradition has established a concept of justice which gives each individual a perception of his rights and obligations. I am sure this conference will further question whether the existing dispute resolutions systems can live up to the expectations raised by these perceptions.

Financially strong organisations are commonly in dispute with much weaker organisations and it is sometimes considered that the financial staying power of a defending larger organisation prevents a smaller organisation obtaining a result which would be expected on the merits of the case. The constant delay and deliberate procrastination increasing the "costs" of the arbitration or litigation can easily create a situation where the investment in the "costs" may become sizeable in relation to the sum originally at issue. The smaller company may be in danger of being taken out of its financial depth in circumstances where it is already financially stretched. This is trial by ordeal, not my concept of justice.

If against such power play a sense of realism causes the smaller organisation to settle outwith an objective consideration of what would be the result if the trial had proceeded, the dispute management procedure will have, in effect, stacked the odds in favour of the larger organisation, at the expense of the smaller organisation, so creating a platform which will in due course affect the shape of the industry.

My point again is that we should not be looking at two parties solving their problems but how our dispute management procedures affect the culture (ways of working) of our construction industry.

We may accept that the larger organisation should succeed at the expense of the smaller, or we may not. Whatever our view, the dispute resolution management procedure which develops will, in some way, also affect the shape and composition of the industry.

## 3 Skills to resolve conflict

In the personal counselling world **Carl Rogers** was perhaps the best known of the **person-centred psychotherapists** writing from the 1940's till his recent death. Maslov, a contemporary, is equally well known for his **hierarchy of social needs** and is often quoted in management education.

Whilst **Maslov** saw man's goal as **self actualisation**, **Rogers** saw **high self-esteem as underpinning man's potential for development**. Both saw an openness to new experience, a flexibility of approach and an ability to see conflict as a challenge and opportunity rather than as a threat, necessitating the use of rigid defence mechanisms as essential goals of human development.

To resolve a conflict both would say each party needs:-

1 Knowledge of the issue.
2 A sense of personal adequacy to the job.
3 A respect for the skills of the other party.

They were not writing about groups or organisations but my experience suggests their conclusions are equally applicable, and it is interesting to see how they reflect the ideals of our legal tradition.

Perhaps those at the Bar represent these ideals but are they reflected in the actions of the silent majority involved in construction conflicts who never use formal dispute procedures?

In my experience, I have seen construction conflict regarded as a threat leading to the use of rigid defence mechanisms. On more occasions, I have seen parties arguing without adequate knowledge of the issues and behaviour being adopted, which will not lead to better mutual understanding.

I have seen parties, who on each side are disturbed by the circumstances they have found their own organisation in and who are preoccupied with possible consequences. A recognition that this frame of mind is not the optimum one from which to find a resolution, may be valuable. They need objective help and support.

Lastly, I have seen many parties who have low respect for the skills of the other party. I may even venture to suggest that this seems a particular feature of the U.K. construction scene. Parties who see each other in a shallow way, as stereotypes, will not have much mutual understanding. Parties who deny they are in a functional conflict but maintain they are in a dysfunctional conflict will also have difficulty e.g. "It's the other person's personality that is preventing a solution, not the quality of the functional arguments submitted on our behalf".

Much of my work is directed towards assisting companies to resolve problems without the need to institute any formal procedures. When companies and individuals are stressed and need to take difficult decisions, what the decision maker may benefit from is being empowered to make decisions and skilled expertise to help him is valuable. The concepts established by Carl Rogers and others like him have proved very helpful even though the actual issues under discussion may be very technical in nature.

## 4 Management ideology, fact or fiction?

Many years ago, I heard John Child, the Professor of Management Studies at the University of Aston in Birmingham, introduce his book "British Management Thought", in which he pointed out the lack of fundamental content in much of the management teaching and its fashionable nature related to the perceived problem in society at a given moment in time.

If he were looking at the Construction Industry today, I feel sure he would identify a similar pattern of belief regarding the efficacy of many traditional, as well as novel (to the UK), contractual procedures and their associated formal dispute provisions.

What would he say about traditional **JCT Public** Forms of Contract? Many years ago I studied the total construction cost per unit area of houses built by a speculative house builder operating at the top end of the market. I compared this with similar costs of houses built in the public sector using appointed professionals and arms length JCT contracting procedures, the market price of which was less, but which had cost more. The cost of management across the commercial interfaces, or as I described earlier, functional conflict created the difference. I am not suggesting that this example has general applications but it demonstrates the need for alertness and facts.

What would he say about Management Contracting or Major Design and Build projects? My experience suggests functional conflict (the inevitable consequence part) exists without significant change from traditional procedures although it was considered it would be reduced. Dysfunctional conflict and its financial consequences, particularly arising from the failure of either or both of the parties to have adequate knowledge of the underlying rules which should govern their relationships, appears to be no less than the traditional procedures they have replaced and some would say more.

What about **Litigation** and **Arbitration?** We can all see the work carried out by the Official Referee's Court and the role it has taken in progressing construction litigation in the interest of the whole construction community.

We all have a vision of what the original ideal of arbitration was and the practical realities of construction arbitrations today.

Because the community assumes arbitration is a consensual arrangement between two parties, no public policy is established which would lead to professional intrusion to keep costs down and in the mind of the common man, make justice affordable for all.

My company is one of those closely identifying with CEDR and I look forward to hearing speakers from that organisation explain the contribution they have made in ADR.

I hope this conference encourages research to get at the facts **inside** the procurement or contracting procedures we use, to address the questions raised in this paper. Facts need to be separated from fiction by accurate research.

## 5 Conclusion

I hope the concepts I have introduced, and even more those that subsequent speakers will develop, will clarify how professional intrusion can be achieved to reduce the cost of conflict (where it is an extra cost) to the construction community, using contributions from sociology, psychotherapy, production engineering and law and any other applicable areas of learning.

Most of all I hope this conference assists in establishing that conflict resolution is not a matter to be left to the parties alone but something that affects our whole construction community and therefore needs similar factual study and a public determination to make improvements for the benefit of the community.

## 6 References

Bendell, T. (1991) **The Quality Gurus : What can they do for your Company?** Department of Trade and Industry, London. Introductions to Crosby, Deming and Juran are given and their publications listed.

Child, J. (1969) **British Management Thought**. George Allen and Urwin Ltd, London.

Coser, L. (1957) **The Functions of Social Conflict**, Routledge and Kegan Paul Ltd, London.

Jung, C.C. (1983) **Memories Dream Reflections,** Flemingo, London

National Economic Development Office (1991) **Partnering: Contracting Without Conflict** NEDO, London.

Rogers, C.R. (1961) **On Becoming a Person: A therapist's view of psychotherapy**, Constable and Co Ltd, London.

Rogers, C.R. (1951) **Client Centred Therapy,** Constable & Co. Ltd., London.

Rollo, (1992) **The Art of Conselling**, Souvenir Press, London.

## Contacts

The Chartered Institute of Arbitrators
24 Angel Gate,
City Road,
London,
EC1V 2RS.
Tel: 071 837 4483

# CONSTRUCTION CONFLICT - MANAGEMENT AND RESOLUTION

RON BADEN HELLARD
Chief Executive, Polycon Group of Consultants, London, England

Abstract
"Construction has a built-in recipe for conflict but good management is the preventive medicine of dispute".

This keynote paper will analyze the fundamental reasons for conflict in the three principal phases of a construction project

**Establishing the Brief** and the interface between client requirements and the possible design alternatives
**During Design and Detailing** - between technical factors and specialist engineering disciplines
**Contract Construction** - between the demands of production and the requirements of the controlling authorities, both within and without the contractual nexus.

It also suggests how management practices having regard to human motivation and teamwork through all the preliminary phases can be sustained during construction by resolution techniques based on social engineering rather than legal confrontation. Nothing less than total quality management of the overall process will be adequate.
<u>Keywords:</u> Overall Construction Process, Project Quality Management Procedures, Contract Management Adjudication.

## 1 Introduction

The modern practices of Arbitration grew from two principal sectors of the world economy - Shipping and Construction. Clearly this results from the unique features involved in a shipping charterparty which, like a construction project, is one-off in time, place, activity and people relationship and perhaps also in terms of legal or bye-law jurisdiction.

Legal precedent was less significant than the technical factors involved. In no area other than construction does the resolution of conflict have an industry all to itself. One where teams of professionals and experts are employed by several parties in major construction projects to make and argue claims for more money or better performance.

The organisation of the Construction Industry today has a built-in recipe for conflict. Each group of professionals, contractors and sub-contractors have developed customs and practices which frequently continue when the building "team" carries out what is a combined operation for essentially prototype construction. Frequently the building owner is the only "non-expert" in the team yet it is he who has to make the key project decisions.

It is this background that led to Construction being one of the leaders in the development of arbitration as an alternative to the courts in resolving disputes arising from unique construction contracts.

Disputes arise between men even if they result from problems with materials, machines, methods and money - the resources of management. The use of management techniques arising from a study of management principles and practices are more likely to be conducive to a satisfactory resolution of such disputes than the practice of the courts - but they should embrace some of the excellent concepts of arbitration.

## 2 Establishing the Brief

Let us examine first why conflict is a particular feature in construction.

Every construction project has four frequently conflicting elements which must be established in "the brief". These four can be classified by the code word FACT, which collectively defines the Quality requirements

**Function** - all the technical and physical requirements: space, servicing, internal relationship between the parts, access, egress and the like.
**Aesthetic** - that is, the satisfaction of all the human and subjective aspects that will be enshrined in the end result. The modern equivalent of "commodity, firmness and delight". But today there are always requirements as to:
**Cost** - both capital and running costs. Perhaps better expressed as lifetime cost of the project.
**Time** - the logistic requirements for commercial completion and occupation which in some cases, for example a short-term exhibition project, can be the most critical requirement in the client's brief.

Sometimes the requirements under each of these headings can be given by the client but sometimes they must be established from the authority that can exercise a modifying or even controlling influence over the project. The influence of such external authorities is rarely complete or black and white. Even when the authority, as with town planning, has statutory powers, negotiation can produce solutions to seemingly impossible conflicts in requirements.

In these situations the client's project manager or design team leader, can still bring about a satisfactory solution by negotiation, and the design may be the better for the challenge presented by these conflicts.

It can be seen that when so many people are involved in providing the criteria for the brief and so many technologies are involved in satisfying the requirements in a design solution, even before work begins on site the whole situation is one where conflict between requirements and resources abound.

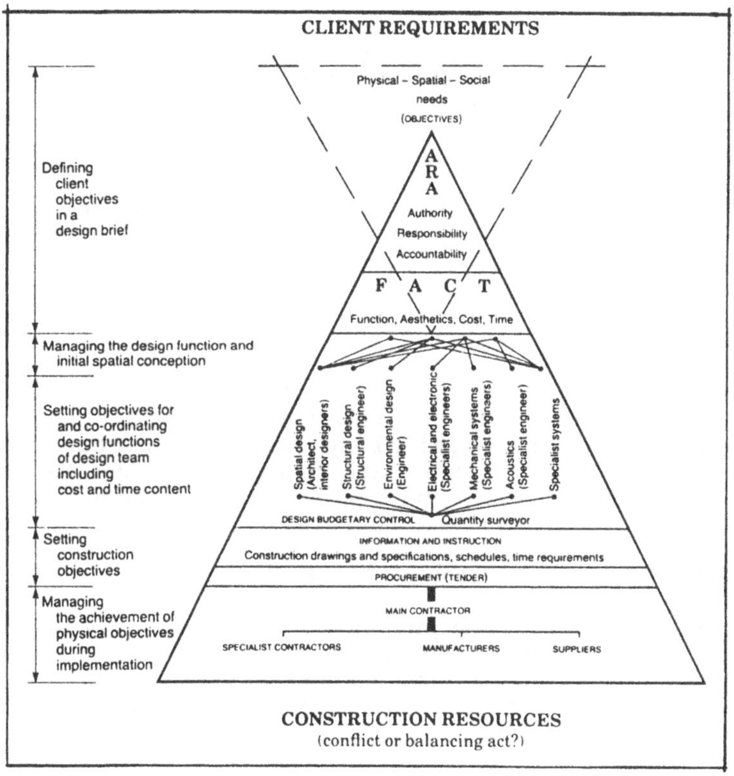


**Figure 1 - Management responsibility pyramid - building design and construction functions.**

It should not be forgotten that within the client organisation too there will always be conflicts in requirements ranging from open enthusiasm (or reluctance) to a keen desire to have a particular aspect fulfilled to their complete satisfaction. A capital building project represents a client's major investment and perhaps a situation that will be experienced by the managers or those involved once only in a lifetime, and each will want to see their own particular interest given satisfaction and priority.

Often these conflicts are not resolved until the project comes out of the ground and a physical presence illustrates, sometimes alarmingly, what were, until then, merely mis-conceptions or perceptions.

If all those involved wish to perform their own tasks to the best of their ability and establish their own programmes there is the need for a great deal of negotiation, which in turn demands excellent communication ability from those responsible for finalising the design brief.

Many of these negotiations will continue over weeks or months and be interactive as between the client's stated requirements and the constraints imposed by the external parties - many in conflict with each other.

### 3 Detail Design

Yet, at some point the design must be frozen to enable the second stage to proceed - the detailing of materials with their consequent quality and cost implications - so that a tender can be obtained for implementing the design by construction.

But before this, conflicts in design requirements between the different technical disciplines must have been resolved. Here the conflicts are technical as designers seek to express their own expertise to the best advantage. In the physical sphere the Structural Engineer may want columns and beams where the Services Engineer needs ducts, and the Architect would like to see an unsupported glass wall anyway!

### 4 Tendering Procedures & Construction

At some stage the design requirements must be tendered for as a construction contract.

The degree of finality in the price obtained from the various contractors for the work required of them will, or should, depend upon the extent of the firmness of the requirements stated in the invitation to tender. The greater the clarity the less the contractor's risk. Therefore the more realistic and competitive should be the price.

But whatever the contract in terms of time and price, external constraints, if not the client's own situation, introduce the possibility of continuing change. Control by an external authority can result in changes of policy, which in turn produce a different situation which might be to the building owner's disadvantage, or produce a benefit, and so produce further changes for the design requirements, which in turn can reflect upon the contractual terms and produce claims.

All this is perhaps to state the obvious but it shows why building contracts provide for variations within the contract. In principle, every change creates a situation where the cost and time criteria are completely open to renegotiation. The various standard forms of contract have grown in size and number to prescribe what are procedures for dealing with these situations. There are now 94 different "standard" contract forms covering the customs and practices which have been developed within the industry. These reflect the commercial conditions prevailing within the industry more than they do the legal environment outside the industry. It is, however, this contractual situation which must be dealt with in these construction conflicts, but which, if successful, will also provide a big pay-off to all parties in the process.

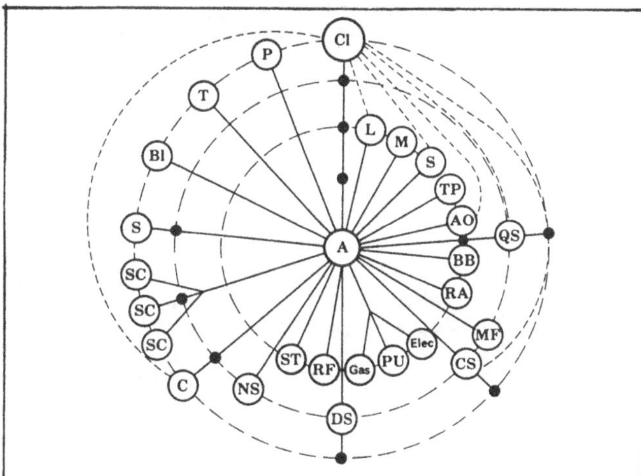


*Note:* Rough guidance on time is given by the distance of the out stations from the centre. The earlier contacts are nearer the centre.

**Group 1:** Cl – client; L – landlord, leaseholders; M – ministries; S – solicitors; TP – town planning authority; AO – adjoining owners; BB – building by law authority; RA – research associations; PU – public utility undertakings – gas, electricity, water; RF – Royal Fine Art Commission. **Group 2:** QS – quantity surveyor; MF – manufacturers; CS – consultants – Services and Structural; DS – designing specialist; NS – nominated specialists and suppliers. **Group 3:** C – main contractor; SC – subcontractors; S – suppliers; BI – Building Inspectors – district surveyors; T – tenants, lessees; P – press (technical and lay).

**Figure 2 THE ORGANISATION NETWORK FOR A BUILDING PROJECT**

## 5 The Client's Project Quality Plan

The project contract through which the legal relationships, promises and procedures are defined has often been completely separate from the technical requirements given in the plans, specifications and component schedules but the project's quality plan should combine all these documents if it is to overcome the conflict that continues during construction. Now it is between the requirements of different trades and sub-contractors who will often want to work in the same place at the same time or may have other demands on their resources from other projects at the precise time they are wanted to perform but at a time not expected when they planned their initial schedule, resulting from earlier changes in programme outside their control.

The changes may be outside the authority and therefore the control of the main project manager or contractor, resulting from the intervention of licensing authorities, public utility undertakings, and building authorities generally, or it may have resulted from unexpected site conditions when sub-soil produces rock where none was predicted or soft areas when firm foundations had been identified. This, together with the vagaries of weather and transport or terrorist disruption combine to produce an ideal climate for conflict during construction.

## 6 Motivation & Behaviour

Conflict, however, is not just the result of a situation created by a series of events. It involves people and human emotions. Motivation is an important but not the only element in human behaviour.

What motivates men? What motivates designers in drawing offices? What motivates men on a construction site? This will depend partly upon the conditions under which they work. There will certainly be different attitudes to working in exposed conditions in the summer with the temperature at 30°C, to that in a similar situation in the winter at -10°C.

But the reaction of human beings, according to Maslow[3], depends fundamentally on the extent of their ascent up their basic pyramid of human needs from physiological at the base through safety and comfort, social, and egotistic to self realisation. The parties and people in a building project can be expected to be at many different personal levels on any one project, and so behave differently.

But behaviour is the individual's total response to **all** motivating forces - only one of which is the particular project situation. Maslow's theory postulates that animal wants are perpetual, and each drive is related to the state of satisfaction or dissatisfaction with the other drives.

Motivation is, however, human - rather than animal-centred and is goal - rather than 'drive'-orientated. All rational human behaviour is caused: we behave as we do because we are responding to forces that have the power to prompt - motivate - us to some manner or form of action. In a sense, therefore, behaviour *per se* can be considered to be an end result - a response to basic forces.

However, behaviour is actually only an intermediate step in a chain of events. Motivating forces lead to some manner or form of behaviour and that behaviour must be directed towards some end. That is to say, there must be some reason why we are responding to the motivating force to satisfy the force motivated as to behaviour in the first place.

Motivating forces are legion and vary in degree, not only from individual to individual but also from time to time. But motivation is not synonymous with behaviour. Motivators are only one determinant of behaviour. Behaviour is almost always biologically, culturally and situationally determined as well. We are, in short, the product of our environment.

**Figure 3**

**Maslow's Hierarchy of Human Needs**

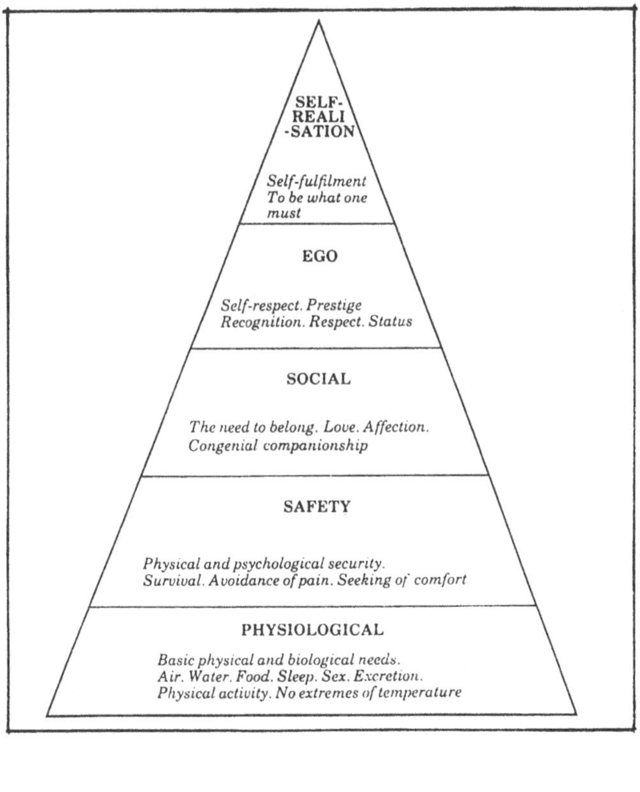

## 7 Theory X, Theory Y, and Self-Motivation[4]

Douglas McGregor takes this behaviour pattern on into what he postulates as Theory 'Y' - that people are self-motivated and will respond to what Drucker called 'management by objectives' in contrast to 'management by control' (Theory 'X') that people are 'directed' to fit the needs of the organisation and without this firm direction people would be passive since they are by nature idle, lacking ambition and resistant to change.

Construction workers, perhaps because of their inherent job satisfaction - the carpenter who lavishes his skills on creating mouldings on doors, frames and staircases, the bricklayer with prowess with an elaborate decorative brick bond, or the paver laying a mosaic floor, all knowing that it will be seen and admired by the generations that pass by the work - appear to fit Theory 'Y' propositions better, and many think that every architect and most designers are all the time engaged upon fulfilling their 'Y' needs at the top of the hierarchical pyramid!

But, because building construction presents a basic situation where every project involves many people whose objectives are widely divergent or on a collision course, the behaviour of the parties may be more the result of these other factors present in conflict situations.

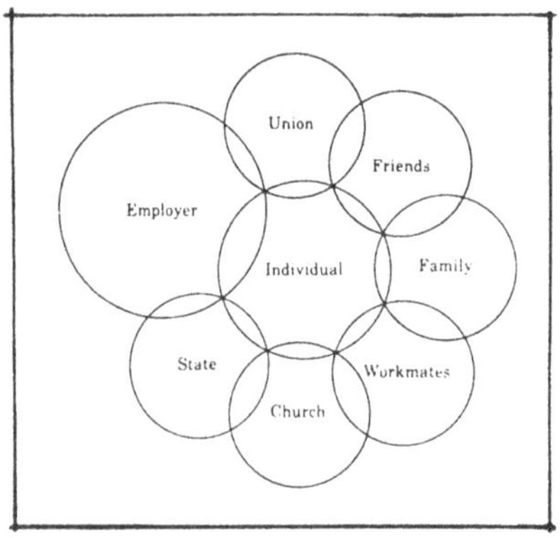


**Fig. 4.A**
**INFLUENCES ON HUMAN BEHAVIOUR IN THE 19TH CENTURY**

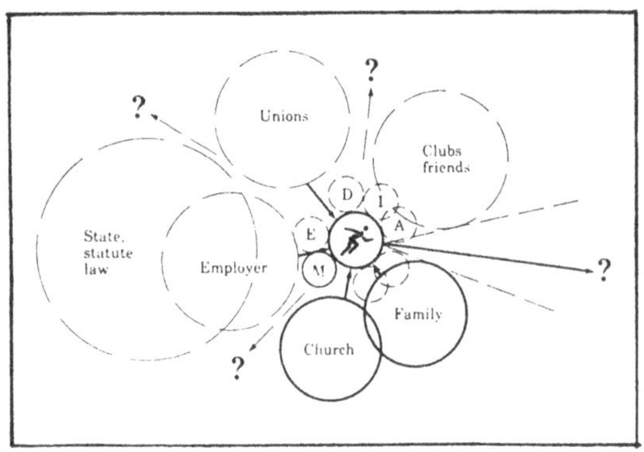


**Fig. 4.B**
**INFLUENCES ON BEHAVIOUR IN THE 20TH CENTURY**

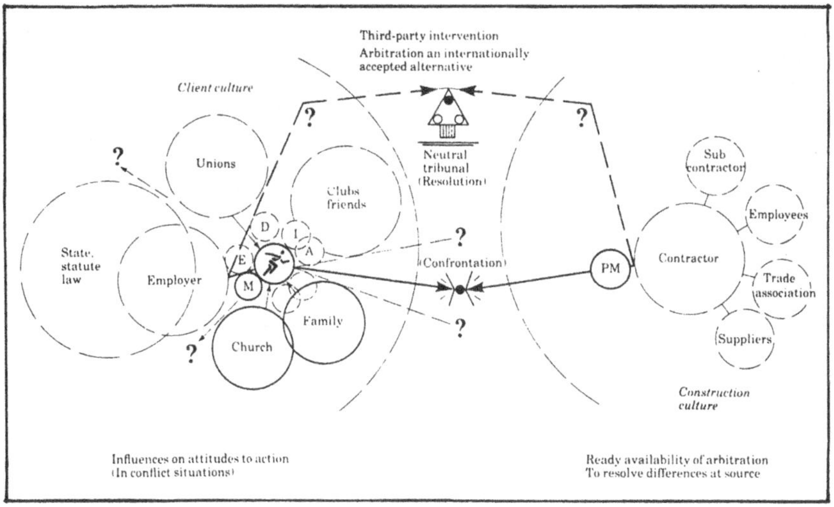


**Figure 4.C - INFLUENCES AND ATTITUDES TO ACTION IN SITUATIONS OF CONSTRUCTION CONFLICT**

Taking this analysis further to the likely implications on the several parties to a dispute, a neutral Adjudication Tribunal is more likely to bring about a solution when it recognises the management culture of construction and studies the contributory elements that caused the situation. This may have flowed from materials, machines, methods or money rather than from weighing legal precedent or the words of the contract defined by lawyers whose interest (apart from winning **their** case) is the law. Above all, construction dispute resolution must take account of the motivation of men.

## 8 The Sports Model

In highly competitive sports arenas we have an excellent model for maintaining the passage of the game, which is, after all, what the teams and spectators alike seek! Here too, conflict certainly exists - in the front row of a rugby scrum, equally between batsmen and fielders in cricket and baseball, no less in professional tennis players with volatile dispositions and many thousand of dollars at stake. All these arenas require and have a referee or umpire available to give an immediate decision on disputed points, without which the game could never end - or prematurely in chaos and conflict!

Thus, whilst teamwork through quality management concepts may, and should, have been developed during the design and planning stages, once work begins on site and enormous sums of money are at stake some suitable form of Adjudication will be needed to provide a firm base for and binding influence over the contractual minefield of potential conflict (and actual dispute).

It may not be long before the referee and umpires in sport are provided with a technological playback to verify their initial decisions. But in providing Adjudication in construction from an instant referee it will certainly be possible, practicable and sensible, to provide for a Final Award after hearing the representations of the parties. This final Adjudication should resolve all conflicts shortly after the completion of the work on site, when the implications of the earlier and interim adjudication decisions can be seen, if necessary remeasured, and the relative responsibilities assessed and a Final Award made. This should preclude the need for legislation at great expense without added value.

## 9 Preventive Measures

Any attempt to resolve conflict expeditiously, economically and effectively should start as early as possible in the chain of events causing the situation. It should not, therefore, ignore the possibilities of prevention rather than cure.

Good management is the preventive medicine of dispute. In my experience of disputes in construction the seeds of the conflict have always been present in the documents that form the technical basis for the construction and the legal framework of the contract. Gaps in the requirements, overlap and conflict between Drawings, Specification, and Schedules, frequently occur due to inadequate understanding, co-ordination and checking between the disciplines and people working in parallel on developing the design.

The current world-wide attention to the concept of Total Quality Management[2] (as an advancement on Quality Assurance and Quality Control) offers the possibility of developing genuine teamwork and unity of purpose for the project if it begins, as it must, at the top of the authority pyramid with the Client's Project Quality Management Plan (Fig. 1). It can then follow down the process chain through Project Managers, Designers and into the Construction zone. Good, strong leadership, and greater participation than is normal by the Client in his project can get the project "team" to the construction start line in the best possible state of morale and technical preparedness.

Unfortunately, the stakes are then so high that the occurrence of difficulties on site referred to in Section

4. above can, and do, frequently shatter the teamwork and unity of purpose of the separate participating firms as they each retreat into their own corner and protect and defend their own individual interests.

## 10 **Contract Management Adjudication**[1]

What will be needed to minimise this effect on what is essentially self-preservation is the provision of an instant impartial umpire - or umpiring tribunal - who by their immediate, if interim, decision on the matter can preserve the parties' position and so maintain their purpose toward the overall objective of a Quality building.

Without this final bridge the chasm between customer satisfaction and construction conflict will remain, and into it will fall too many projects which set out to achieve Quality in all its dimensions.

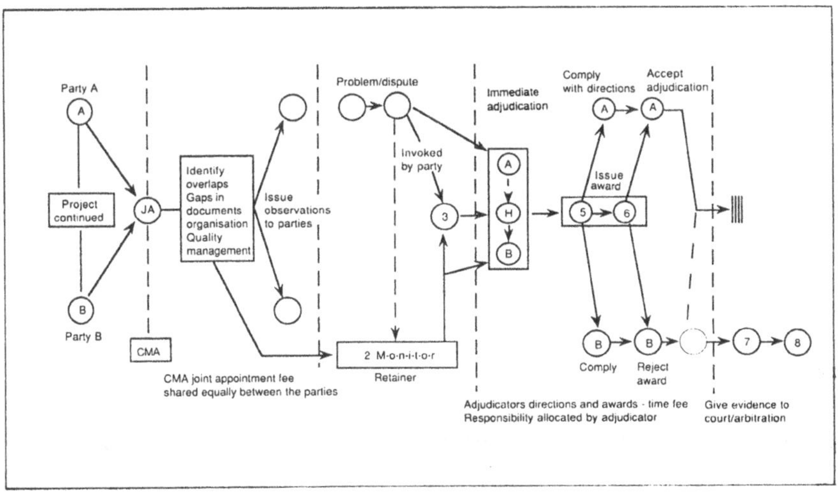


**Figure 5 - CONTRACT MANAGEMENT ADJUDICATION PROCEDURES**

## 11. **Conclusions**

Team building must take place, quality plans must be developed, planning, scheduling and documenting of procedures must take place. All have a part to play and this undoubtably requires leadership, management education and attitude changing. In the construction industry this must include the Customer first, last, and at all stages if he is to receive from the industry what he wants - a quality project. This will need to have regard to all the cultures and contracts through which Quality will be achieved.

It will also need to maintain through the construction phases the discipline and procedures of instant umpiring as the final stage of Total Quality Management.

## References


1. Baden Hellard, R. (1988). **Managing Construction Conflict** Chapter 8. Longman Scientific & Technical.
2. ISO Definition (1989) **Total Quality Management** - below.
3. Maslow, A. H. (1943) **Psychosomatic Medicine** vol. 5
4. McGregor, Douglas M. (Nov. 1957) **The Human Side of Enterprise.** AMA Management Review, Vol 46, No. 11.


## Footnote

The Draft Addendum to ISO 8402 "Quality Vocabulary", at item 3.50 Defines Total Quality Management as:

*"A way of managing an organisation which aims at the continuous participation and cooperation of all is its members in the improvement of:*

*- the quality of its products and services*

*- the quality of its activities*

*- the quality of its goals*

*to achieve customers satisfaction, long term profitability of the organisation and the benefit of its members, in accordance with the requirements of society."*

# MANAGING DISPUTES

C.A. CREE
Coopers & Lybrand Deloitte, London, England

Abstract
This paper describes the need for, and appropriateness of, using good project management techniques to handle the dispute process. It identifies why that need exists and goes on to develop this proposal through analysis of the environments in which disputes take place.
Keywords: Dispute Management, Strategy

## 1 Introduction

A project is something that has a beginning and an end. In construction that beginning and end are readily recognised and can often be described before the project starts. For example when a client decides that he wants a new warehouse we know that the project will start with brief collection and preparation of the financial case. We also know that the project will end as soon as the new warehouse has been set to work and the client has taken it over.

Is the same true of a dispute? It is; there is a beginning and an end of a dispute.

Whilst many of us can pin point likely causes of dispute in advance of their happening we cannot always be precise and one hundred percent certain when a dispute will start. The same cannot be said of how and when a dispute will be settled. So our ability to predict the beginning and end of a dispute is much less certain than for a construction project.

Similarly the processes that we go through during a construction project can be scoped, options tested and uncertainties managed. In any construction project we should go through an hierarchical planning exercise. By this, I mean working from the strategic plan, which is prepared against business and project objectives, then progressively building up more detailed plans, so, for example, planning applications are progressed at the right time in relation to building concept development and site acquisition negotiations etc.

Within any plan there will be points where different options exist for how to progress. The decision on which option to follow cannot be decided until a certain physical event has happened. What should be known and detailed before getting to that decision point is the decision making framework, the risks associated with each option for progressing the work and the impact on the time, cost and quality of the remaining work.

All of these points that you will be very familiar with in planning and managing construction projects apply to a dispute. The fact that a dispute has a beginning and an end has already been established. The fact that there are uncertainties in how it will run cannot be queried. As an example consider how you would decide what the main thrust of the expert witness' evidence should be. A decision cannot be made until the basic facts of the dispute are known, so the key question is when will that occur? Perhaps at initial briefing by the client, perhaps when witness proofs are taken, perhaps not until after discovery or perhaps not until the expert witness has done a large amount of detailed work.

At each of these possible decision points the same sort of decision making framework will need to be considered so that a detailed plan can be developed and the process managed and controlled. For example, at these decision points it will be necessary to consider the following types of questions so that plans can be revised, risks balanced and likely cost commitments calculated: do things look so bad that your client should withdraw or accept an offer of any settlement now before the case deteriorates further? should the expert be instructed to look at specific items relating to the dispute rather than consider all items within his field of expertise? do the matters which now appear to be most significant mean that a different expert is needed? and so on.

So a dispute is just like a construction project. It can be planned within a strategic framework and decision points, decision making frameworks and risks associated with each option for progressing the dispute can all be identified.

And one final point of similarity between a construction project and a dispute: you can guarantee that circumstances will change during its currency so flexibility and forward looking management are vital.

I have made my case for using good project management tools to manage the process of a dispute. I will now look at how that might be done, by examining the fundamentals of project management itself. These are:-

(a) understand your client's objectives;
(b) define the brief;
(c) prepare the project plan;
(d) be forward looking; and
(e) make timely decisions.

## 1.1 Understand your client's objectives

Before embarking on planning the proposed course for a dispute to run it is vital to understand what your client wants to achieve. These objectives could range from wanting to see justice done at any cost; to getting an early settlement; to delay any settlement for as long as possible; or to minimise legal costs and so on.

Obviously each of these objectives would have a different impact on the actions you might take to plan and progress the dispute. Both the outline plan for management of the process and the resources could vary for different objectives. To begin to flesh out the conceptual plans which are available and to differentiate between plans, it will be necessary to delve more deeply into your client's needs and requirements. Only then can a preferred plan can be developed.

## 1.2 Define the brief

Brief taking and definition is vital to the success of construction projects and the same is true for disputes. Ultimately a client wants his project to be successful and to provide value for money. Similarly a happy client knows what is going to happen, when and at what cost. Greater certainty about timing, cost and success can be achieved through good brief taking.

Typically it will be important to establish what experience your client has of disputes. This includes knowing about any inhouse resources that your client wants to devote to the dispute and knowing about their skills and capability to contribute to the resolution of the dispute. It also includes an assessment of what records are available and what the strengths and weaknesses of the arguments on both sides of the dispute are. Finally it will be essential to understand what external commercial pressures affect your client and how these might impact on the management of the dispute.

In seeking answers to these searching questions, the options for handling a dispute will begin to crystallise. Of course, the normal questions about contractual requirements, technical and legal issues have to be formulated as well. Answers to all these questions enable the strategic plan to be mapped out and preliminary costings, time scales and sensitivity analyses calculated. This is a vital stage in project management as it provides, after a very short time, the preliminary scoping for the client in terms that the client will understand and be able to consider from his own commercial view.

Once the brief is defined your client should be able to take business decisions within the framework of the brief. For example, he should be in a position to balance whether the opportunity cost of having his commercial director tied up in preparing witness statements, researching correspondence and giving evidence is acceptable when the commercial director could be out negotiating a new contract.

### 1.3 Prepare the project plan

Once the client brief has been shaped and detailed, the project manager should be able to put together the project plan. Frequently this task will need to be led by the project manager with essential input from the legal team, those who have been involved in the dispute and someone who has a dispassionate view of the strengths and weaknesses of the potential witnesses. If the legal team involved in the original contract drafting have practical experience of disputes they could contribute to the project planning team.

With this team the project plan can be developed through a pooling of knowledge about the various processes that could be involved, through intimate knowledge of the dispute itself and of the client's needs. For each activity (or potential activity) a series of questions will need to be addressed. For example, some of the first questions to be asked will focus on what legal routes could be followed and what commercial routes for settlement exist.

Considering commercial routes: is it possible to apply business pressure to another part of the other side's organisation to get a settlement? If the answer is no then (depending on your client's objective) this branch of the project will finish and pursuit of it will not increase the likely payoff from the dispute. If the answer is yes then this branch of the project plan can be extended by answering questions about the types of pressure that can be applied, to whom and when etc. The likelihood of each action producing a successful result can then be assessed and this can be brought into the overall analysis of the likelihood of getting the right outcome for your client. As we all know times within the construction industry are hard and the pressure caused by the potential loss of goodwill because a dispute is pursued has to be considered. One therefore has to consider: what will the other side's response be if a high profile option is followed? is a successful settlement more likely to be achieved in private through agreement at Chief Executive level, rather than through arbitration, conciliation or through the Official Referee's Court?

Outlined below is a hypothetical example of the structured approach to planning out a part of a dispute. It clearly illustrates that with careful brief-taking and analysis of potential actions, a strategic framework can be set up, within which the dispute can be managed.

The first task in creating the strategic framework is to identify all of the decisions to be taken. This is a tall order but is it important to remember that a project manager's task is not just a "one off" planning task but one that requires repeated attention. The strategic framework is an essential device in ensuring consistency and focussing thoughts. New decision points will arise because there are often unexpected twists during the currency of a dispute. The strategic framework model will help to identify what the sensitive factors are in handling the dispute. Running the model for different scenarios will help decision making.

The terminology used by statisticians in this type of modelling is rather confusing. The exception is the generic title "decision tree" which, for me, describes clearly the overall objectives of pursuing each possible route up through the trunk, along the branches and onward to the ends of the twigs. For example the term "expected monetary value" (EMV) does not necessarily represent an actual monetary value. It is, just, a comparative measure where the higher the value the better. Similarly "utility" is a means by which the decision maker's preference for monetary return as opposed to avoiding risk is measured.

Through the simple example below I demonstrate the power of using decision trees to analyse complex problems where a range of outcomes is possible and preparedness to take risk varies. Contract disputes display all of these features.

The drawing below shows a simple analysis of how to select the best route for pursuing a dispute.

**DECISION TREE**

There are two choices, either to go to arbitration or to go to trial. In taking either of these routes it is possible to get a partial or total win. The probability of success for each basis is shown by p( ). The EMV is calculated for each route working from the end point and the estimated payoff back towards the decision point.

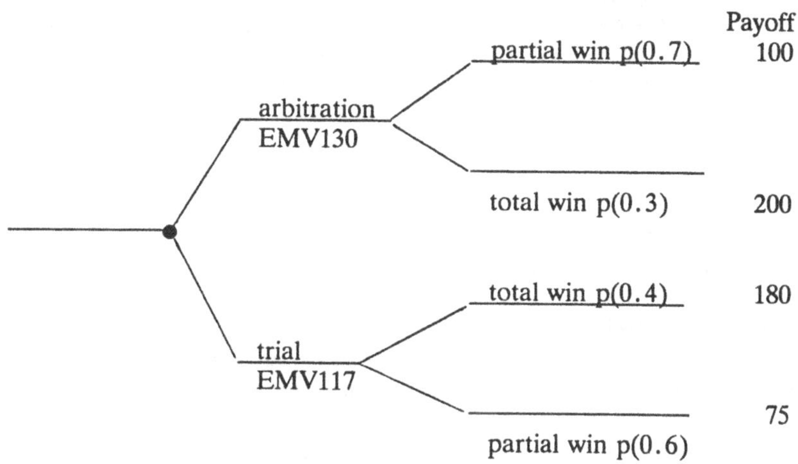


Working through the route marked arbitration the calculation process would be: arbitration EMV = 0.7 x 100 + 0.3 x 200 = 130 similarly, trial EMV = 117. In terms of EMV the preferred route would be to go to arbitration. However to take a decision on whether to go to arbitration or trial, it is also important to consider the range of outcomes because this will help the decision maker assess the risk (or utility) that he will be assuming. Using variance to reflect riskiness, where variance = sum of the probabilities times the square of the difference between the payoff and the EMV, then at arbitration the EMV is slightly higher and the variance is $0.7 \times (100-130)^2 + 0.3 \times (200 - 130)^2 = 2{,}100$ (at trial the variance is 2,646). So if I wanted to assume as little risk as possible I would go to arbitration.

From a high level decision taking framework like this it is easy to move progressively to the next level of detail and so form the whole dispute management plan. This plan will include time, cost and resources analysis.

The preferred option is the one that matches your client's objectives best and the one that should be formalised as the project plan. This is not planning for the sake of it; disputes are inherently complex and uncertain, to have even a reasonable chance of success it is vital to understand the complexities and their impacts so that good negotiating opportunities can be engineered, costs can be managed and risks reduced.

### 1.4 Be forward looking

Things always change; there is the legal equivalent of unforeseen ground conditions, the internal memorandum found at discovery or the failure of the expert to make his inputs on time or the appointment of an arbitrator, who in your view is totally unacceptable. In that you can ask these questions without knowing precise details you can preplan alternative courses of action, and should do so during preparations of the project plan.

As time progresses some possibilities will not have come to fruition, some will, so as you move forward you will have more precise details and be able to review risks and the different likelihoods of being able to reduce them.

By being forward looking and reviewing your plan for a dispute you are able to reduce foreseeable risks before they crystallise and will be able to refine cost, resource and time estimates. This then provides you, at all points in time, with the information needed to manage the dispute process. It also helps you keep your client well informed about costs and progress. This enhanced client handling is particularly valued by my clients.

### 1.5 Make timely decisions

Often decision taking is not the project manager's prerogative. But a good project manager should always create the mechanisms so that recommendations on the best course of action are provided to the decision maker early enough for a rational decision to be taken (if a rational decision is what is required!).

Recommendations should not be simply technical they should include the cost, time and risks associated for each alternative course of action. This data allows the project manager, using structured analysis, to calculate the likely payoff for the dispute under the new conditions. It also allows the client to be forewarned of a deterioration in the likelihood of success and prolongation of the dispute, or a change in emphasis or a new opportunity to develop a new negotiated settlement. With project planning which is sufficiently detailed and analytical it will be extremely difficult for the other side to catch the project manager, by surprise, and for costs to escalate without early warning. Also it will be a lot easier to think through the other side's possible strategies, to carry out "what if" and scenario analyses so that you can optimise your own responses to moves made by the other side without panicking or being left short of time.

## 2 Summary

In summary, I have demonstrated that a dispute is analogous to a construction project and that the application of good project management will increase opportunities for success and minimise uncertainty for your client. It provides the client with the potential to save considerable sums of money both directly, through reduction of opportunity costs and through ability to predict and preempt the other side's strategy.

To maximise the advantages for my clients I work closely with the legal team because, just as in construction projects, it is true that no individual specialist is likely to be as effective as a team of necessary specialists, motivated and directed by a project manager to achieve the client's objective.

# SUCCESSFUL CONFLICT MANAGEMENT

ROBERT V ZIKMANN
Zikmann and Associates, Sydney, Australia

Abstract
The management of conflict is an inescapable part of a Project Manager's responsibility. The ability to identify and effectively respond to conflicts can significantly affect a manager's overall success in managing a building project. This paper discusses the importance of effectively responding to conflict. Typical forms of conflict are identified and the implications of adopting different responses to these conflicts are evaluated in terms of their likely impact on relationships and the project. Strategies for responding effectively to conflict are included in the paper.
Keywords: Conflict Responses, Conflict Implications, Conflict Resolution Strategies.

## 1 The importance of understanding conflict

The building industry poses unique challenges to those working in it. Traditional industry training, economic necessity, modern procurement methods and a heavy reliance on the subcontract system have produced an industry which is extremely fragmented.

In this environment, project managers are required to establish and manage intricate relationship networks for projects of limited life and budget. The very nature of such projects often provides little incentive for the establishment of long term working relationships between project participants.

Short term financial concerns often overshadow the potential benefits of developing and maintaining relationships beyond the limits of the project duration.

The result is often the development of an aggressive 'winner takes all' project mentality.

Use of threats, financial manipulation and other forms of coercion almost inevitably become an established part of the project environment.

In these circumstances, several forms of conflict commonly occur.

These include:

- Interest conflicts
- Structural conflicts
- Value conflicts
- Relationship conflicts
- Data conflicts

Unless project managers are alert and have the skills to manage the levels of conflict effectively, relationships between project participants can deteriorate to such an extent that the original project goals become impossible to achieve.

## 2 Responses to conflict

Project managers adopt different styles in responding to conflict. These styles are usually a combination of individual personality, training and past experience.

Different responses bring with them not only implications for the specific conflict but also for the project.

Effectively managed conflicts can help identify previously undetected problems and attitudes. They can also help clarify uncertainties and improve overall cooperation.

Poorly managed conflicts can conversely create a pool of further unresolved issues, frustration and resentment. This may result in subsequent and often escalated conflicts.

### 2.1 Passive responses

Passive responses occur in the following forms:

- Conflict denial
- Conflict avoidance
- Capitulation

When parties adopt a passive response to conflict, their needs or the needs of others inevitably go unmet.

Denial of the existence of conflict (when unresolved issue do in fact exist) also inevitably leads to increased tension. This can result in concealed hostility and the cultivation of a false sense of security.

In these circumstances, issues of real importance to everyone involved are seldom adequately addressed, often resulting in frustration and a gradual withdrawal of cooperation.

A desire to maintain the peace or to avoid 'rocking the boat' at all costs can also have undesirable long term consequences.

Failure to adequately confront and deal with problems can result in the creation of 'no go' areas and encourage shallow commitment to project goals.

Capitulation to the demands and threats of other project participants also often brings with it an incorrect perception that a conflict has been resolved when in fact it has only been unwillingly suppressed.

## 2.2 Active responses

Active responses to conflict can take several forms. These include:

- Domination
- Distributive bargaining
- Compromise
- Integrative bargaining

Active responses are normally either aggressive or creative in nature.

### 2.2.1 Aggressive responses

Aggressive responses include attempts to dominate others (particularly perceived weaker parties). This can occur when unreasonable demands are made or 'one sided solutions' are imposed on others.

The undesired consequences of domination can often be the stifling of future initiative, reduced creativity and the creation of an environment where poor future decisions are allowed to go unchallenged.

Many managers in the construction industry pride themselves on being 'hard nosed' and capable of 'driving a hard bargain'. As a result their responses to conflict are usually characterised by distributive bargaining. This response is usually accompanied by the use of threats, manipulation, the cultivation of power bases and the defence of adopted positions.

An obsession with 'winning' by one side often only results in the opposing party withdrawing cooperation and setting about defending its adopted position. This is hardly the environment in which workable solutions are easily identified.

Other managers in the industry subscribe to the view that compromise is the best response as it normally ensures that the needs of all parties are at least 'partially met'.

The disadvantage of always adopting this response is that an environment can be created in which parties regularly 'over inflate' initial demands.

Significant time and resources can be wasted in arriving at solutions which are at best only partially acceptable to the parties.

### 2.2.2 Creative responses

In sharp contrast to the other active responses, creative responses are those that are usually characterised by integrative bargaining.

In this process both parties are encouraged to cooperate in joint problem solving. The emphasis is on identifying creative and workable solutions which can satisfy the needs and dispel the fears of the parties involved.

This response contrasts sharply with the other aggressive responses where maximum effort is usually directed towards persuading or forcing some of the parties to modify their adopted positions.

Instead of directing resources and effort towards the defence of positions, the parties concentrate on developing a wide range of possible solutions (i.e lateral thinking).

If both parties can be satisfied that their needs can eventually be met, it is far more likely that they will be prepared to modify their adopted positions.

Even if it eventually becomes clear that the needs of all parties cannot be fully met, a cooperative climate will have been established. In this environment the probability of achieving a satisfactory resolution to the conflict will have been greatly increased.

## 3 Adopting an appropriate response

The adoption of an appropriate response is crucial if the project manager wishes to effectively resolve a conflict with the minimum disruption to the relationships or the project.

Most major conflicts develop from relatively insignificant issues which were not identified and/or correctly responded to when they first occurred.

For this reason passive responses such as denial, avoidance or premature capitulation should be avoided and actively discouraged by project managers.

Similarly aggressive responses such as domination and distributive bargaining should be discouraged wherever they could be detrimental to relationships or the project.

Informal resolution processes such as negotiation and/or mediation should preferably be the resolution processes initially adopted. Where possible these processes should be encouraged as a mechanism for integrative bargaining rather than a search for compromise solutions.

Providing they are constructively used, significant areas of common agreement can usually be identified. This can have the effect of defusing much of the hostility and tension that has accumulated.

If areas of the conflict remain that cannot be resolved through further negotiation, these can often be settled if the parties will agree to abide by the findings of a mutually agreed independent expert.

There is little point in undertaking protracted negotiations in situations where large power imbalances exist or some of the parties obviously have no desire or interest in resolving the conflict.

In such cases formalised processes such as arbitration or even litigation may eventually be required to settle conflict issues.

It should however be appreciated that settlement of a conflict in this manner is not usually synonymous with conflict resolution and often results in the total destruction of any longer term working relationship.

## 4 Conclusions

The ability to effectively identify and respond to conflict is a crucial requirement for successful project management.

Ultimate success or failure in achieving project goals can often depend on a project manager's ability to identify the causes and to respond appropriately.

## 5 References

Boulton, R. (1982), **People Skills**, Simon Shuster, Sydney.
Cornelius, H. (1989), **Everyone Can Win**, Simon Shuster, Sydney.
De Bono, E. (1982), **Conflicts, a better way to resolve them**, Harrap, London.
Fisher, R. (1981), **Getting to Yes**, Business Books, London.
Zikmann, R (1990), **Conflict Management in the Commercial Environment**, University of New South Wales Report.

# CONSTRUCTION CONFLICT - THE SPECIALIST CONTRACTORS VIEW

RONALD S. DAVIES
Federation of Associations of Specialists & Sub-contractors (FASS), Stevenage, England

Abstract
Umbrella organisations are in a prime position to be able to assess the issues that face construction professionals, since they see all sides of the problem at one time or another. Hence this paper attempts to trace the development of the construction process from the pre 1939/45 period through the present day and to look to possible future developments, particularly so far as both contractual issues and the settlement of disputes is concerned. Whilst looking at the position within the industry created for Alternate Dispute Resolution, it seeks to suggest how such procedure could be encompassed within a future framework.
Keywords: Deterioration, Professionalism, Education, Conflict, Settlement.

## 1 Introduction

FASS, being an umbrella organisation representing a significant number of specialists in the construction industry, is in a prime position to comment upon the current, and future, role of the specialist contractor in the present day construction process. Our industry has moved from a position where the architect controlled the whole process, through various developments to the present day when perhaps the architect is the person least suited to have ultimate control of the contractual scene.

## 2 Pre-world war two to the present day

The industry has moved over the years from a position where considerable mutual respect between professionals, contractors and sub-contractors resulted in well designed buildings being constructed to a relatively high standard and without conflict, to a position where the complexities of design, not only of the structure but of the services within it, has led to many designers being involved, to specialist contractors taking on all manner of additional responsibilities so far as design is concerned, whilst planning and programming techniques have developed considerably to establish control and progress of the works which had hitherto not been considered necessary. All this has let to a proliferation of working parties, working groups, committees and sub-committees, all with the same aim, that of preserving the interests of particular sector groups, be it professionals, employers, contractors or sub-contractors.

The method of execution of the works has also altered considerably from the position of the contractor who employed directly almost all of the tradesmen necessary to complete the works to the present day position where, as is now generally accepted, something over 90% of the construction process is carried out by specialists or sub-contractors. Indeed, the contractors site staff are often confined to merely operatives involved in unloading and generally servicing the specialist trades.

Recent years have seen changes such as were not even contemplated by our predecessors, hence the proliferation of contract systems that we have encountered over recent years. Prime Cost, Design & Build, Management Contracting and now Construction Management. Each brought their own contract forms, some structured for contracts with quantities, some without. Of course, following in the wake of all these sub-contracts were all the sub-contract documents and yet more complications added in the shape of nomination, naming, pre-naming, to identify but a few.

But even the foregoing was not enough, for in addition others, who for their own reason, be it dissatisfaction with the present forms or merely that they thought they could do better, sought to confuse the issue by the introduction of their own "pet" forms and so we saw the publication of yet further documentation in the form of contracts designed and published by the British Property Federation, The Association of Consultant Architects, The Faculty of Architects and Surveyors and so the list of available options grew.

The matter was made yet more complex by the actions of contractors in recent years, for despite the fact that Joint Contracts Tribunal had produced, after much deliberation and negotiation, forms of sub-contract to cater for particular instances and that the specialist umbrella organisations, in conjunction with contractors, had produced standard sub-contract forms, the contractors by and large decided that these forms were not to their liking and spurred on by the strength they had gained in the path of the ensuing recession in construction, they in the main decided to ignore standard forms and publish their own individual forms, or alternatively produce numerous pages of amendments to those standard forms which rendered them unrecognisable.

And so the scene was set by the early 80's for the arrival in strength of the legal profession. Since the mid 80's it seems that the industry has been in the grip of both lawyers and accountants and in many companies the "professional" contractor has been forced to take a back seat.

### 3 How does conflict arise?

It seems therefore that in the early 80's the industry was ready to welcome the lawyers. Everything about the contractual scene seemed to be ready for them, there was a proliferation of contract and sub-contract documents of all shapes and sizes and only the Joint Contract Tribunal documents had, in the main, been "tried and tested". Case law already abounded, but there was little or nothing 'tried' so far as the non-tribunal forms were concerned. In addition the latter part of the 80's saw a take off in the requirement for Bonds of various types and more latterly the collateral warranty hit the contractual scene.

Gone, finally, were the days when the Quantity Surveyor and the Architect could complete the task of mediation between the feuding parties to the contract, the quasi judicial role of the architect seemed in the main to have slipped quietly away.

To repeat therefore, how then does conflict arise in the modern building contract? The reasons seem to be many, but the following are probably the principal causes of problems.

Firstly, in the traditional type of contractual set up there is now little time for proper consideration to be given in the design stage to provide sufficient detail to permit the contractors to properly assess the implications of design etc upon their programme and thereby upon the ultimate financial outcome.

The effect of the current economic situation has been to cause two significant problems.

1 To cause designers, of whatever persuasion, to restrict the extent of their work at pre-tender stage, thus fairly positively ensuring the failure of SMM7 which, in the main, has just not worked.
2 To cause contractors, be they main or sub-contractors, to be encouraged to submit prices which cannot but cause either financial problems when the

final outcome is known, or to bring additional financial muscle to bear on specialist and sub-contractors to reduce margins to a dangerously low level.

Secondly, the training in both the professional and commercial sectors has severe shortcomings. In the professional sector training is concentrated in areas which largely ignore the contractual and management spheres and an architect who has effectively little formal training in the managerial scene, is suddenly expected to control perhaps a multi-million pound contract.

In the commercial scene of construction there are four principal training areas namely:

Architectural
Engineering
Quantity Surveying
Accountancy

and with various sub-divisions in each category. There is however too much specialisation in practice. This means that within construction the real leaders tend to emerge from either the engineering or quantity surveying areas.

## 4 Future development

Having witnessed failure from the various systems experienced over previous years it seems likely that the current atmosphere lends itself to the latest and most interesting of contractual systems, that of Construction Management. This area will not however develop without considerable thought and input from the various sides of industry.

In this regard we have to firstly decide from where this new animal, the Construction Manager, is likely to emerge. Certainly the training of such a person will have to be developed to provide the right knowledge and ability to properly administer contracts. It is unlikely that such an administrator would come from the architectural scene and even less likely that he will will be an accountant.

Training needs will have to be thought out very carefully and further thought should be given to the background knowledge that will be necessary for such a task. It may be that the additional skills would be obtained form a second degree.

The knowledge required will encompass, construction, contractual procedures and law relating to building contracts, ability to appreciate the financial implications upon client and contractor, ability to programme and effectively control not only the construction process, but also the total project from initial conception to settlement of the final account. Finally, leadership qualities of a very high degree will be essential in such a person.

## 5 The Contractual position

It would be our view that the contractual position would be one that should exist as between the client and specialist contractors executing the various works packages.

It is essential that the contract between these parties should be fair and even handed, unlike some of the contracts currently imposed upon specialist contractors. Careful attention should be given to the management of disputes throughout the duration of the contract. There is nothing new about such a concept for quantity surveyors in general, and indeed architects and contractors, have been in many instances practicing Alternate Dispute Resolution without the assistance of the legal profession for many years.

The contract should therefore include means of managing and settling disputes as they arise and it is clear that adjudication should play a significant part in such a contract. It should not be restricted to set-off problems as is presently the situation in various sub-contracts, but it should be enlarged to encompass all the problems that can be envisaged as arising, within a contract, in which adjudication could be of assistance.

Finally, it is essential that the activities relating to adjudication should be designed in such a way as to keep the activities of lawyers to an absolute minimum. For what we are proposing is very akin to an Alternate Dispute Resolution, but if one looks at the present ADR scene it is disturbing to see a predominance of lawyers yet again and what is more likely to ensure the failure of a scheme than that.

# CONTINGENCY MANAGEMENT OF CONFLICT: ANALYSIS OF CONTRACT INTERFACES

D.A. LANGFORD, P. KENNEDY and J. SOMMERVILLE
University of Strathclyde, Glasgow Polytechnic,
Glasgow College of Building and Printing, Scotland

Abstract
This paper discusses the sources of conflict which have been found to exist in a range of procurement methods in current use within the construction industry. It considers the relationships between between main contractors and sub-contractors and between separate trades contractors. The variables which may influence the contracting organisations and the key players engaged in the execution of construction projects are identified. A model is proposed which may indicate the relationships between the variables identified. This model may be used as a diagnostic tool to evaluate interfaces which historically have given rise to conflict and forecast possible future conflict zones in novel procurement methods.
Keywords: Sources of Conflict, Procurement Methods, Conflict Variables, Sub-contractors.

## 1 Background

Conflict between contracting companies may be seen as an inevitable by-product of organisational activity. Each contractor has conflicting aims. The management of every contracting company has to consider the short and long term goals of their company against the objectives of the management teams of each of the projects they are servicing.

In the last ten years, the construction industry has been undergoing profound change in the way in which it procures work. These changes have been ushered on predominantly by clients who have been dissatisfied by their experience of the building process. These very changes have resulted in greater fluidity in the roles undertaken by construction organisations. For example, the large traditional main contractor has been drawn to provide management services for clients whilst the smaller medium sized contractor may be identified as a sub-contractor or specialist contractor dependant upon the procurement system chosen on large projects.

On large projects, medium sized companies, used to being main contractors in their own right, may be called upon to perform as the civil engineering sub-contractor or the builderwork sub-contractor. In Construction Management contracts the usual configuration is for every contractor to be engaged directly by the Client. In large complex projects where the services element is substantial, there may be a strong case for the services contractor to be the main contractor and the building trades contractor to be merely a sub-contractor constructing a building in which to house the services.

Contractors will be required increasingly to act in a capacity which is outside their traditional role and the means by which they previously managed the conflict between themselves and others engaged in the building project may no longer be appropriate or effective.

A study by Dodd and Langford (1990) of a major building project which was procured by means of Construction Management has shown some likely areas of conflict which arise due to this ambiguity of role. This ambiguity, it must be said, was not due to poor drafting of contractual arrangements but due to the perception of roles played by the individuals and the contracting firms themselves. Initially, trades contractors who normally performed as main contractors felt the urge to manage activities outwith their own packages of work; specialist contractors who normally worked as sub-contractors adopted a passive mode as if waiting to be managed. It was also noted that the Construction Managers, with staff drawn from a main-contracting background, had to 'unlearn' habits of a lifetime and not allow themselves to be drawn into managing in detail the work of the trades contractors. Hence the new procurement methods have provided fresh arenas where conflict may be expressed. This paper seeks to identify the sources of this conflict and proposes a research model to analyse conflict.

## 2 Sources of conflict in traditional procurement routes

Before identifying the variables influencing conflict, it is worth examining the way in which conflict is managed under traditional contractual arrangements. Conflict resolution by 'dominance' is a well tried and tested method exercised by many main contractors but it relies on the use of power; the power to deny or delay payments, the power to withhold new contracts, the power to levy damages, etc. When the status of a traditional trades contractor changes in innovative contracts and with it the balance of power between contractors, then dominance may no longer be a viable strategy and other means must be sought to manage the conflict. Effective management of any project requires the formation and development of teamwork. In traditional contracts the concept of teamwork may be somewhat constrained by the contractual and legal framework by which the participants are bound together.

Conflict is relatively common between the main contractor and his domestic sub-contractor (Harding, 1991), usually related to disagreement over the payment of extra items, late payment of interim instalments, etc. These disputes can be acrimonious and are normally solved by the dominance of the main contractor or by the legal process. This is a situation which is increasingly being experienced by sub-contractors working for Management Contractors.

The relationship between nominated sub-contractors and main contractors tends to be of a different hue. The zones of conflict tend not to involve late payment, because the sub-contractor has some protection by virtue of his special relationship through nomination, but conflict can arise when a main contractor attempts to compensate for his own lack of progress by shifting the responsibility for delay to the nominated sub-contractor and thereby claim an extension of time.

Conflict between domestic sub-contractors is of considerable interest because it involves parties of relatively equal stature who have no contractual relationship with each other. Dominance as a method of resolving conflict is of little use in this situation and therefore the parties are required to negotiate, compromise and generally communicate at site and director level to resolve difficulties.

It would appear that a form of teamwork is experienced by domestic sub-contractors in a traditional contract, but this is adhoc and unstructured. The only structure which may exist is where the main contractor actively attempts to co-ordinate their activities. A similar pattern of co-operation was observed by Dodd and Langford (1990) between trades and specialist contractors operating under Construction Management. Here however, a distinction could be drawn between trades contractors and specialist contractors. The trades saw their roles as being diminished and found co-operation more difficult whereas the specialists, being able to make an enhanced contribution to decision-making, were more positive and showed themselves to be more adept at handling these **'democratic'** relationships.

Teamwork therefore is a function of the contractual option which is being utilised by the Client (Baden Hellard, 1988). Some options foster a more adversarial relationship others a more co-operative one (Grout, 1991).

## 3 Variables influencing conflict in innovative procurement methods

The changing procurement scene of the last ten years has redefined the arenas for conflict. Research carried out by Harding (1991) and Dodd and Langford (1990), have identified three variables influencing the form and extent of conflict.

The variables are:

Ambiguity of role
Interpersonal skills of key players
Responsiveness to change

## 4 Ambiguity of role

Uncertainty in a construction project tends to lead to conflict. Mention has been made above to the unfamiliar role which some contractors find themselves having to resolve, but this they can learn with experience. A major source of conflict between contractors or between sub-contractors is at the boundaries of the work packages. Sometimes it is the gap between packages which causes uncertainty when it becomes apparent that, due to an oversight by the Construction Manager or the Management Contractor, vital work is not being carried out because it was not specifically included in a work package. Such problems tend to be more prevalent in the fast-track forms of procurement.

Attitudes to responsibility are subject to change with newer forms of procurement. In traditional main contractor/sub-contractor relationships the sub-contractor sees no need to identify with the Client - the sub-contractor has no contract or communication lines and consequently the responsibility to the project, in terms of progress and quality, only goes as far as the fulfilment of obligations to the main contractor. If, on the other hand the sub-contracting organisation is directly employed by the Client it is more likely to identify with him and as Dodd and Langford found it takes on a much higher level of responsibility, perceiving the Client's goals as its own. If each of the package contractors does likewise the motivational effect of this identification of common goals by all concerned would eradicate a significant source of conflict.

## 5 Interpersonal skills of key players

People are the principal resource on any project and the dynamics of the interpersonal relationships should be considered. The two main factors which influence this relationship are personality and environment. Psychological tests such as FIRO-B and the Myers-Briggs Type Indicators(Myers and McCauley, 1985) may be used to gain some insight into an individual's personality. Dodd and Langford (1990) and Langford and McDermott (1984) used the FIRO-B test with construction managers.

The FIRO-B uses three dimensions of personality **Inclusion** - that is how the desire to be included and include others in a group; **Control** - how controlling was the construction manager's behaviour: and **Affection** - how the manager seeks out or expresses affection.

Each of these dimensions can be measured as High, Moderate or Low. From a study of construction managers a typical personality profile is shown in Fig.1

| | INCLUSION(I) | CONTROL(C) | AFFECTION(A) |
|---|---|---|---|
| Expressed behaviour(Ex) | M | H | M |
| Wanted from others (Wa) | L | L | M |

M - moderate score
L - low score
H - high score

Fig.1. Typical personality profile of construction manager

From the point of view of conflict management the most interesting characteristic relates to **control**. Two individuals with high expressed control scores are extremely competitively incompatible on the originator compatibility scale. If the managers of trade or specialist contractors are required to interact with each other and both parties to the interaction have high expressed control scores there is the basis of a power struggle and hence conflict. If, however, one has a 'democratic' profile, ie. with a moderate control score for expressed and wanted behaviour then the two parties are more likely to be able to interact without conflict. A mixture of these personality types would help to reduce conflict on sites but bringing this situation about poses some practical difficulties. Assuming it were possible to establish the personality profile of each manager at each hierarchical level of each contracting company it is unlikely that a contractor would respond positively to the suggestion that, because the Construction Manager has a 'autocratic' personality type (high control score) the contractor should appoint a more 'democratic' (moderate - low expressed control score) manager to help reduce friction.

The understanding of these human relationships is useful in helping to identify possible areas of conflict between individuals and either re-allocate the conflicting personnel before conflict arises or change the environment in which they operate.

## 6 Responsiveness to change

Managers in the construction industry with their baggage of attitudes and behaviours are products of the industry in which they have been nurtured (Dorch, 1989). When a company's role changes from that of main contractor to sub-contractor or an equal contractor with many others under Construction Management, these attitudes and behaviours may pose a barrier to effective relationships within the new environment. It is necessary for the managers to unlearn these habits and indeed for company cultures to change to enable companies to survive in modern procurement options. This unlearning may be more difficult for large well established trades contractors with a main contracting history for much of their perceived status derives from pride in the list of impressive projects for which they claim total credit for constructing and a belief in themselves that they are better than their competitors. Indeed these qualities are still important as they bid for work under certain procurement options, so a schizophrenic outlook may be useful for these contractors.

Specialist sub-contractors need less re-education as they have less historical baggage to carry and are used to working with work packages. They are also skilled at working in a cooperative manner with other specialists at the boundaries of their packages without the use of contractual power.

Communication is the core of the working environment. Communication is primarily a function of organisation;

> '...the pattern of relationships and the divisions of responsibility in any building team have much more effect on the way communications function than have any particular aspects of the techniques of communication themselves.' Higgin and Jessop (1963).

If a framework for effective communication has been planned in advance of commencement on site then conflict should be reduced.

Some argue that the end result intended by communication is behavioural change, ie. getting something done. Techniques of communication which may be adopted to bring about this objective involve direct interaction. It is particularly important therefore, in fast-track projects or work undertaken using modern procurement options where several trade and specialist contractors are mutually reliant on each other but not bound together contractually, that meetings form the base of the communication system. Apart from the obvious use of meetings to pass information, co-ordinate activities, etc. it allows for the use of group dynamics to apply pressure on participants to conform to goals set by the group.

## 7 Modelling the variables

The dependant variable is the amount of conflict generated and, in earlier work, Langford and Bradley (1987) created an index of conflict which scored the quantity and level of hostility related to communications about construction claims.

Using the same format, the future research could test the relationship between the variables identified and the level of conflict experienced in projects using innovative procurement methods.

It may be theorised that the dimensions of role definition, social skill and responsiveness to change be modelled as in Fig. 2.

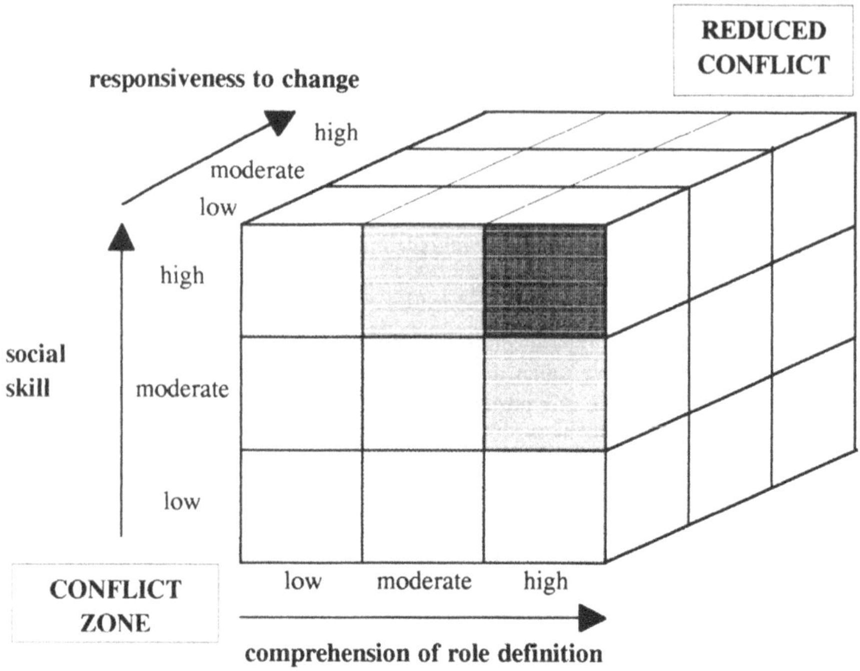


Fig.2. Matrix of conflict variables

The three dimensional matrix would contain 27 cells and one may hypothesise that highly conflictual projects would fall into low social skill, low comprehension of role definition and unresponsiveness to change cell whilst more adaptive organisations staffed by highly skilled managers with a clear view of their role, would avoid conflict.

Case studies will need to be carried out to test these polar positions.

## 8 Conclusion

Conflict between contracting companies may be inevitable and it can take different forms depending on whether the firm is trade or specialist. It is possible for informed project management to anticipate zones of conflict; between contract management and separate contractors, between main contractor and sub-contractors, between sub-contractors or between individuals critical to effective progress. Having identified conflict zones managers must increasingly turn to methods other than dominance to resolve these conflicts. It is clearly better to avoid conflict by planning for it rather than react to the consequences of it. It would appear that teamwork may be an effective way forward but only if the members are bound together by mutually set, internalised goals rather than by contractual arrangements alone.

## 9 References


Baden Hellard, R. (1988) **Managing Construction Conflict**, Longman, Essex, 10-12.

Dodd, J. and Langford, D.A. (1990) Construction management on one large project in London: A case study. **Construction Management and Economics**, 8, 385-398.

Dorch, T.C. (1989) Job Person Match, **Personnel Journal**, June 1989, 49-57.

Grout, B. (1991), Battle Front, **New Builder**, 21 November 1991, 26-27.

Harding, C. (1991), Building Without Conflict, **Building**, November 1991

Higgin, G. and Jessop, N. (1963) **Communications in the Building Industry, The Report of a Pilot Study**, Tavistock Publications, 37.

Langford, D.A. and Bradley, P. (1987), Contractors Claims, **Building Technology and Management**, June/July 1987.

Langford, D.A. and McDermott, P. (1984), **The sources, causes and effects of variations in Building contracts**, unpublished report, Brunel University.

Myers, I.B. and McCauley, M. (1985) **Manual: A guide to the development and use of the MBTI**, Consulting Psychologists Press, London.

# AVOIDING CONFLICT BY RISK MANAGEMENT - THE ROLE OF THE CLIENT'S PROJECT MANAGER

J. LEWIS, D. W. CHEETHAM, D. J. CARTER
School of Architecture & Building Engineering, University of Liverpool, England

Abstract
This paper discusses a source of conflict in construction - the failure by a party to a project to make adequate allowance for the risks involved. The methodology of Risk Management is briefly outlined, and the central role of the client's Project Manager in applying these techniques is discussed. The hypothesis that Risk Management is the primary role of the Project Manager is proposed. This is illustrated by two case studies.
Keywords: Risk Management, Project Management, Conflict Avoidance.

## 1 Introduction

Risk and uncertainty is inherent in all construction projects. Should these risks actually occur, they may have an impact on on the cost of the project, its duration or the quality of the finished product (or, indeed a combination of them). All parties to a project, be they clients, consultants, contractors, sub-contractors, financiers or tenants, clearly have an interest in the successful completion of a project in a manner satisfactory to each of them.

Each of the parties to a project will take a different view of risks. Although this paper considers risk and its management from the client's point of view, the identification and consideration of risk early in the life of project will lead to a more satisfactory project for all parties. Risks may not affect all of the parties - the design team or contractor would not normally be penalised for a developer's failure to let a building on completion. The ability of parties to control the likelihood and impact of risk will vary - nobody can control the weather, but the contractor is able to take steps to enable production to continue during inclement weather and, may be able to recoup costs or claim extensions of time from the client.

Methodologies for dealing with risk and uncertainty have become known as Risk Management. The authors' definition of Risk Management is given below.

> 'Risk Management involves the identification of the particular, significant risks which may impair the performance of a specific project. It requires the assessment of the effect of these risks on the project and the establishment of policies for dealing with them. These policies may include: clear allocation of risk to the various parties, determination of appropriate time, cost and quality allowances for risks ,whether transferred or retained, and steps to reduce the likelihood, magnitude and impact of risks.'

The process and techniques of Risk Management will be discussed in more detail below, but it is sufficient here to assert that a project on which Risk Management has been effectively applied will be one where the potential for conflict is small. This paper will support this view by presenting case studies illustrating the role of the Project Manager in managing risk.

## 2 Risk and conflict

### 2.1 Risks in construction

The risks inherent in any project will arise from a variety of sources and vary both in likelihood of occurrence and in potential impact on the success of the project. For example, it is virtually certain that inflation will increase the cost of a project during its procurement, but, currently, the effects of inflation are relatively low and predictable. On the other hand, loss or extensive damage to works during construction caused by fire, will have a major effect on the project (even though the cost of reinstatement may be covered by insurance, the duration will be increased), but is mercifully rare.

Seven broad categories of risk may be identified - physical, construction, design, political, financial, legal and environmental. A sample of typical risks in each category is given in Table 1. A more complete listing of risks may be found in Perry and Hayes (1985a). It will become apparent from consideration of the categorisation, that it is arbitrary and that there is considerable overlap between some categories. Nonetheless, it provides a framework within which the risks associated with a given project may be considered.

### 2.2 Contractual complexity

Building construction management is concerned with providing acceptable quality buildings which satisfy client's requirements, on time and at the right price. Those engaged in this endeavour are all concerned with some

Table 1. Typical risks affecting construction projects

| Risk category | Typical risks |
|---|---|
| Physical | Fire, flood, earthquake... |
| Construction | Weather, industrial action, ground conditions, quality and availability of labour/materials, site safety ... |
| Design | Buildability, quality of brief, innovative applications... |
| Political | New legislation, war, embargo... |
| Financial | Insolvency of client, contractor sub-contractor or supplier, inflation, inability to let or sell on completion... |
| Legal | Liability to third parties, legal differences between countries, liquidated damages... |
| Environmental | Delay through public enquiry, pollution, environmental damage... |

aspect of the three criteria for judging project success - time, cost and quality.

The legal and organisational relationships on most projects are extremely complex. A typical set of relationships for a 'traditional' contract is shown in Figure 1. Despite the growth in use in recent years of alternative methods, the traditional procurement route still predominates in the U.K. building industry [Franks (1990)]. Such complex relationships mean that communication links between some parties are tenuous or non-existent. Furthermore, this complexity does little to foster a cooperative, 'team' attitude to the production of buildings. In fact, some would argue that it causes and sustains the adversarial nature of construction projects. Parties seek to gain advantage (financially or otherwise) at the expense of others, with the supposedly common cause of building production merely providing the medium for such struggles.

One method by which parties seek to gain advantage is by amending 'standard' forms of contract to include onerous conditions. Few groups can claim to be innocent in this respect; clients pass inequitable risk onto contractors, contractors enforce 'pay when paid' clauses on sub-contractors etc. A consequence of onerous contract conditions is that some risks tend to be passed, without consideration, further and further down the line, often

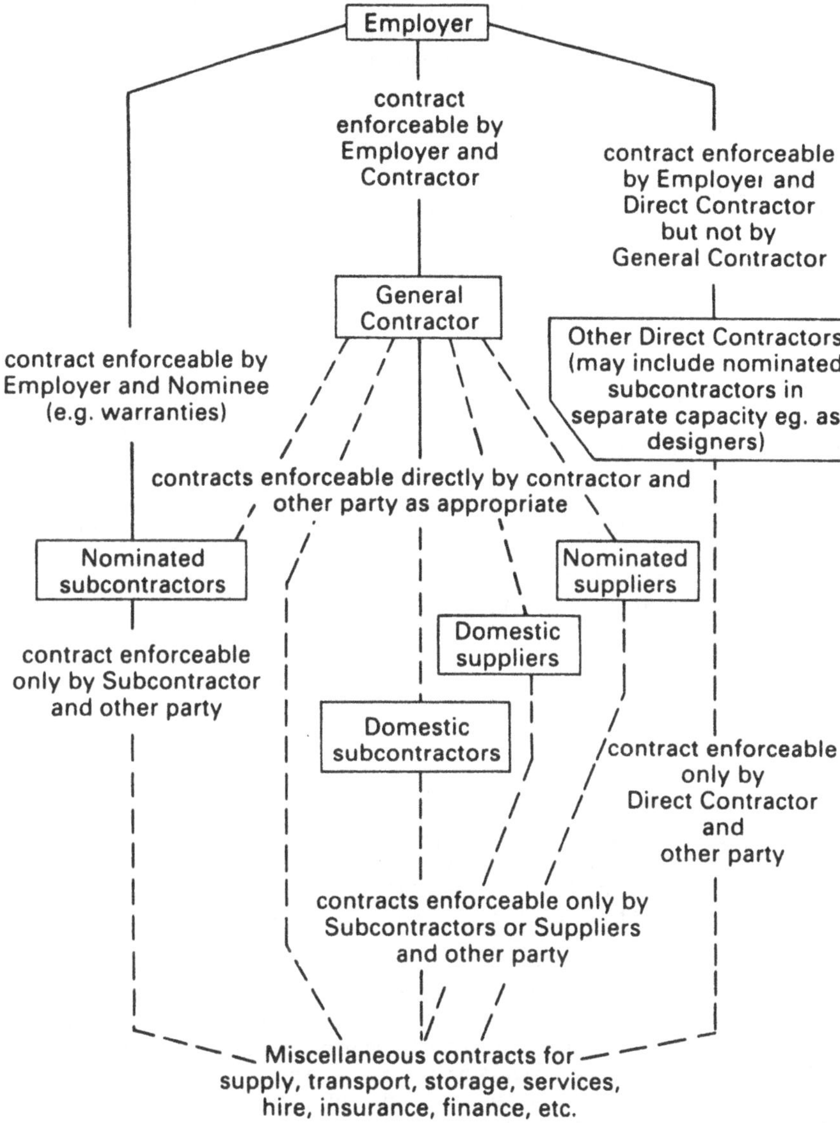


Fig. 1. A typical pattern of contractual relationships

resting with the smallest parties involved in the project. From an ethical standpoint this may be unfair, less altruistically, it may be unwise. The smaller parties to a project (sub-contractors, material suppliers or even individuals) may well be the least able to appreciate the magnitude of the risks they are accepting or least able to control them, certainly they will be the least able to withstand their effects. Yet the financial failure of a small sub-contractor, due to inequitable risk loading, can cause considerable difficulties and loss to parties higher up the contractual chain.

Contractual complexity can make the application of Risk Management more difficult. Decisions as to the equitable allocation of risks and which parties can most effectively control the likelihood and impact of risk become much more awkward when the client has no formal relationships with many of the parties. Accordingly, Project Management must place much emphasis on the choice of an appropriate contract and procurement strategy as well as the application of Risk Management methodology.

## 2.3 The causes of conflict

Risk and uncertainty can result in conflict between the parties to a project when the following conditions arise:-

a) one of the potential risk events occurs,**and**
b) one or more of the parties suffers some loss as a result of it,**and either**
c) the damaged party had not identified the risk as relevant to the project,**or**
d) the risk was identified but insufficient steps were taken to mitigate its effects,**or**
e) the allocation of risks between the various parties to the contract was not clearly established in the first place.

In these circumstances the damaged party will seek to redress their loss, and the result will often be conflict and dispute. Claims and disputes can damage all parties to the construction process, in the words of the report 'Building Britain 2001' [CSSC(1988)]: '...claims have attacked British industry like a cancer.' The link between risks and contractual disputes is also supported by other workers findings [e.g Perry(1986)].

Clearly, poorly managed risks are not the only cause of claims and disputes. For instance, it is widely accepted that in times of low workload, when tender prices are reduced in order to win work, contractors will be more 'claims conscious'. Some will devote considerable management effort to identifying and pursuing claims. This is unsatisfactory for it creates uncertainty in both the client's expenditure and the contractor's income. However, irrespective of the causes of claims, Risk Management, with its emphasis on the early identification of potential problems and their solutions, can reduce the magnitude and number of claims.

It should not be imagined that problems from claims and disputes will only afflict other people's projects, or that they only occur on jobs which have little management effort expended on them. Carter et al (1990) refer to claims and disputes suffered by Regional Health Authorities (RHA)on large hospital building projects. RHAs are experienced building procurers, with well established systems for the management of building work - this did not prevent them from falling victim. More recently, the delays and disputes on the channel tunnel project [John (1992)] provide further illustration that even the most prestigious and intensively managed projects suffer from disputes.

## 3 Project Management and risk

### 3.1 Project management

Major investigations into the performance of the U.K. construction industry [NEDO (1975), NEDO (1983), NEDO (1988)] have advocated the use an integrated management system for building procurement - Project Management - as a means of improving time, cost and quality performance. Such improvements also serving to reduce the level of claims and disputes.

The definition of Project Management provided by the Chartered Institute of Building [CIOB (1988)] is considered most apt:

> 'The overall planning, control and coordination of a project from inception to completion aimed at meeting a client's requirements and ensuring completion on time, within cost and to required quality standards.'

It should be noted that the role described in this definition goes far beyond that of contractors' site and contract managers who are commonly termed 'project managers', to embrace the entire building procurement process.

### 3.2 Risk management in project management

The duties and responsibilities of the client's Project Manager have been well described by a number of authors [e.g.: CIOB (1988), Rougvie (1987)]. The roles of the Project Manager in planning, co-ordination, communication, contract administration and leadership are clearly identified. Systems for fulfilling these roles have been described by the many authors [e.g.Bennett [1985], O'Neill [1989]). However, little consideration has been given to the responsibilities of the Project Manager for the assessment and control of risk and uncertainty.

General texts on the procurement process and contract

administration [e.g.Aqua Group (1990a) & (1990b) & Franks (1990)] acknowledge, explicitly or implicitly, that the management and apportionment of risk between parties is important, especially when considering contract conditions. However, they devote little space to methods of managing risk or the responsibility for applying them.

This paper argues that the majority of the Project Manager's duties fall within the definition of Risk Management, and as such, Risk Management is the most important of the Project Manager's duties. Furthermore, because of its importance, a deeper study of Risk Management will enable Project Managers to perform their duties more effectively and discharge their responsibilities to their clients by minimising conflict in the projects that they administer.

## 4 Risk management methodology

### 4.1 When to apply risk management

Risk Management techniques should be applied as early as possible in the course of a project when the ability to manage and control is greatest. Also, because the type and magnitude of risks and the nature of the project itself may change throughout its life, regular reviews of risk should be carried out.

Perry and Hayes (1986) confirm the above, citing the initial project appraisal stage as an important time for the identification of risks and the development of broad policies for risk response. They proceed to identify two further stages in the life of a project at which Risk Management can play an important role. Firstly, the time at which proposals are submitted for the client's approval to proceed. At this stage they cite the development of a contract strategy and the consideration of technical responses to risks as being of importance. Secondly, at the receipt of tenders, though the major decisions on risk control and allocation have already been made, Risk Management principles can still be used to improve the selection process. It is unlikely that all of the contractors tendering will have identified and assessed all of the risks involved. Additionally, the need for changes to the project may affect individual tenders differently. The consequence of these two factors is that the lowest tender may not necessarily result in the lowest final cost to the client. Due allowance when formulating budgets and judging tenders must be made for these effects.

### 4.2 The client's role

Uncertainty will be greatest at the inception of a project

and decisions made then will be subject to most risk. Furthermore, it is at the inception of a project that policies and management structures to control risk can be most easily and effectively established. Since the client is responsible for project inception, it is the client who must be the prime mover in the adoption of Risk Management on any project. The effective application of Risk Management to a construction project requires extensive knowledge of the techniques and organisation of construction. Therefore the client will be reliant on his or her principal advisors - typically this role will be filled by the Project Manager.

### 4.3 The phases of risk management

The process of Risk Management has generally been considered to consist of three phases - Identification, Analysis and Response [Perry & Hayes (1985b) and Hayes et al(1986)]. Though Fellows (1989b) adds a fourth stage - Allocation - which other authors have included within Response.

#### 4.3.1 Risk identification

This phase involves the consideration of the specific nature and circumstances of a project in order to assess the uncertainties which may result in the inability to achieve optimum performance. The principal task is to establish which risks represent the most serious threat to the successful performance of the project. Clearly, each project is unique, but 'standard' lists of risks provide valuable sources of reference and prompt a more profound consideration of the project. Also, identification of risks implies that some assessment, at least qualitative, is made of the likelihood and potential impact of each risk.

Risk identification is the most important of the three phases [Perry & Hayes (1986)] for Project Managers. The subsequent stages will only be effective if the relevant risks have been identified. Identification provides the opportunity to think through the project and consider problems and how they may be solved. A rigorous analysis with numerical assessment is usually only performed for large and complex projects.

#### 4.3.2 Risk analysis

Risk analysis involves the quantification of the effects of the identified risks on the project. Several techniques, of varying degrees of sophistication, are available for this purpose. The following paragraphs give an outline of some of the techniques, more detailed discussions of these techniques may be found in the references cited in section 4.3 above.

The simplest technique is sensitivity analysis. In this

technique, a range of values for each risk variable is estimated and the effect on some performance parameter of the project (e.g. capital cost, rate of return on capital, duration etc) is assessed across the range. Sensitivity analysis is mathematically straightforward and focuses management attention on the important risks. However, the technique has some deficiencies: each risk variable is treated separately and no attempt is made to examine the effects of interacting or combining risks or to take account of the different likelihoods of occurrence of risks.

Probability analysis seeks to address these defects. In this technique probabilities of occurrence are assigned across the range of each risk variable (e.g. by defining a normal distribution). A random selection is made from each distribution to determine a level of each risk. The impact of each risk variable on one of the performance parameters is then assessed. The effects of each risk can then be summed to produce one possible total outcome for the parameter. It is necessary to repeat the process several times, with different random selections, in order to determine the range and distribution of the possible outcomes for the parameter. The large number of numerical operations required in this technique means that it can only practically be performed using a computer.

Given the above, it is perhaps not surprising that several researchers have produced specific software for probabilistic risk analysis of construction projects. CASPAR [Hayes et al(1986) and Thompson(1986)] and Construction Project Simulator [Bennett(1985)] are two such programs. Both are based upon a time network of a project onto which cost and financial risks may be grafted. Importantly, these programs can produce an holistic view of the effects of risk on a project. For example, since some costs are time related (e.g. preliminaries), time delays can themselves result in cost increases.

Other techniques which have been proposed for risk analysis are decision trees and 'Delphi' analysis. The former method has had limited application in construction; a brief description and sources may be found in Hayes et al(1986). The latter technique [Fellows(1989a)] involves the iterative consultation of experts for their opinion of project outcome (and is hence named after the Delphic oracle).

### 4.3.3 Risk response

The response to identified and quantified risks can take many different forms. On any single project several different responses may be adopted, depending on the type of the risks identified. When deciding between alternative forms of response two basic decisions must be kept in mind: whether to avoid or reduce risk and whether to retain or transfer risk.

In the extreme, complete avoidance of risk may mean the abandonment of a project. More probably, however, risk avoidance and the reduction of risk probability and impact will require a technical response. Such a response may include a change in design or further development of the design, or the acquisition of more complete information, e.g. by undertaking or extending the site investigation.

The principal medium for transferring and allocating risks is the building contract. The standard forms of building contract in common use [e.g. JCT(1980)] involve both explicit and implicit allocations of risk between parties. It is not surprising that researchers in construction risk management have devoted much space to the choice of procurement route and formulation of contract documents [e.g. Hayes et al(1986)]. The general conclusions of these investigations are as follows:

a) Contracts should allow the explicit allocation of specific identified risks between parties.
b) Contracts should include incentives for risk control and sound management practice.
c) Risks should only be allocated to the parties who are best able to control them and/or sustain their effects.

Another common method of risk transfer is through insurance. This insurance may cover the physical works during construction (e.g. the insurance required by most forms of contract covering the works for loss or damage), the building in use (e.g. Latent Defects Insurance) or the performance of the design team (e.g. the Professional Indemnity Insurance which most clients require their consultants to carry).

Whether the choice is to retain or transfer risk, there will be a cost implication. This cost may take the form of an insurance premium, a payment to the contractor for risks accepted or a cost allowance for risks retained. Whatever the option chosen for a particular risk, the process will have quantified the likelihood and effects of that risk, allowing the adequate and equitable assessment of payments and allowances.

## 5 Case studies of risk management

The two projects discussed in the case studies presented below vary widely in nature and size. The first case study covers the construction of a vicarage - essentially a large domestic property. The second discusses the development of a large 'out of town' retail complex involving some £20,000,000 of construction work.

Despite these wide disparities in scale, the risks to the organisations and individuals involved were, relative to their size and resources, large in both cases. In fact, it might be argued that the risks were relatively higher on the smaller project. In both cases the early identification of important risks and formulation of a strategy for dealing with them proved beneficial to the project.

## 5.1 The construction of a vicarage

### 5.1.1 Introduction

One of the authors has served as vice chairman of the council of an ecumenical church situated in a North West New Town. The members of the team ministry lived in houses rented from the Development Corporation but with the mellowing of the town, the foreseeable abolition of the Development Corporation and formal creation of an Anglican Parish the opportunity arose to make proposals to the Diocesan Board of Finance for the provision of a purpose built vicarage for the Church of England member of the team. The author acted as unpaid project manager for the Church Council and helped in the purchase of a vicarage at a substantial saving compared to normal prices for such buildings.

### 5.1.2 The usual procedure for procuring vicarages

Most Anglican parish churches have vicarages nearby, on land owned by the church. New vicarages are normally purpose designed for their sites by consultant architects in accordance with clear guidelines laid down by the Church Commissioners. The design being approved by the incumbent, church council and the Parsonages Committee of the Diocesan Board of Finance. Consultant Quantity Surveyors prepare bills of quantities and selected small and medium sized building contractors invited to tender. A standard JCT contract provides the legal agreement between the Diocesan Board of Finance and the builder. Typically at the time vicarages were costing £80,000 to £90,000 plus professional fees on land owned by the church. There would normally be little change from £100,000. The parsonages committee had a responsibility to provide a vicarage and proved amenable to an unconventional procurement route.

### 5.1.3 Building Land in New Towns

Land in a new town is all owned, through compulsory purchase, by the appointed Development Corporation. When the Ecumenical Centre was initially built, the Development Corporation had given the site and some adjoining land for a vicarage to promote this important community development. Changes to the plans for the town centre following that early decision had made the vicarage site unsuitable.

Where should the new vicarage be sited? The Development Corporation had a few building plots for sale in the most expensive part of the town. Not only were these in an inconvenient location but they would isolate the incumbent from his parishioners, most of whom live in rented accommodation.

### 5.1.4 The site

Some ten years earlier a large estate of high density low rise housing had been commenced on the site. For a variety of reasons, two main contractors abandoned the site whilst a third went bankrupt. This resulted in there being several hundred derelict, partially completed houses on the site. A decision was made by the development corporation to demolish them but to leave the roads and sewers in place. The site, part of which had not been built on, was subsequently taken over by a speculative house builder, who built and sold small, conventional semi detached and terraced houses and small blocks of four one and two bedroom flats.

An initial approach to the builder's regional office ascertained their interest in providing a vicarage. The novelty appealed to the regional director and a location adjoining an existing road identified. There was neither surface evidence nor recollection that housing had previously stood there. The firm offered to cost a design proposed by the Church and, subject to agreement on cost, to build the vicarage.

### 5.1.5 Risk identification and avoidance

The church council voted to support the provision of the vicarage and authorised the author to act on its behalf. While supporting the scheme, the author realised that not only was he taking on a voluntary undertaking requiring a considerable commitment of his limited 'spare time' but that he was also potentially exposing himself to claims for negligence if anything went wrong! Without professional indemnity insurance for risks that as vice chairman are hard to define at the very least refuting any alleged negligence could be both time consuming and expensive. Though professionally qualified as a chartered builder his competence as a designer could be challenged for he does not possess a design qualification. Similarly the question of supervision of the works on site and authorisation of stage payments to the contractor arose. For attendance at site to inspect work at major stages such as trench bottoms before concreting, or to attend in order to value work completed to comply with contract requirements for stage payment might be difficult.

The duty of care owed to the church council and indirectly to the Diocesan Board of Finance could be onerous should one of either party chose, in the event of mishap with the project, to claim that such a duty existed. Hedley Byrne v Heller (1963) suggests that a duty of care cannot be avoided by the use of phrases such as 'without responsibility' and Uff [Uff(1991)] when discussing the case suggests.

> 'Thus persons such as engineers and architects must be on their guard when making statements to their clients even concerning matters in which they are not directly instructed **And a duty of care may equally arise when giving gratuitous advice to strangers if the circumstances are such that there is an implied undertaking of responsibility.**' (authors' emphasis)

Although the decision as to whether or not negligence has occurred is ultimately the responsibility of a judge in the particular circumstances of a case the author had no desire to be cited in legal text books!

### 5.1.6 Development of the design

Neither the speculative builder, the author or the vicar had previously been involved in building a vicarage. Two Saturdays were spent looking at a recently completed vicarage in the North West and at the large detached houses built by the builder on more prestigious sites. These generally sold for £100,000 to £150,000. A booklet on design criteria for parsonage houses [Church Commissioners(1982)] was obtained and the vicar and author set to work with squared paper. The design criteria specify aspects of the design which enable vicarages to be used both as a residence and a work place.

By adapting a standard detached house layout the basis of a design was created. Access to the study was from the family hall rather than by separate entrance and access to the dining room through the kitchen. Both these variations to the standard design criteria were acceptable to the Church Commissioners in this particular case. In fact in dimension, only the staircase remained as a standard house type but the proposed design was carefully developed to the same vertical dimensions and same range of horizontal dimensions found in the company's other houses to facilitate use of standard components such as roof trusses and partitioning. The design was developed with the practicability of economic construction and minimising circulation space as prime criteria. It was intended to blend with other properties, appear totally conventional, and avoid ostentation.

### 5.1.7 Responsibility for the design

A schedule of requirements, cross referenced to both those of the Church Commissioners and the National House Building Council was prepared. This also required the builder to obtain all legal and planning approvals and agreements for supply of electricity, gas, water, drainage, telephone and communal aerial T.V. A 'U' value of 0.6 W/sq m.K was sought for walls, roof and ground floor (these being of a higher standard than required by building regulations at that time).

The sketch plans for a 160 square metre house were approved by the Church Council and the Parsonages Committee. The Parsonages Committee included among its membership an architect in private practice who served in a voluntary capacity. The sketch plans were costed by the builder and the offer price found to be within the Committee's budget.

Sketch plans were developed by the builder's architect into a design for detailed planning permission at 1:100 and 1:20. The design used the same components, windows, doors, roof tiles, bricks, bathroom and kitchen fittings as elsewhere on the site. It complied with NHBC requirements and carried their warranty. The builder's architect assumed that the site had not previously been built on for all the foundations being dug adjoining the proposed vicarage site were into undisturbed soil.

### 5.1.8 Inspecting work and authorising payments

The suggestion was made to the secretary of the Parsonages Committee that the consultant architect, who served as a member of the board, be invited to act on behalf of the board (and church council) in supervising the works. This was accepted by the committee and the architect appointed as supervising officer. The contractor was informed by letter from the vicar of the Dioceses' intention to purchase the vicarage and he signed the standard house purchaser's agreement and an additional agreement to make a 10% deposit payable at the time of execution of the agreement and stage payments when the property reached slab level and when watertight. The Church Commissioners undertook the legal work associated with the conveyance of the property.

As the project developed the builders tried to introduce methods of construction which did not satisfy the initial schedule of requirements. The supervising architect ensured adherence to both the preliminary schedule of requirements and to the builder's own drawings. As the working drawings were not prepared until after the price had been offered and accepted, the supervising architect also ensured that the working drawings complied with the initial specification. The supervising architect issued snagging lists and authorised payment of the builders final account.

With hindsight a risk that was not identified was that the contractor might have accepted a higher offer from another purchaser before payment of the initial deposit was made and accepted.

### 5.1.9 The construction phase

The enthusiasm of the site foreman for the rather unusual nature of the project contributed to the success of the scheme. He enjoyed the idea of building a vicarage and the workers were made well aware of the prestige nature of the project.

During excavation for foundations a small tree trunk and main branches about 6 metres long was encountered horizontal about 1.5 metres below the surface. It was not as assumed a virgin site but the ground had been made up when the road had been constructed 10 or so years earlier. This necessitated taking foundations to some 2 to 2.5 metres deep, sides of trenches were battered to avoid the need for planking and structuring and a considerable area within the curtiledge excavated to 2 metres to remove all the branches. The extra costs of excavation, backfill with hardcore, concrete and brickwork in footings was borne by the building contractor. Under the usual system for procuring vicarages a variation would have been claimed and extra expense incurred.

Otherwise, the work proceeded well with only minor variations. On completion a good sized family home with many features such as coloured bathroom suite and close boarded private garden was provided at a price of some 55% of the normal construction cost of vicarages, excluding provision of the site.

### 5.1.10 Observations

Unconventional contractual solutions can work satisfactorily when there is goodwill by all participants to the procurement process and the client states their requirements in an unambiguous way at the beginning. This is particularly helped when the 'lay' participant recognises the risks involved. The fixed priced contract benefitted the client. The builder accepted responsibility for the design decisions of his consultant architect whose failure to undertake a site investigation for the plot led to extra expense being incurred.

The responsibilities of the consultant architect appointed by the Parsonages Committee were clearly identified and limited to general advice, inspection of works and authorising completion. No Quantity Surveyors fees were incurred.

A hidden factor in the project may have been any loss that the builder may have been prepared to carry in order to enhance sales of adjoining properties. A detached house and vicar as neighbour would not detract from the estate. This, if present, is an element never present under regular contractual arrangements for vicarages.

It is doubtful that the builder knew how much vicarages

normally cost. Once the builder was committed to the low fixed price all financial risk was carried by him.

## 5.2 A retail development

### 5.2.1 Introduction

This case study concerns the speculative development of a large, multi-tenant retail complex by a commercial development company ('the Developer'). The work was to be carried out under a Development Agreement with the local Development Corporation who would be providing the land and constructing a new interchange and access road on the adjacent dual carriageway 'A' road. The development, whose original estimated construction cost was £20,000,000, was to take place on a green field site on the outskirts of a large town in Northern England.

The project principally comprised twenty four retail units, providing some 40,000 square metres of retail space, and associated external and infrastructure works. In addition, one area of the site was to be cleared, levelled and provided with roads and services for direct sale to a retailer who would construct a 7,500 square metre food store. The Developer employed consultants to produce a sketch design for the retail park, sufficient to obtain outline planning permission and to provide the basis for competitive tendering.

The form of construction proposed for the retail units was single-storey, steel portal frame with half height external cavity walls and coated, lined and insulated steel cladding panels for the upper walls and roof. Because of the soft, alluvial sub-soil on the site, all of the portal frames had piled foundations. In addition, it was considered necessary to pile the ground floor slabs in half of the units.

Tenders were sought on a two stage basis for a design and build contract with an established 'guaranteed maximum cost', savings below the maximum cost to be split equally between client and contractor. In stage one of the tendering process a number of contractors submitted details of preliminaries, overheads and profit. On the basis of these costs one contractor was selected to proceed to stage two. In the second stage the Contractor was required to develop 'contractors proposals' and to obtain competitive quotes for the work packages necessary to complete the project. These quotes and the stage one costs were used as a basis for the negotiation of the maximum cost figure with the Developer's Quantity Surveyor. In parallel with this activity, the Developer, by means of letters of intent, had authorised various enabling works to be carried out by the Contractor and principal sub-contractors prior to the contract being let.

During this second stage the Developer was acquired by

a competitor. One of the first actions of the new owners (hereafter referred to as 'the Client') was to suspend all work and negotiations on the project and to employ a firm of consultant project managers ('the Project Manager') to review the status of the job, to assist in the decision on whether to continue with the project or not and to develop a strategy for completion if the decision was in the affirmative. The selection of the Project Manager was made on the basis of a successful working relationship on a number of previous projects.

The compilation of the report on the status of the project, which consisted mainly of a factual review, proved to be straightforward. However, the decision to proceed and the development of a plan for completion were seen to require much careful consideration. The identification and analysis (albeit qualitative) of the principal risks faced by the Client were an important source of input to the above decision and the development of a strategy for continuing the project was, in essence, an exercise in risk response.

**5.2.2 Risk identification - financial viability**

The decision on whether or not to proceed with the project was based primarily on an assessment of the financial viability of the project. Based on the initial information presented to the Client it seemed that the project would not be financially viable. In such circumstances, the Client would have been willing to accept the loss arising from the aborted enabling works and professional fees etc., amounting to some £300,000, and withdraw. However, the appraisal and analysis carried out by the Project Manager indicated a number of areas in which the viability might be improved and that the project merited further investigation.

Clearly, the financial viability of an individual project directly influences the overall profitability of a development company. Other issues must also be considered. In the event of resuming the project, the Client would wish to obtain the majority of the finance by means of a bank loan. In order to obtain such a loan, a bank would also need to be convinced of the viability of the project. Furthermore, it would be unlikely that the Client would wish to retain the completed development. A successful sale of the development, to a pension fund or property company, would also depend on an adequate demonstration of its viability.

A number of factors and their associated risks were identified by the Client and Project Manager as important influences on the viability: the amount and timing of construction costs; the level and timing of rental and sale income; the value and saleability of the completed project; payments to be made to the Development

Corporation on signing the Development Agreement and potential damages payments which could be claimed by the Development Corporation in the event of late completion of the project. The risks identified often affected a number of the above factors and covered financial, contractual and technical aspects of the project.

Under the maximum cost/incentive scheme previously proposed, the final construction cost was uncertain and accordingly, the assessment of viability would need to be based on the guaranteed maximum cost. Since the higher the maximum cost, the greater the payment to the Contractor, there was no incentive for the Contractor to obtain keen prices for the work packages during the second stage of the tendering process. The maximum cost might therefore be unnecessarily high and would certainly be higher than could be obtained through a more conventional fixed price approach. The maximum cost approach would put at risk the ability to demonstrate the viability of the project both within the Client's own organisation and to external institutions.

The Project Manager was also able to identify a further factor causing increases in construction cost. The Development Corporation where using the project as a 'flagship' development and had required the original developer to provide various enhancements to the design. These enhancements, which mainly affected the external appearance of the buildings and grounds, resulted in a design which was to a higher standard and more expensive than would normally be provided in such a commercial retail development and therefore further prejudiced the financial viability of the project.

One risk which all speculative developers face is the possibility that they will not be able to find tenants to fill the development. This project was no different, clearly the higher the rental levels obtainable and the sooner this income started to come in, the more financially viable would be the project. At the time of the takeover, no tenants had signed Agreements to Lease. The number and prestige the agreed tenants would strongly influence potential financiers and potential purchasers of the completed development. Clearly, this risk would have a vital effect on the success of the project.

In addition, Project Manager also identified some aspects of the design which might further put at risk the acquisition of tenants, the procurement of interim funding and the long-term sale of the development. These aspects concerned mainly the external cladding and rainwater drainage system, and would have the effect of making maintenance of these items both difficult and expensive.

The consideration of all of these (and other) issues and the formulation of a strategy for addressing them could not be carried out at leisure, there were significant time pressures. Even if work proceeded, the Client would be liable for some £200,000 additional costs arising from delays in proceeding with the works and in connection with the enabling works previously instructed. Since a large part of these costs were due to interest and storage charges, they were increasing as the delay in restarting work increased. More importantly, the Development Agreement, which though not yet signed was substantially agreed, included a obligation on the Client to complete the Development by a given date. The longer the delay in recommencing works, the shorter the construction period available and the greater the risk of the client defaulting.

### 5.2.3 The response

Firstly, the procurement method was addressed. A 'Design and Build' approach was still favoured as this provided a single point of contact and responsibility and thus provided a clear allocation of risk between the parties. However, the maximum cost/incentive approach to payment was not considered appropriate. Accordingly, the Project Manager was able to re-negotiate with the Contractor to arrive at a fixed price, lump sum significantly below the contractor's original maximum cost. The greater certainty of final price was considered to be of more importance than the potential for cost savings in the long term.

Whilst these negotiations were taking place assurances were given to the the contractor, the major sub-contractors and the statutory authorities that work would proceed. In return for this assurance sub-contractors were willing to drop their claims for abortive work and delays and enabled them to reinstate the project in their work programmes. The statutory authorities were able to recommence planning for servicing the site and proceed to arrange for the wayleaves required to pass through adjacent land. These steps were taken to try to avoid delays on restarting the project and thus reduce the risk of over-running the development period.

Negotiations also took place with the Development Corporation regarding the design of the retail units. As a result of these negotiations, some of the design enhancements required by the Corporation were modified and additionally, the Development Corporation agreed to make a contribution to the extra costs arising from the remaining enhancements. Simultaneously, these changes reduced the concerns regarding the practicality and cost of maintenance. Further control over the repair and maintenance aspects would be applied by the Project Manager during the detail development of the design.

In order to increase the speed of lettings on the development, the Client employed a London-based estate agency practice, in addition to the local practice employed by the previous Developer. It was anticipated that this would improve access to national companies and provide the impetus to finalise agreements on a number of tenancies which were in negotiation. Ultimately, however, it was accepted by the client that lettings were affected by many factors outside the control of the project team, e.g. general economic and market conditions. The risk of being unable to let the units had to be retained by the Client.

In order to improve the cash-flow for the project, the Project Manager developed a scheme for phasing the works to enable early generation of income from the development. This scheme allowed the early completion and sale of the site for the food store. Also, since agreement then had been reached with a retailer for the sale of a completed unit of some 6,000 square metres, it was possible to include this unit and associated infrastructure in the first phase works.

The review and analysis resulted in the conclusion that the risks could be reduced to acceptable levels, rendering the project viable. On this basis the client decided to continue with the project.

**5.2.4 Observations**

The procedure which the Project Manager and Client followed has, without great difficulty, been fitted, by the authors, into the Risk Management methodology of Identification, Analysis and Response. The fact that the Project Manager was not familiar with the term 'Risk Manager' does not alter the fact that, in performing what he perceived to be his duties to his client, he was filling this role.

The result of the Risk Management exercise was that a project which initially seemed non-viable, and could have been abandoned, was brought to an acceptable level of viability (not all risks can be avoided) and undertaken. The complete process of appraisal, strategy formulation, re-negotiation and re-mobilisation was performed under time pressures and allowed work to re-commence around two months after the takeover.

The Risk Management was carried out on behalf of (and partly by) the Client, but all parties to the project benefitted from its result. Few contractors, designers or suppliers can afford to lose the income from a £20,000,000 contract in the current market conditions. Also, the analysis and re-negotiation of contracts between the principal parties resulted in a clarification of responsibility and a clearer allocation of risk. As has been proposed above, unclear allocation of risks and inadequate response to them can lead to disputes.

The pause in the progress of the project inadvertently provided the opportunity to follow one the tenets of Risk

Management - a review and amendment of strategy to suit changes in the project's internal and external conditions. Too often projects of all kinds take on a great momentum; progress towards some previously defined goal becomes the over-riding imperative without recognition that the position of the goal might have changed.

This case study also illustrates the need for Project Managers to work across conventional disciplinary boundaries and display an understanding of areas (economics, marketing, estate agency) not normally covered by construction professionals.

## 6 Discussion and conclusions

Risk management methodology places much emphasis on the choice of an appropriate contract and procurement strategy. This is illustrated in both case studies where the adoption of a design and build approach resulted in simplification of contractual relationships. In the first case study, clear allocation of financial risk prevented the contractor claiming for for the failure to conduct a ground investigation. In the second case study, the re-negotiation of the contract to a fixed price basis allowed a lower initial price to be obtained for use in viability assessments.

Although Risk Management involves the consideration of uncertainties which not only affect the cost, but also the duration and/or quality of a project, most risks have a financial implication. For example, in the second case study, the risk that the design would fail to achieve required standards was identified. This can clearly be categorised as a risk to quality, but the concern of the Client and Project Manager was ultimately that defects in design would prejudice the sale of the development and therefore its financial success. In both of the case studies it was the financial implications of risks that were of the greatest interest to the clients.

In neither of the studies presented were quantitative risk analysis techniques applied. The analysis was qualitative, intended to provide a rank ordering of risks. Nonetheless, in both cases, the sequence of identification, analysis and response was followed. Responses adopted included both the allocation of risks and the reduction of likelihood and impact of risks to the benefit of both clients. In both cases contractors understood the risks they were expected to bear and to make appropriate allowances in their prices.

Success in a project can most easily be achieved if there is good communication and cooperation between the contributing parties. A lack of communication and cooperation will often manifest itself in the form of disputes. The role of the Project Manager in leading and coordinating the project team, coupled with the production of a clear recognition of each party's needs and responsibilities can reduce the possibility for disputes.

Project Managers may not be familiar with the term 'Risk Management', or the more esoteric aspects of the theory and techniques which underlie it. Nonetheless, the management of risk is implicit in the performance of many of a Project Manager's duties. Risk Management is the core of good project management and an acknowledgement of this fact and the dissemination of the methodology could improve the performance of Project Managers.

## 7 References


Aqua Group (1990a) **Contract Administration for the Building Team.** BSP Professional Books, Oxford.

Aqua Group (1990b) **Tenders and Contracts for Building**. BSP Professional Books, Oxford.

Bennett, J. (1985) **Construction Project Management.** Butterworths, London.

Carter, D. J., Cheetham, D.W. & Lewis, J. (1990) A Systems Approach to Information Flows in Construction Management. **Habitat International**, 14, 137-156.

Chapman, C.B., Ward, S.C., McDonald, M. & Curtis, B. (1989) **Roles, Responsibilities and Risks in Management Contracts**, SERC Research Report, University of Southampton.

Church Commissioners (1982) **Today's Parsonage**.The Church Commissioners, London.

C.I.O.B. (1988) **Project Management in Building**. The Chartered Institute of Building, Ascot.

CSSC - Centre for Strategic Studies in Construction - (1988) **Building Britain 2001**, University of Reading, Reading.

Fellows, R.F. (1989a) **The Management of Risk**,CIOB Technical Information Service, No. 111.

Fellows, R.F. (1989b) Risk in Management Contracting, **Chartered Builder**, December 1989, 32-35.

Franks, J. (1990) **Building Procurement Systems.** The Chartered Institute of Building, Ascot.

Hayes, R. W., Perry, J. G., Thompson, P. A., Willmer, G. (1986) **Risk Management in Engineering Construction.** S.E.R.C. Project Report, Thomas Telford, London, 1986.

JCT (1980) **The Standard Form of Building Contract With Quantities.** Joint Contracts Tribunal, 1980.

John, D. (1992) Bickering flares as Eurotunnel delay sparks fears of three month set-back. **The Guardian**, 11th February 1992, 11.

NEDO (National Economic Development Office) (1975) **The Public Client and the Construction Industry**. Her

Majesty's Stationery Office.
NEDO (1983) **Faster Building for Industry**, Her Majesty's Stationery Office.
NEDO (1988) **Faster Building for Commerce**, Her Majesty's Stationery Office.
O'Neill, J. J. (1989) **Management of Industrial Construction Projects**. Heinemann Newnes, London.
Perry, J. G. (1986) Dealing with risk in contracts. **Building Technology and Management**, April 1986, 23-26.
Perry, J. G. & Hayes, R. W. (1986) Risk Management for Project Managers. **Building Technology and Management**, Aug/Sept 1986, 8-11.
Perry, J. G. & Hayes, R. W. (1985a) Construction Projects - Know the Risks. **Chartered Mechanical Engineer**, February 1985, 42-45.
Perry, J. G. & Hayes, R. W. (1985b) Risk and its Management in Construction Projects. **Proceedings of the Institution of Civil Engineers**, Part 1, 78, 499-521.
Rougvie, A. (1987) **Project Evaluation and Development**. Mitchell, London.
Thompson, P. (1986) The Application of Risk Analysis to the Management of Engineering Construction. **A.R.C.O.M. Conference**, Building Research Establishment, February 1986.
Uff, J. (1991) **Construction Law**.Sweet and Maxwell, London.

# RESOLVING CONFLICT IN THE FORMULATION OF BUILDING DESIGN OBJECTIVES

S. D. GREEN
Department of Construction Management & Engineering, University of Reading, England

Abstract
This paper describes the benefits of developing a formal decision model during the briefing and outline design stages of a new laboratory building. A case study illustrates how understanding and communication between the design team and client can be considerably improved by the formulation of a requisite decision model based on the simple multi-attribute rating technique (SMART). The terminology of value management is considered to provide a useful way of introducing decision modelling to unsophisticated clients. However, notions of 'design optimisation' and 'value maximisation' are rejected as being unrealistic.
Keywords: Building Design Objectives, Value Management, Decision Theory, Soft Systems Methodology.

## 1 Introduction

The purpose of this paper is to describe how decision modelling techniques can be used to clarify and communicate building design objectives during briefing and outline design. The proposed methodology is based on the simple multi-attribute rating technique (SMART) as proposed by Edwards (1977). It will be demonstrated that SMART provides a suitable basis for the development of a requisite decision model within the context of a multi-disciplined building design team. It is important to emphasise that the purpose of developing such a model is to establish a shared understanding of the design objectives and to encourage new insights into how they may be achieved (Green, 1992a). The model is therefore intended to facilitate a **learning** approach which will continue throughout the design process. Attempts to implement an optimising approach will inevitably fail due to the limitations of bounded rationality (Simon, 1957) and the labile nature of human values.

The development of a requisite decision model is directly comparable to Checkland's (1981, 1989) soft systems methodology in that the existence of a clearly defined objective is not taken for granted. An important part of the exercise is to achieve an understanding of what the objectives actually are, or, more realistically, what the objectives are perceived to be.

The concept of a requisite decision model, as developed by Phillips (1984), is concerned with helping a group of decision-makers develop a common **understanding** of a decision situation. Unlike the 'Olympian' subjective expected utility model (Savage, 1954), a requisite decision model makes no pretence of producing a 'right' solution.

This paper also draws upon the terminology of value management as a useful way of 'selling' the techniques of decision analysis to unsophisticated clients. However, the traditional approach to value management is clearly a 'hard systems' approach in that it assumes that the functions required of a design solution (i.e. the design objectives) are static over time. Furthermore, the literature on value management (e.g. Dell'Isola, 1982; Green and Popper, 1990) invariably assumes the existence of a coherent group of decision-makers whose values are both transitive and consistent. Once these assumptions are weakened, the notion of value maximisation becomes meaningless. In the final analysis, value for money has more to do with psychological comfort than it does with objective economics.

## 2 A SMART decision model

The suggested methodology is based on the 'simple multi-attribute rating technique' (SMART), as originally described by Edwards (1977) and subsequently revised by Edwards and Newman (1982) and Edwards *et al* (1988). The following brief overview is drawn from these three sources. The suggested approach involves the creation of a decision model which represents a shared perception of the design problem at a particular point in time. It is therefore important that the key project **stakeholders** are actively involved in the development of the model. The model would normally be constructed within the context of a one or two-day decision conference. Chairmanship would be provided by a facilitator who is skilled in the techniques of decision modelling.

The first step of the SMART methodology is to structure the design objectives into the form of a hierarchy, otherwise known as value tree. The value tree for the design of a health centre is shown in Table 1. It is important that the value tree is produced by group consensus and that each participant feels involved. Whilst this process might result in disagreement, it is preferable that any conflict regarding the design objectives is made explicit at an early stage. Judgement is obviously required with respect to how far the objectives should be subdivided. In the final analysis it is only the lower-order objectives which are carried forward to the next stage. It is the 'twigs' of the value tree which provide the attributes against which the decision options are evaluated. The required level of breakdown is therefore dictated by the need to compromise between the ease with which the attributes can be measured and the number of attributes with which the model can cope.

Table 1. Value tree for health centre
(adapted from Edwards and Newman, 1986)

A. Good staff conditions
  - AA. Office size
  - AB. Convenience of staff commuting
  - AC. Office attractiveness
  - AD. Office privacy
  - AE. Availability of parking

B. Easy access for patients
  - BA. Closeness to patient's homes
  - BB. Access to public transport

C. Suitability of space
  - CA. No. of consulting rooms
  - CB. Suitability of reception area
  - CC. Storage space

D. Administrative convenience
  - DA. Adequacy of space for secretaries
  - DB. Flexibility of layout

The next stage is to allocate an importance weight to each of the lower-order attributes. Once again, it is critical that this process is performed on the basis of consensus. However, the literature does describe a number of procedures which can be used for resolving conflicts. The SMART approach to weight elicitation is based upon the sum of the weights being equal to one for each group of attributes which derive from a single node. Each group is dealt with in turn and the final weights for the lower-order twigs are obtained by 'multiplying through the tree' (see Table 2). The relative weights for each group are determined by the ratio method. Attributes are initially listed in order of perceived importance and the least important is awarded an arbitrary weight of ten. It is then necessary to allocate weights to the other attributes on the basis of their relative importance. The weights are then summed and each is normalised such that the total weight for the group adds up to one. It is important that the facilitator regularly checks for the inevitable inconsistencies. The process is therefore likely to involve a good deal of iteration before the elicitated weights are fully consistent and the group is fully comfortable with the final result.

The third stage of SMART is concerned with assessing each decision option against the attributes identified at the lowest level of the value tree. The assessment is performed on the basis of a single-dimension utility measurement. Whilst the majority of scores will be allocated on a subjective basis, it is important that objective measures are used where possible.

Table 2. Value tree for health centre with importance weights (adapted from Edwards and Newman, 1986)

| Node code | | Normalised weights | Twig weights |
| --- | --- | --- | --- |
| A. Good conditions for staff | .43 | | |
| AA. | | .39 | .17 |
| AB. | | .21 | .09 |
| AC. | | .14 | .06 |
| AD. | | .14 | .06 |
| AE. | | .12 | .05 |
| B. Easy access for patients | .24 | | |
| BA. | | .50 | .12 |
| BB. | | .50 | .12 |
| C. Suitability of space | .19 | | |
| CA. | | .52 | .10 |
| CB. | | .32 | .06 |
| CC. | | .16 | .03 |
| D. Administrative convenience | .14 | | |
| DA. | | .64 | .09 |
| DB. | | .36 | .05 |

Subjective attributes are measured on an arbitrary scale of 0-100, where 0 represents the minimum acceptable standard and 100 represents the maximum which is achievable. For attributes which can be assessed objectively it is necessary to convert the measurements so that they are also represented on a 0-100 scale. It is usual to assume that each utility function is linear, thereby ignoring the possibility of diminishing marginal utility.

The fourth step of the SMART methodology is to aggregate the weighted utilities for each decision option. In mathematical notation:

$$U_i=\sum_j w_j u_{ij} \tag{1}$$

where $U_i$ is the aggregated utility for the *i*th decision option, $w_j$ is the normalised importance weight for the *j*th attribute and $u_{ij}$ is the re-scaled utility measure for the *i*th decision option assessed against the *j*th attribute.

Within the spirit of a requisite decision model it is not necessary to worry too much about the theoretical validity of the above additive utility function. The objective of the exercise is to provide a structured framework for thinking rather than a normative idealization.

The next stage of the process is to perform a sensitivity analysis to test how sensitive the outcome of the rating process is to marginal changes in the key variables. Particular attention should be given to any importance weights or utility scores about which members of the group had expressed discomfort. It may well be thought necessary to adjust the structure of the value tree. The model would continue to be revised until it was considered to be representative of the group's value system.

Edwards (1977) refers to a special case where one of the attributes possesses an upper limit. Within the context of building design the capital-cost attribute will often be subject to a budget constraint. In this situation, it is recommended that capital cost is omitted from the initial decision model. An additional step would then calculate the ratio $U_i/C_i$ where $C_i$ is the estimated capital cost of the *i*th attribute. It could then be argued that the highest value of $U_i/C_i$ would dictate rational choice. However, if the decision-makers were relatively indifferent to the value of $C_i$ provided that it was below the budget limit $C^*$, then the highest value of $U_i$ could be chosen provided that $C_i$ is less than $C^*$.

## 3 Case study

### 3.1 Background

The following case study describes the application of the SMART methodology to the design of a research laboratory for an international pharmaceutical company. The new building was to be located on a confined site in close proximity to several existing buildings. The complex was an established research centre with an international reputation. The identities of the client and other parties are suppressed for the purposes of confidentiality. The case study has been simplified in order to emphasise the SMART methodology.

The client was initially attracted to the concept of value management by publicity in the technical press. The client's project team was especially keen to demonstrate to its head office that the new building represented value for money. The notion of achieving the required functions at least cost was considered to be particularly attractive. The author was engaged as a value management consultant to chair an initial one-day study during the development of the conceptual design. This presented an ideal opportunity to test the SMART methodology in practice.

A multi-disciplinary design practice had been commissioned to develop the conceptual design and a firm of quantity surveyors were appointed to act as independent cost consultants. The date of the value management study was fixed several weeks in advance by a process of negotiation involving all parties. Whilst the designers were somewhat sceptical of the benefits of the study, they were persuaded into cooperating by the client.

During the weeks prior to the value management study the performance of the design team was increasingly causing concern. In particular, the client was critical of the perceived failure of the designers to communicate effectively with each other. There was also concern that the client's requirements were not being interpreted correctly. The members of the design team were becoming equally frustrated in their efforts to respond to the client's needs. They were of the view that different members of the client's team were stating different requirements. The designers were also finding that the statements given by the client's project coordinator were alarmingly inconsistent. The relationship between the client and the designers was consequentially deteriorating rapidly. The objectives of the value management study had therefore taken on an additional dimension: to re-establish trust between the client and the design team.

**3.2 Value management at conceptual design**

The value management team consisted of four members of the design team, five representatives of the client and one quantity surveyor. The agenda for the study was structured around the traditional phases of value management. It is convenient to describe the study in accordance with the stages defined on the agenda.

Stage 1: Information

After the initial introductions, each team member was asked to make a brief presentation on what were perceived to be the key objectives. It soon became apparent that there were many conflicting views. Not only did the designers fail to agree with the client, different representatives of the client's team disagreed amongst themselves. After a good deal of rigorous discussion the following design objectives were finally agreed:

1. provide a 'world class' research facility;
2. provide a new focal point for the entire site;
3. facilitate good communication;
4. offer flexibility and adaptability;
5. provide a sound M & E strategy;
6. cater for extendability;
7. ensure maintainability;
8. provide pleasant working environment;
9. provide a safe and healthy working environment;
10. achieve operational efficiency;
11. ensure financial approval.

The architect completed the information stage by presenting the three alternative design concepts which had been produced to date. The client's team had serious reservations about each of these options and a heated debate soon developed. This was brought to a close by the value management facilitator who insisted that the agenda was followed.

Table 3. Value tree at conceptual design

---

A. Good conditions for staff
  AA. Safe and healthy environment
  AB. Pleasant working environment
B. Attract customers
  BA. Focal point for site
  BB. Visibility within building
  BC. Functional image
C. Promote teamwork
  CA. Encourage interaction
  CB. Ensure communication
D. Provision for future change
  DA. Extendability
  DB. Flexibility and adaptability
E. Operational efficiency
  EA. Sound M&E strategy
  EB. Low running cost
  EC. Maintainability

---

Stage 2: Functional Analysis
A value hierarchy was constructed on the basis of the above list of objectives. The construction of the value tree involved a considerable amount of iteration before all the team members were satisfied with the outcome. The resultant debate also caused the identification of a number of additional objectives. The issue of capital cost was omitted from the value tree as it was considered to be a constraint rather than an objective. The final value tree is shown in Table 3. It is of note that the architect was no longer sceptical of the value management process and now felt that he was gaining valuable information from the client which had not previously been forthcoming. Of particular importance was the emphasis which was now being given to the need for interaction within the building.

Stage 3: Speculation
One of the criticisms which had been levelled at the design concepts produced by the architect related to site utilisation. There was also a particular concern that the new building would have a very low impact from the main road running through the site. An initial brainstorming session therefore addressed

the three issues of building shape, building location and its relationship with the existing buildings. Many interesting ideas were generated, some of which were clearly not feasible. However, the session did produce a number of ideas which were both innovative and practical.

A second brainstorming session addressed the issue of how the building could be designed to promote teamwork amongst the researchers. Once again numerous ideas were produced, some of which were impractical and some of which seemed to have potential for further development.

Stage 4: Evaluation

Each of the ideas produced in the previous stage were evaluated in turn. Those which were obviously nonsensical were deleted from the list and those which were considered worthy of further consideration were highlighted. The process of evaluation generated considerable discussion and there was an increasing feeling that real progress was being made.

The value management study was concluded by listing the design options which were to be developed further. Specific actions were allocated and a timescale was established. It was agreed that the results of the development work would be discussed during the next design team meeting. An anonymous questionnaire was then distributed which asked the participants a number of questions about how useful they had found the exercise. The responses were extremely positive, not only from the client representatives, but also from the designers. Several respondents considered the process of constructing the value tree to have been particularly beneficial in clarifying the design objectives. The study was also perceived to have made a significant contribution to team building and conflict reduction.

The client was so pleased with the outcome that he subsequently commissioned a further study. The value management consultant recommended that this should take place towards the end of the outline design stage when the client would be required to make a decision regarding the choice of outline proposal.

### 3.3 Value management at outline design

The duration of the second study was once again limited to one day. The date was agreed several weeks in advance. The design team had developed five distinct outline design options and were anxious for the client to make a decision so that the scheme design could proceed. The client, however, did not want to be rushed into a decision. Furthermore, he wanted to be able to demonstrate that the outline design had been chosen on the basis of rigorous value-for-money criteria. The composition of the value management team was the same as it had been previously. A full agenda was circulated in advance; however, on this occasion, the traditional value management stage descriptions were not used. The agenda was structured around the following titles:

Stage 1: Information: Re-formulation of Objectives

The study was commenced by a brief statement from the value management consultant outlining the purpose of the exercise and explaining the procedure which would be adopted. Each member of the value management team was then invited to comment on the extent to which the value tree constructed during the previous study remained valid. It was generally agreed that the value tree was still representative of the design objectives, although it was emphasised by the client that the need to meet the overall budget figure of £13,000,000 was even more paramount.

A summary presentation of each of the five outline design options was then made by the architect. This led to a free-ranging discussion regarding their relative merits. This discussion was allowed to continue for fifteen minutes before being curtailed by the value management facilitator.

Stage 2: Definition of Attributes

The purpose of this stage was to 'prune' the value tree in order to produce a manageable list of assessment attributes. Each lower-order objective on the value tree was considered in turn and the question asked 'should this attribute be used to assess the relative merits of the design options?'. The following lower-order objectives were therefore eliminated from consideration:

**Safe and healthy working environment** - eliminated on the basis that this was a fundamental requirement for all feasible design options.
**Functional image** - this was considered to be an issue which was more relevant to detailed design.
**Sound M & E strategy** - this was seen to be achievable irrespective of the choice of outline design.
**Maintainability** - the choice of outline design was not considered to have any significant maintenance implications.
**Ensure communication** - it was considered that this should be combined with 'encourage interaction', otherwise the same features would be taken into account twice.

The elimination of the above branches produced the revised value tree shown in Table 4. The lower-order objectives on this simplified value tree were adopted as the assessment attributes for the choice of outline design option.

Stage 3: Assigning Importance Weights

The ratio method was then applied in order to produce the importance weights for each attribute. The relative importance of the following attributes was initially considered:

pleasant working environment;
attract customers;
interaction & communication;

provision for future change;
low running costs.

Table 4. Value tree at outline design

A. Pleasant working environment
B. Attract customers
   BA. Focal point for site
   BB. Visibility within building
C. Interaction and communication
D. Provision for future change
   DA. Extendability
   DB. Flexibility and adaptability
E. Low running cost

These were listed in rank order and the least important was assigned an arbitrary weight of 10. The other attributes were then compared against this baseline and an appropriate weighting allocated. The weights were checked for consistency before being normalised. The results are shown in Table 5. The relative importance weights were then applied to the lower-level attributes as shown in Table 6. The final weights were produced by 'multiplying through the tree'.

Stage 4: Utility Assessment

The five design options were then assessed in accordance with the seven lower-order attributes from the value tree. Each design was given a score for each attribute on a scale of 0-100. The scoring process was entirely subjective for all the attributes, with the exception of running cost, where some quantitative annual costs had been calculated on the basis of energy losses. The scores were entered into an analysis matrix and multiplied by the appropriate importance weights. The weighted scores were then summed in order to produce a utility rating for each design option.

The value management consultant then made a particular point of emphasising that the scores produced in the analysis matrix were only as good as the assumptions upon which they were based. He also asked all the team members individually if they had any reservations regarding the assumptions which had been made.

Stage 5: Sensitivity Analysis

A number of team members stated that some of the weights and utility scores which had been used in the decision model did not really reflect their own opinions. Several separate adjustments were then made, but the resultant utility scores still favoured the same design option. The additional discussion regarding the assumptions which had been questioned also served to ease the doubts which had been raised. A general consensus was therefore obtained that the model did indeed represent the values of the assembled team.

Table 5. Case study: elicitation of importance weights

| Attributes | Weights | Normalised Weights |
|---|---|---|
| A. Pleasant working environment | 40 | 0.32 |
| C. Interaction and communication | 40 | 0.32 |
| E. Low running cost | 20 | 0.16 |
| D. Provision for future change | 15 | 0.12 |
| B. Attract customers | 10 | 0.08 |
| TOTALS | 125 | 1.00 |

Table 6. Value tree at outline design with importance weights

| Node code | | Normalised weights | Twig weights |
|---|---|---|---|
| A. Pleasant working environment | .32 | | 0.32 |
| B. Attract customers | .08 | | |
| BA. Focal point for site | | .75 | 0.06 |
| BB. Visibility within building | | .25 | 0.02 |
| C. Interaction and communication | .32 | | 0.32 |
| D. Provision for future change | .12 | | |
| DA. Extendability | | .25 | 0.03 |
| DB. Flexibility and adaptability | | .75 | 0.09 |
| E. Low running cost | .16 | | 0.16 |

Stage 6: Cost/Value Reconciliation
The one issue which had been omitted from the decision model was that of capital cost. The utility rating for each design option was now compared to the estimated capital costs produced by the quantity surveyor. Unfortunately, the estimated cost of the favoured option was £300,000 higher than the budget figure. However, it was felt that this difference could be reduced by minor design changes without affecting the overall performance.

Stage 7: Search for Areas of Marginal Improvement
A brainstorming session was then held to generate ideas of how the cost of the chosen option could be reduced without effecting the level of performance. Several interesting possibilities were identified and the designers felt confident that the projected cost could be reduced by the required £300,000. The client was therefore confident that the design could proceed on the basis of the chosen option, provided that the identified economy measures were introduced.

An anonymous feedback questionnaire was distributed at the end of the study and the participants were once again very positive about the outcome. There was a general feeling of satisfaction that an important decision had been taken which would now allow the design process to continue with confidence. The client representatives felt that the decision had been made on the basis of defined criteria. Furthermore, the decision was accountable in that the value management report recorded the basis upon which it had been made.

## 4 Discussion of case study

The above case study has illustrated the application of the SMART methodology to value management during briefing and outline design. It has also been demonstrated that designers and client representatives feel comfortable with the SMART approach. It is of note that minimum use was made of the mathematical formulations and terminology of multi-attribute utility theory. Indeed, for the study at the briefing stage the traditional labels of traditional value management were adhered to in order to satisfy the client's expectations.

The use of value trees for the determination and structuring of the design objectives proved to be particularly successful. It is clear that the value tree did not represent any sort of underlying truth. Indeed, the very process of constructing the value tree caused the team members to revise their perception of the design objectives. Had the discussion developed along different lines, the resultant value tree may well have looked very different.

The use of the decision analysis matrix for the assignment of a utility rating to each design option also worked well as a means of structuring discussion. The very process of determining the importance weights and utility scores for each attribute went some way towards satisfying the client's desire for the decision-making process to be rational and explicit. The real benefit of this approach is that it ensured that the team **thought** about all the issues which had been identified. The adjustments which were made during the sensitivity analysis ensured that the decision model was **requisite**. It is important to stress that the SMART methodology does not replace the professional judgement of the team members. Poor judgement and a lack of expertise will inevitably produce poor decisions, irrespective of whether a formal decision model is used or not. The advantage of SMART is that it encourages professional judgement to be applied within the context of rigorous framework.

The validity of the SMART decision model produced in the case study was also dependent upon the composition of the value management team. Had all the major stakeholders within the client's organisation not been involved then the value tree would clearly not have been representative of the **client's** objectives. Whilst the designers also had a significant influence on the structure of the decision model this should not be interpreted as being detrimental. The traditional process of briefing depends upon an interactive process between client and designers. Indeed, helping the client to articulate his own requirements has always been a recognised function of the architect.

The two reports which were produced following the value management studies were referred to throughout the remainder of the project. It is important that value management reports fully record the design objectives and assessment attributes which were established at the time of the study. Whilst there is no reason to assume that these will remain constant over time, at least the above procedure will record what they are at the briefing and outline design stages. It should be recognised that any subsequent post-occupancy evaluation may well apply different criteria of assessment. However, it is surely beneficial to be able to distinguish between evolving objectives and a failure to design in accordance with the objectives as they were understood at the time.

Whilst six out the seven assessment attributes used in the case study were entirely subjective, this will not necessarily always be the case. It is easy to envisage how, in some projects, 'objective' attributes, such as internal rate of return, net present value, net/gross ratio and capital gearing might be used to assess the extent to which financial objectives are achieved.

In the final analysis, there is no sure way of knowing if the development of a formal decision model has led to an improved building design. The only meaningful way to assess the success of the exercise is to determine how useful it was to the participants. Based on this criterion the method used in the case study was clearly a success.

## 5 Conclusion

This paper has demonstrated the benefits which can be attained by the formal construction of a SMART decision model during building design development. However, it is important to emphasise that the use of a decision model is not recommended from a normative point of view. Attempts to 'optimise design' or 'maximise value' are seen to be entirely unrealistic. The primary objective of the exercise should be interpreted in terms of establishing a shared understanding of the design objectives and the relative benefits of competing designs options.

Different clients are likely to build different decision rules into their model. Although the SMART multi-attribute decision model provides a suitable basis for general application, in many cases it is possible to develop a requisite decision model which is based on a single criterion. For example, a commercial client may feel that his objectives can be represented by the single-attribute decision rules commonly used in investment appraisal. In this case, a requisite decision model could be based solely on some measure of financial performance (i.e. net present value). However, if the client wishes to take other objectives into account, then some form of multi-attribute decision model must be used. The benefit in making the design objectives (and their relative importance) explicit is that any conflict amongst the project stakeholders is brought into the open. This is considered to be a worthwhile end in itself.

## 6 Acknowledgements

The case study described in this paper has been adapted from Green (1992b), which also describes the theoretical origins of SMART and the wider assumptions upon which multi-attribute utility theory is based.

## 7 References

Checkland, P. B. (1981) **Systems Thinking, Systems Practice**, Wiley, Chichester.

Checkland, P. B. (1989) Soft systems methodology, in **Rational analysis for a problematic world: problem structuring techniques for complexity, uncertainty, and conflict** (ed. J. Rosenhead), Wiley, Chichester.

Dell'Isola, A. (1982) **Value Engineering in the Construction Industry,** 3rd edn., Van Nostrand Reinhold, New York.

Edwards, W. (1977) Use of multiattribute utility measurement for social decision making, **in Conflicting Objectives in Decisions** (eds D. E. Bell, R. L. Keeney and H. Raiffa), Wiley-Interscience, Chichester, 247-276.

Edwards, W. and Newman, J. R. (1982) **Multiattribute Evaluation**, Sage, Beverley Hills, CA.

Edwards, W., von Winterfeldt, D. and Moody, D. L. (1988) Simplicity in decision analysis: an example and a discussion, in **Decision Making: Descriptive, Normative and Prescriptive Interactions** (eds. D. E. Bell, H. Raiffa and A. Tversky), Cambridge University Press, Cambridge, 443-464.

Green, S. D. (1992a) Value engineering during design: new directions in research, in **Architectural Management** (ed. P. Nicholson), Spon, London, 256-267.

Green, S. D. (1992b) **A SMART Methodology for Value Management**, Occasional Paper No. 47, Chartered Institute of Building, Ascot.
Green, S. D. and Popper, P. A. (1990) **Value Engineering: the Search for Unnecessary Cost,** Occasional Paper No. 39, Chartered Institute of Building, Ascot.
Phillips, L. D. (1984) A theory of requisite decision models, **Acta Psychologica,** 56, 29-48.
Savage, L. J. (1954) **The Foundations of Statistics**, Wiley, New York.
Simon, H. A. (1957) **Administrative Behaviour,** 2nd edn., MacMillan, New York.

# THE RELATIONSHIP BETWEEN CONFLICT, CHANGE AND PROJECT MANAGEMENT STRATEGY

P.D. GARDINER[1] and J.E.L. SIMMONS[2]
[1]School of Engineering and Computer Science, University of Durham, England
[2]Department of Mechanical Engineering, Heriot-Watt University, Edinburgh, Scotland

Abstract
The management of conflict and change in construction projects is an important factor in determining the success of a project and the satisfaction of the client. Examples of project conflict and change have been collected in structured interviews within a series of organisations which are clients of construction activities. Five stages of conflict have been described and used as a basis to classify examples of conflict in particular projects. Conflict occurs to some extent in all projects and the outcome of this can be dysfunctional or functional to the project. A model is proposed for modifying project management strategies in the light of conflict and change.
Keywords: Conflict, Change, Project Management, Client, Strategy.

## 1 Introduction

Conflict and change in construction projects have become a subject of considerable interest and importance. A contributing factor is the increasing number of specialised participants in many projects. Research in construction management has so far failed to address the subject of conflict head on, although it has been an important sub-theme for many years. Socio-technical work carried out by the Tavistock Institute of Human Relations focussed attention on the relationship between the social, technical and administrative functions in construction organisations and brought to light some examples of conflict in the industry (Higgin and Jessop, 1965; Crichton, 1966). It has also been proposed (Cherns and Bryant, 1984) that many of the problems concerning design changes, delays and difficulties during the construction phase have their origins in unresolved conflicts within the client organisation which remained unresolved when the decision to build was taken.

To some extent, conflict between groups in organisations is inevitable (Bowditch and Buono, 1990). Armed with an understanding of project conflict and change it should be possible to reduce the occurrence and limit the damage caused by dysfunctional conflict and at the same time provide conditions which encourage 'controlled' functional conflict and change of benefit to the client. The course of conflict is open to influence even under the most unfavourable circumstances (Deutsch, 1969).

The purpose of this paper is to present some results of a current research programme whose objectives are to: (i) identify and classify examples of conflict occurring on a number of selected construction projects; (ii) investigate the extent that conflict is inherent in construction project environments; and, (iii) investigate the relationship between conflict, change and project management strategy.

## 2 Conflict and change

Conflict is defined in this paper as:

> any divergence of interests, objectives or priorities between individuals, groups, or organisations; or nonconformance to requirements of a task, activity or process,

and change as:

> any alteration (by modification, omission, or addition) to a project document, design, process, or method previously approved or accepted.

Handy (1983) suggests five different situations in which conflict can arise. These are listed below together with examples in construction.

*1. Formal objectives overlap.* A consulting engineer may have as one of his objectives the long term safety and stability of a bridge or building. Whilst these are also important to a construction firm, the firm is primarily concerned with cost and profit margins. The engineer who requests unplanned changes at various stages of the project to keep safety tolerances high, perhaps responding to new information, can meet with resistance from the main contractor and subcontractors in line with their objectives.

*2. Role definitions overlap.* This can lead to conflicting objectives. For example, in the installation of building services a conflict may arise over who has responsibility for quality assurance. The site agent working for the main contractor may regard it as the clerk of work's responsibility, who may in turn regard it as the site agent's, insisting that his own role is not to 'inspect in' quality but rather to verify quality has been achieved.

*3. The contractual relationship is unclear.* Is the allegiance of the clerk of works to the client or the design leader, or is he the servant of the site manager?

*4. Roles are simultaneous.* An organisation or an individual may provide both services and coordination. This is typical of the architect's dual role as lead designer and project manager. It may not always be clear to other participants which is the current role.

*5. There are hidden objectives.* The architect may be looking towards future design awards, rather than only the client's needs; or the main contractor may put in an unrealistically low bid to secure a contract during a period of very poor business.

A certain level of conflict in an organisation is not only inevitable but desirable, for conflict is both a cause and effect of change (McGivering, 1983). It has also been reported (Pondy, 1967) that an organisation's success hinges to a great extent on its ability to set up and operate appropriate mechanisms for dealing with a variety of conflict phenomena or sequence of interlocking conflict episodes. Pondy (1967) distinguished five stages of conflict (see also Bowditch and Buono, 1990).

*1. Latent conflict.* Refers to the source of a conflict, such as role conflict or competition over scarce resources. The assumption is that due to certain antecedent conditions conflict 'should' occur. The situations outlined above are examples of latent conflict.

*2. Perceived conflict.* The realisation that there is a conflict, but neither party is upset about it. Perceived conflict may accompany latent conflict or be present when there is no latent conflict.

*3. Felt conflict.* Conflict which grieves the parties involved, but which neither would normally do anything about. Stress and tension are usual outcomes of felt conflict.

*4. Manifest conflict.* Involves openly aggressive behaviours ranging from mild passive resistance through sabotage to actual physical conflict. It is that behaviour which, in the mind of the actor, frustrates the goals of at least some of the other participants.

*5. Conflict aftermath.* The response to, and outcome of, conflict which may involve change. There may be no 'active' response but there will be an outcome, even if it is sustained chronic conflict (continuous, high-level conflict). If a conflict is actually resolved this can lead to greater satisfaction among the participants. If a conflict is not resolved then what appears to be a satisfactory resolution may only be a reversion to a prior level of conflict.

## 3 Project systems

### 3.1 Process system

Construction procurement is a process in that it has a start point and an end point, and there is input and transformation of resources between the two. Project management serves to manage this process, or rather to manage the sum of all the component subprocesses that together constitute the project, including inception, briefing, design, tendering, and construction.

Conflict can occur within single subprocesses, such as design or site investigation, or between subprocesses, such as marketing and design (important for developers), or design and construction. Change takes place to remedy conflict. The nature of the change will depend upon the severity, in cost and time terms, of the conflict. Usually, the longer it takes to discover a conflict in a subprocess, the greater the cost of the remedy. The concept of conflict in a project subprocess is similar to the concept of nonconformity in quality parlance. Design control procedures should ensure that nonconforming design work is recognised and changed before being utilised. Change in this sense includes the 'do nothing' solution which is not to ignore a conflict, but to recognise its presence and approve 'doing nothing' as the best course of action under the circumstances. The key points are: (i) early recognition of conflict within and between subprocesses; and (ii) the mechanism to enact a rapid and appropriate response.

### 3.2 Organisation system

Models of project organisations show the relationships between the individuals and groups involved in construction projects. The following model has been proposed as a result of the present research and focuses on the relationships between the separate component organisations.

*1. Client system (CS).* This term includes all the organisations which satisfy one or more of the following criteria: (i) has the authority to approve expenditure on the project; (ii) has the authority to approve the form the project has to take and its timing; (iii) will be the owner of the project; (iv) will be a major tenant or user; (v) will administer or manage the project upon completion (Walker, 1988).

*2. Project organisation (PO).* The temporary multi-organisation established for the limited and finite purpose of bringing the project into being from inception to completion, and which consists of parts of several separate and diverse organisations drawn from the project participants (including the client system), and whose members will eventually disperse, going back to their own organisations or on to some new project (Cherns and Bryant, 1984).

*3. Client project organisation (CPO).* The intersection of the project organisation and the client system; that part of the client system designated or assumed as having project responsibility.

*4. Project management (PM).* A subset of the project organisation whose responsibility includes one or more of the following management functions: boundary control, monitoring and maintenance activities (in connection with the activities of the project organisation), project recommendation and approval powers (Walker, 1988).

Examples of how these organisations relate to one another are given in Fig. 1.

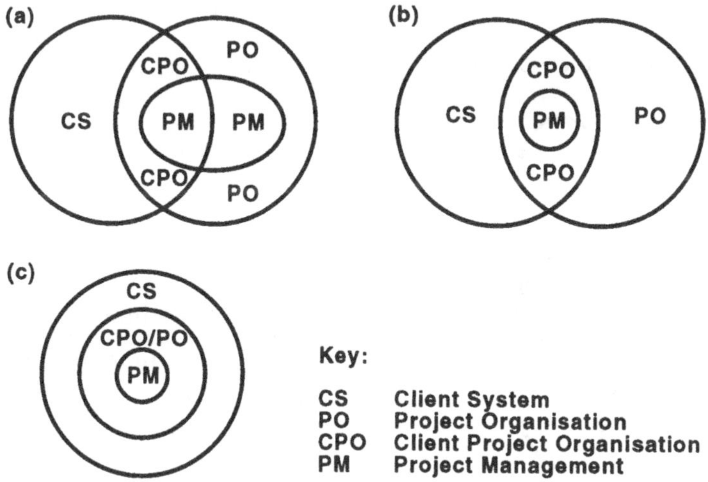


Fig.1. Model of the organisation system.

Fig. 1 (a) is typical of the majority of projects. The role of project management is shared between the client project organisation and at least one other member of the project organisation. In Fig. 1 (b) the project management function is contained entirely within the client project organisation, e.g. many property developers. The situation in Fig. 1 (c) is less usual; the entire project organisation is a subset of the client system, for instance the house development arm of a construction firm.

Organisational conflicts may originate in one person, or in one group in which case they are called *intra*personal, or *intra*group conflicts. Or they may reflect incompatible actions of two or more persons, or groups in which case they are called *inter*personal, or *inter*group. The characteristics and culture of a project organisation are important in determining: (i) the relative frequency of conflicts; (ii) the ability of an organisation to resolve conflicts; and (iii) the likelihood of achieving a productive or dysfunctional outcome.

**3.3 Quality and control system**

A quality management system (QMS) provides a method of working which constantly checks the validity of each operation or activity against identified requirements, highlighting modifications and changes that need to be carried out to continue meeting and to meet more closely those requirements. The presence of a QMS in a construction project can help to secure the benefits of conflict for the client and also control and regulate system changes; document change control is a major component of quality management systems (BSI, 1990). As a client turns its attention to the introduction of a QMS, some of the issues surrounding conflict and change are brought to light and dealt with. This is highlighted by the experience of one of the collaborating clients: "As a result of introducing BS 5750, project control has had to be applied in cases where before it was very much up to the project manager."

Each project subprocess (normally involving two or more participants) and the interfaces between them need to be managed and controlled for the duration of the project. It is unrealistic to expect a QMS to embrace more than one organisation because of administration difficulties, conflicting objectives, and differing organisational structures. Nevertheless, in the event of each of the participants of a

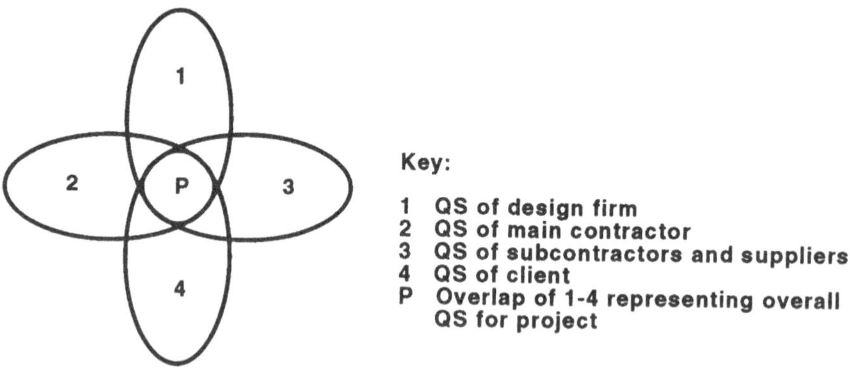


Fig.2. Project quality system.

construction project operating their own QMS, part of each of these separate quality systems would inevitably overlap. Fig. 2 shows the overlap required by the individual participants' quality management systems, including the contribution, if any, required by the client. This concept has been represented to some extent in practice by bringing together the 'quality plan' of each participant of a particular project; sometimes collectively referred to as a 'project quality plan' (Cornick, 1990). Most quality systems include periodic quality audits to review existing procedures. Applied to the formulation of project management strategy, this practice would benefit project managers in situations involving conflict or change.

### 3.4 Contract system

There are two main causes of contractual conflict. The first is the use of inappropriate procurement routes, conditions of contract or contract condition modifications for the project being undertaken. The second arises out of a failure to interpret correctly (by ignorance or wilful intent) the contract conditions in use resulting in misunderstandings, negligence or misrepresentation. Contracts normally include some procedures for resolving disputes and manifest conflict between the parties. They rarely include detailed information regarding the prevention of conflict, nor are they applicable to situations relating to the internal affairs of the separate companies involved for which a quality system is more suitable. There has already been a substantial amount of work done on the selection of appropriate contracts to use (e.g. Skitmore and Marsden, 1988) and the procedure for construction contract claims (e.g. Kim and Adams, 1989) and it is not the purpose of this paper to reiterate this, but to point out that certain types of conflict exist in this area.

## 4 Research methodology

A number of previous research workers have collected information from companies within the construction industry, for example Cherns and Bryant (1984), Fisher (1984), and Bresnen and Haslam (1991). In the present work a new procedure was followed which built on the information gathering techniques of these investigators and added to them a novel method of data processing and reduction.

The collection and reduction of the data formed a five tiered hierarchy consisting of: (i) raw data - tapes of recorded interviews, documentation, letters, etc.; (ii) transcribed interviews; (iii) primary level processing - collation and sorting of all contributions relating to conflicts within a single project as (a) general information and personnel, (b) individual conflict episodes; (iv) secondary level processing - cross checking of information, reduction into summary form, and critical analysis of conflicts; and (v) tertiary level processing - assignment of attributes and statistical analyses.

A preliminary investigation was carried out to verify the methodology which is reported elsewhere (Gardiner and Simmons, 1992). A major study followed the preliminary investigation and some results of this are given later in this paper. A more complete report will be published in due course.

## 5 Analysis of preliminary conflict and change data

In the preliminary investigation, 130 data entries were collected from sixteen recorded interviews across six client organisations. Fig. 3 shows the distribution of conflict and change across the four project systems: process output*, organisation, quality and control, and contract. The system having the greatest number of conflicts is *organisation* (42 occurrences) of which 11 also include change. Fig. 4 shows the distribution of conflict and change across the main subprocesses of a construction project. The subprocess having the most occurrences of conflict and change is *design* with 53 and 47 examples of conflict and change respectively. The success of the construction subprocess depends to a large extent upon the success of the design subprocess, which depends upon the quality of the briefing subprocess, which in turn depends upon the inception subprocess and the reasons behind the decision to build. A latent conflict originating at inception, perhaps between the concept architect, a user representative and the client's projects officer which leads to a change once the

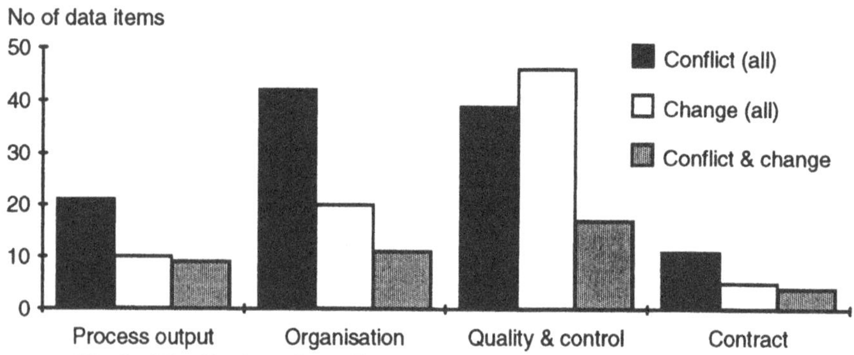

Fig.3. Distribution of conflict and change across project systems

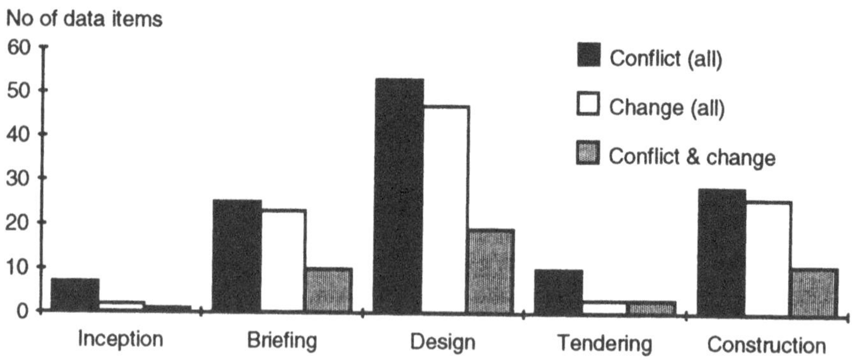

Fig.4. Distribution of conflict and change across project subprocesses.

* Since all the data relate to one or more subprocesses only those involving conflict or change in the finished product or output of a subrocess are included here.

construction subprocess is under way, is likely to have serious implications of cost and time. Table 1 shows the distribution of conflict and change for the subprocesses briefing, design, and construction within the context of process output. It is perhaps surprising that the highest proportion of conflicts leading to change in this system should be in the construction phase, when the cost of change is probably at its highest.

Many situations were identified in which conflict and/or change occurred. A selection of these are listed in Table 2.

Table 1. Distribution of conflict and change across subprocess outputs

| Subprocess | Conflict | Change | Conflict and change |
|---|---|---|---|
| Briefing | 2 | 1 | 1 |
| Design | 17 | 7 | 6 |
| Construction | 8 | 5 | 4 |

## 6 Conflict, change and strategy

During the main investigation three projects from each of the six client organisations were used as case studies giving a total of eighteen projects studied in detail. Using the methodology developed earlier, many interviews were undertaken, both face-to-face and across the telephone, to collect data to investigate interactions between conflict, change, and project management strategies (PMS).

Systematic error was avoided as far as possible by cross checking conflict episodes with several participants. Examination of the data gathered suggests there is a link between the effects of conflict and change and the management strategies adopted. This inference has applications in project management for many organisations and should certainly be of interest to clients. For regular or sophisticated clients it was noticed that some aspects of strategy already evolve from project to project; documented project procedures (Gardiner, 1991) are updated from time to time to incorporate new knowledge and experience.

A better understanding of the relationship between conflict, change and PMS would enable project managers to make greater and more effective use of 'feedback' information in the formulation of their PMS. Fig. 5 illustrates a model of the situation in flow chart form. The model describes the decision making procedure for modifying PMS using conflict and change information from recent projects. Each stage highlights an important concept to be understood or task to be carried out.

A central theme is the hypothesis that PMS can influence the latency of project conflicts in terms of inducing, avoiding, exposing or hiding effects. 'Bad' project management strategy can *induce* conflict in one or more of the four project systems outlined. It can also *hide* the early stages of conflict so that when conflict does erupt

Table 2. Locations of conflict and change

| | |
|---|---|
| **A. Inception/Briefing/Tendering** | |
| Client or users lack experience | Getting a consensus view from users |
| Briefing procedure | Coordination of information |
| User committees: low recognition | User committees: lack of authority |
| **B. Design** | |
| Design error | Insufficient/late design information |
| Design not meeting specification | Getting written approval from users |
| Interpretation of drawings by client | |
| **C. Construction/Operation** | |
| Construction: failure to meet design | Site: quality of work |
| Cost overrun | Running late |
| Operational faults | |
| **D. Project management/Other** | |
| Client's internal politics | Lack of agreement between users and PM |
| By-passing a 'single' point of contact | Conflict of loyalty (e.g. clerk of works) |
| Different levels of change control | Changing emphasis on project change |
| Maintaining interfaces to serve the client | Use/misuse of a quality system |
| Contract condition modifications | |

its effects are more damaging and longer term. 'Good' project management strategy, on the other hand, generally *avoids* conflict but where latent conflict does exist, tends to *expose* it at an early stage when its damaging effects are less or when it may even bring some benefit to the client.

The model also highlights the importance of change control and the need for techniques to evaluate conflict and project outcomes. Scope and design changes occur in many different project situations. What is important here is the control of change. A system which allowed anybody to change anything would be catastrophic. Similarly, a system which prevented the consideration of any changes could result in a building quite unsuitable for its client, or even the discontinuation of a project. Changes themselves are not usually a problem. Most problems arise when there is no suitable mechanism to recognise required changes and enable them to occur promptly with

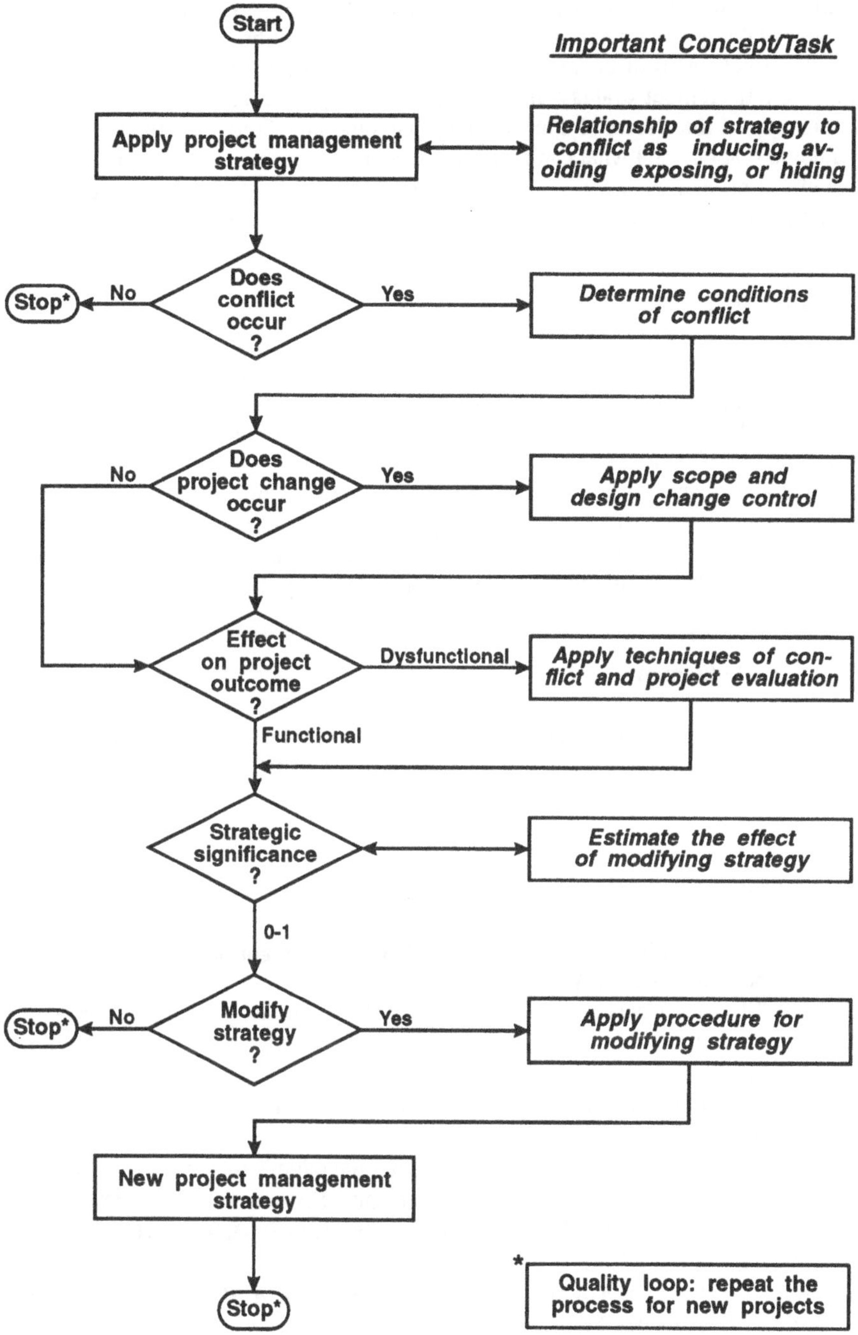


Fig.5. Modification of project management strategy.

minimum disruption. The same mechanism should also help to prevent unnecessary changes taking place without creating additional conflict between the parties.

The final stages of the model draw attention to its capability as a modifier of project management strategy. In this respect the important concepts/tasks are the evaluation of the effects of conflict on project outcome, how these relate to current strategic practice and what can be done to 'capture' this new information for the benefit of project managers of future projects. The model should not be operated in isolation but as a continuous cycle, being reapplied with each new project.

## 7 Conclusion

The investigation suggests that conflict, in one or more of its forms, occurs to some extent in all projects and that this frequently precipitates project change. Certain factors such as the experience of the client, the degree of competition involved in the appointment of participants, and the amount of uncertainty in a project influence the number and severity of conflicts. The relationship between conflict, change and project management strategies can be exploited by project managers as shown in the model outlined, to harness the functional outcome of conflict, resulting in project change for the better, and to limit the damage caused by dysfunctional conflict.

## 8 Acknowledgements

We should like to thank all the contributing organisations for their help and time in the preparation of this paper. We are also grateful to the SERC for the research studentship held by one of the authors (PDG).

## 9 References

Bowditch, J.L. and Buono, A.F. (1990) **A Primer on Organizational Behaviour** (second edition). John Wiley & Sons, New York.

Bresnen, M.J. and Haslam, C.O. (1991) Construction industry clients: a survey of their attributes and project management practices. **Construction Management and Economics**, 9, 327-342.

BSI (1990) Quality Systems BS 5750 : Part 4 : 1990. British Standards Institution, Milton Keynes.

Cherns, A.B. and Bryant, D.T. (1984) Studying the client's role in construction management. **Construction Management and Economics**, 2, 177-184.

Cornick, T. (1990) **Quality Management for Building Design - A Model for Process Control**. Butterworth and Co., London.

Crichton, C. (1966) **Interdependence and Uncertainty: A Study of the Building Industry**. Tavistock Publications, London.

Deutsch, M. (1969) Conflicts: productive and destructive. **J. Social Issues**, 25, 1, 7-41.

Fisher, G.N. (1984) Towards a general model of project monitoring and control systems - as used by broadly similar and successful building contractors, in **Proceedings CIB w-65**, 4, 1351-1362.

Gardiner, P.D. (1991) Example project control procedures of the University of Durham. Unpublished report, University of Durham.

Gardiner, P.D. and Simmons, J.E.L. (1992) The analysis of conflict and change in construction projects. Accepted for publication in **Construction Management and Economics**, 10, Part 4.

Handy, C.B. (1983) **Understanding Organizations**. Penguin Books Ltd., Harmondsworth.

Higgin, G. and Jessop, N. (1965) **Communications in the Building Industry**. Tavistock Publications, London.

Kim, M.P. and Adams, K. (1989) An expert system for construction contract claims. **Construction Management and Economics**, 7, 249-262.

McGivering, I.C. (1983) Conflict. In **A Handbook of Management** (ed. T. Kempner), pp.94-95. Penguin Books Ltd., Harmondsworth.

Pondy, L.R. (1967) Organizational conflict: concepts and models. **Administrative Science Quarterly**, 12, 296-320.

Skitmore, R.M. and Marsden, D.E. (1988) Which procurement system? Towards a universal procurement selection technique. **Construction Management and Economics**, 6, 71-89.

Walker, A. (1988) **Project Management in Construction** (second edition). Blackwell Scientific, Oxford.

# KARMING CONFLICT

R.F.FELLOWS
Construction Study Unit, School of Architecture and Building Engineering, University of Bath, Bath, England.

Abstract
This paper contends that seeding more efficient and effective ways of resolving disputes in the construction industry, although valuable, is a means of tackling symptoms rather than causes and as such, may serve to foster complacency. It is argued that the crucial goal should be to prevent conflicts and disputes arising (or, at least, reducing their incidence to a minimum). In so aiming, considerations of attitudes of participants and the cultural contexts of projects are reviewed and imperatives for changes discussed. A central tenet is that the traditional ethos of the merits of competition amongst groups of players leading to bidding for work (most obviously and commonly through price competitive tendering) is a prime factor in the generation of conflicts and disputes. Alternative approaches have been initiated and have proved successful demonstrating the credibility and advantages of effecting the changes necessary.
Keywords: Conflicts, Disputes, Tendering, Competition, Peoplism, Karma, Financing, Satisficing.

## 1 Introduction

Fenn (1991) discussed the incidence of disputes in the UK building industry. He found that over a period of some twenty years (to 1986) although the incidence of initiating legal action had increased exponentially (from 200 in 1960 to 1150 in 1986), the number of cases coming before the courts was amazingly stable at around 150 per annum.

Several issues emerge:

Is the UK getting more litigious (and following the USA in so doing)?
Is there more, and increasing, conflict and incidence of disputes in the industry?
Are more 'claims' being made spuriously (as 'try-ons')?
Are on-site provisions for resolution of disputes less appropriate/efficient/satisfactory?
Are the parties less competent/flexible/tolerant/forgiving?
Are pressures to pursue every potentially profitable avenue increasing?
Is arbitration less available/appropriate?

The list is far from comprehensive! Clearly, the potential causes are many and varied. The effects include higher insurance premia, more claims specialists in all their guises, increased risks and uncertainties and higher project costs and prices - in the long term industrial context if not otherwise too.

Generally, responses to the issues are rather cosmetic in that they seek to tackle the symptoms - making resolution of the disputes quicker, cheaper and easier to obtain. In so doing, such solutions may, themselves, engender further conflict and disputes.

It is rare for the 'consequential costs' of conflict to be considered. Good relations and communications between the parties to a project foster good performance. Hence, disputes give rise to hidden, consequential costs which arise from reduced performance which supplement the obvious costs - those which frequently form (a major component of) the subject of the dispute. 'Direct' economic loss is difficult to recover (if at all); so what price recovery, or even quantification, of consequential costs?

A somewhat contrary view is that some incidence or 'threat' of disputes can be beneficial in that the consequent heightened awareness of and care exercised by the parties enhances overall project efficiency. However, such a situation, it is submitted, is quite dangerous in that minor disputes could, themselves, escalate into counterproductive conflicts and a number of minor disputes could have a similarly counterproductive consequence due to their synergy.

If the approach that some small level of conflict enhances efficiency is valid, it seems highly likely that a ceiling exists above which counterproductive effects/consequences will dominate.

## 2 Illustrations

In simple, neo-classical economic terms, people and organisations are believed to behave as maximisers; in particular, people endeavour to maximise satisfaction whilst organisations (persona, the entrepreneur) attempt to maximise profits. Even for monopolies - in a free market situation rather than enshrined by legal protection - the extent of profits is restrained by the spectrum of competition from new entrant producers should the profits be regarded by such potential producers as, at least, sufficient to compensate them for the costs of entering the industry. Such a constraint is not very far removed from Baumol's (1982) discussion of contestable markets.

Although monopolies (and close associates) may not be bad for consumers always, even in bastions of free enterprise such as USA and UK legislation to provide some restrictions on the activities of 'monopolies' exists.

Through the divorcing of ownership and management, a dichotomy has been identified between the objectives of owners and managers. The owners, as investors, are concerned with yield on their investment; managers are concerned with turnover. Institutional investors, whether via equities or loans, may have a relatively sophisticated approach to the organisations in which they invest but their 'bottom line' is contribution of the investment to their own profitability (in a portfolio context). Hence, organisations must reconcile the desires for turnover with the need for profitability and although a 'power play' will determine outcomes, the concept of normal profit as an operational imperative remains intact. Even accepting that, for some, construction is a way of life, the necessity for an organisation to earn some minimum profit in the long period, if not universally, is obvious.

## 3 Work allocation

Marketing has been given many definitions. Presently, a popular view is that marketing involves identifying actual and potential customers, determining their needs, gearing the organisation to produce items conveniently for the customers and securing adequate recompense for such provisions. One view, Ohmae (1990), implies that effective and efficient marketing makes the notions of competition virtually irrelevant/redundant. In this context especially, differentiation between marketing and selling is crucial.

Clearly, emphasis is shifting away from price competitive approaches to non-price competition. In construction, recent attention to time performance on projects and the current vogue for quality provision through quality assurance is acknowledgement of the trend. It is regrettable that standard procedures for allocation of work remain focused firmly on price competition as the final/primary factor. In contrast to the marketing philosophy, price competition in construction is increasingly widespread - most notably due to fee bidding amongst design consultants. However, changes are afoot which accord more closely with developments in marketing. Although sub-contractor selection remains locked into price competition, even the infamous 'Dutch-auction', mechanisms for letting work to main contractors is moving towards multi-stage tendering. Simply expressed, the steps in single stage selective tendering are followed but two or three tenderers are selected, usually on price, to give presentations to the client and primary designers of the team, construction methods and management procedures which the contractor will use on the project. Thus, time, personnel, control, quality and other non-price factors re-emerge as the criteria which determine which contractor is awarded the work.

Unfortunately, although such an approach is common for design and build and management contracting, the majority of projects are let more traditionally with price competition's remaining the factor which finally determines the successful contractor. Indeed, it is easy to establish that cost to the client remains the sole factor in not only contractor selection but in project selection too. The argument contends that, due to the capitalistic operating criteria required of most organisations (most obviously expressed as 'profit seeking'), the non-price factors are evaluated through their impact in overall cost and then combined with the tenders to select the best bargain. So, back to price competition but on a global basis. Organisational operating criteria govern behaviour.

## 4 Sunrise in the east

Fellows (1991) discusses the appropriateness of the law of Karma to bidding in the construction industry. It is contended that hard, bad, unfair etc bargains have 'circular flows' which produce results detrimental to the health of the industry through a cause and effect spiral. Given that organisations must earn at least 'normal profits' to survive, being on the receiving end of a hard bargain must produce claims, and other consequences, such as pressuring sub-contractors and suppliers prices downwards, suppressing wages etc to obtain the necessary profits through cost reductions. The likely detrimental effects on relationships, quality and other performance measures are well known. Traditionally, major advances in industrial productivity have been secured by increases in the use to plant and via technological developments. Such advances require investment and, although the construction industry in the UK has changed its methods, approaches and plant-based technologies fairly rapidly over recent years, the funding for the investments increasingly has come from overseas. The rise of overseas, albeit European, investment in UK-based construction companies may be viewed as unwelcome by all apart from those whose jobs are saved by those investments; UK investors being reluctant to put their funds to such uses - possibly due to the period required to yield a return as well as the likely level of the return! However, in comparison with their counterparts in continental Europe, UK construction companies profitability is approximately double.

In many countries, particularly UK, the extension of free-market raw capitalism has been marked. Japan's economic miracle has been extolled as a shining example of the success of such a system which many have sought to emulate. The Japanese model is different. Trades unions, life time employment in one company, attention to quality and product/service performance in general, just-in-time inventory control - all are Japanese - led arrangements which still distinguish that economy. Far from representing raw capitalism, the Japanese approach has been termed peoplism. Peoplism involves developing employees, links investors and bankers to the company and organises markets in order that long term involvements and relationships prosper. The Japanese system does not make shareholders sovereign; companies organise as social organisations in which work is regarded as a mechanism through which people (express their needs to) build social relations.

Whilst UK companies have to sell shares on equal terms to any purchaser, Japanese companies can, and do, sell shares to preferred banks, suppliers etc cheaply thereby creating a favourable and highly supportive ownership through a network of helpful relationships.

Many Japanese companies have over half their shares owned by 'supporters', the resulting financial structure provides security to enable managers to work on the basis of the company's having perpetual life (a similar assumption but with a different basis, level of security and performance requirements from UK) and so can organise and involve employees in the company's future. The upshot is more sharing of information on joint decision taking; groups compete with each other but from a base of extensive internal co-operation.

Japanese legal and financial structures allow external debt to be used by companies to finance investment together with mechanisms to nurture young companies, to protect them from competition and to promote acquisition of skills by people (long term need). High gearing, via bank-based debt financing is common for launches of new products etc. Japanese companies are renowned for aggressive pricing to win market share (even dominance) and cost reductions secured by constantly innovating. The relatively high level of funding for R&D by Japanese companies is well known - many times that of UK construction companies.

In the global context, and in UK, Japan's construction industry appears to be following hard on the heels of its automotive, electronics and optics industries. The reputation includes good, reliable products and hard bargaining; high incidence of disputes and conflict does not feature - Japanese face fosters trustworthiness.

As the UK construction industry is subject to an exponentially increasing incidence of disputes, even if at a fairly constant level of ultimate conflict, the disparities in objectives between parties become highly pronounced. Disparities are accentuated through many work allocation mechanisms, especially those which focus on bidding as the crucial stage, and in forms of contract which, all too commonly, accent and foster the adversarial approach. (Interestingly, it is that approach which is enshrined in the mechanism of litigation in UK.)

The law of Karma, with its counterparts in almost every culture, demonstrates that the consequences of someone's actions, sooner or later, 'come home to roost'.

A clients' forcing down contractors (and consultants) bids on a project is likely to have a quite extensive ripple effect; if sufficient clients behave in such a manner, even simple market theory asserts that supply and demand will attempt to restore equilibrium by liquidations of contractors etc and higher prices from the remainder.

Fortunately, sophisticated, repeating clients, whilst glad of low bids, generally do adopt an approach which is commensurate with the long term 'health' of the industry - any less responsible approach would be detrimental to their own future projects. So, recognising that all involved are in business, with operational imperatives which have some degree(s) of commonality, is likely to reduce disputes through recognition of others requirements and, thence, removing some of the major causes.

## 5 Conclusions

The notion of peoplism is complimentary to an approach which accords with the law of Karma. By fostering a wide network of loyalties on a sound financial base which has a long term focus as well as shorter term requirements (in contrast to the popular belief that UK 'financiers' are concerned with short term performance only - and increasingly shorter terms at that), cost reductions, technological developments and loyal staff are encouraged. All should benefit. Simon (1960) noted that an organisation had to adopt sub-optimal performance against individual criteria in order to obtain acceptable performance over the spectrum of criteria, some of which were likely to be in conflict with each other! If that 'satisficing' approach could be employed by all parties on construction projects, perhaps driven by the possible advantages noted in this paper, it is contended that, as noted the introduction, the industry and its clients would benefit and profit through fewer disputes and conflicts - those direct and indirect costs consequent upon disputes occurring would be eliminated. Pity those poor, unemployed and thin claims lawyers which would result!!

## 6 References

Baumol, W J, Panzar, J C, Willig, R D (1982), Contestable Markets and the Theory of Industrial Structure, Harcourt Brace Jovanovich.

Fellows, R F (1991) Karma Bidding, Habitat International 14, Issue 2/3 December, 115-118.

Fenn P (1991) Managing Corporate Conflict and Resolving Disputes on Construction Projects, Proceedings Annual Conference of the Association of Researchers in Construction Management (ARCOM).

Ohmae,K (1990) How to Win Without Competing, Best of Business International, Whittle Communications International (UK) Ltd, 58-63.

Simon, H A (1960) The New Science of Management Decision, Harper and Row.

# CONTRACTS CAUSE CONFLICTS

STEWART R. CLEGG
University of St. Andrews, St. Andrews, Scotland.

Abstract
This paper draws on field work materials collected from project meetings on construction sites. Far from minimizing or eliminating conflict, contracts cause them. Contracts need to be understood as the key to the institutional framework which constitutes the social relations on the site. The contract seeks to specify the key variables concerning construction. What it cannot specify is its own 'indexicality': that is, how it will be read, interpreted and used by occupationally and organizationally self-interested parties to the contractual relationship. Members of the site organization are oriented to the contractual documents less as a blueprint, however fallible, more as an opportunity in the occasion of site-meetings for re-negotiating to their advantage what they understand the contract to mean. Contracts, as any set of rules, can never provide for their own interpretation. Conflicts are inherent to any situation of complex organization in which are vested inter-organizational and inter-occupational interests, the reconciliation of which proceeds through the formal mechanism of a contract which always requires substantive understanding within a competitive profit-oriented context.

The argument is illustrated with examples drawn from the industry, such as 'normal clay' which are shown in their naturally occurring form - as transcriptions of actual site-meetings.
Keywords: Construction sites, contracts, conflict, indexicality.

## 1 Introduction

In the early 1970s I conducted field-work on construction sites in the north of England, studying one in particular. I was actually doing a PhD at the time, at The University of Bradford Management Centre, into the topic of 'Power in Organizations'. I ended up on the sites because a part of the process of gaining the PhD was to collect organizational data. I had once worked in the construction industry, as a joiner's labourer. Finding that access was difficult to the media organizations which I had initially wanted to study, I turned to a setting that I knew well from this past work: construction sites.

I had wanted to study media organizations because I had become interested in what in the literature on power were called 'non-issues and 'non-decisions'. These were the things which did not happen because they were prevented from happening by the present disposition of power in a setting. I reasoned that if I were to study the ways in which the myriad of things which get to be the news, and, more importantly, didn't get to be the news, were constituted in the organization, then I would have some empirical purchase on a key issue in the current theoretical literature.

When I switched to construction sites it was with no such well-rehearsed idea in mind. Here my approach was far more exploratory, far more the typical field-worker with no real a priori sense of what would be, and what would not be, interesting. However, I knew that I still had to produce a thesis on 'Power in Organizations', so, combined with this very exploratory data-collecting approach there was a strong sense of analytic focus.

As a methodology I had resolved to adopt an approach which was relatively novel at the time. Instead of preparing a formal questionnaire I carried a small portable tape recorder and tape-recorded both what are referred to as 'naturally occurring conversations', that is discussions and conversations which I was not party to and which would have transpired in my absence, as well as discussions and conversations which I initiated. It was a rather laborious research method, particularly when it came to transcribing the data, but it threw up a rich seam of accidental data. By 'accidental data' I mean to suggest that this data was not something which, a priori, I would have even have thought of as significant. Yet, it proved to be. I doubt that I would have ended up thinking about power, contracts and construction sites in quite the way that I did, without this 'accidental data'.

## 2 The site

The people on site knew that I had the tape recorder and knew that I had permission to use it, although they rapidly forgot that both I and it were present in most instances. Only once did it come up as an issue. My reason for having the tape recorder, I said, was because it was important for my research that I taped the actually occurring conversations which comprised managerial work in the organization. Rather than speak to ideal or theoretical models of what management was, I would tape it , in real-time, and thus have a sounder data-base for subsequent work.

The site was a contract to build a multistorey car park and bus station for the Local Authority Borough Corporation of a Northern Town. I joined the site at an early stage in its development. Much of the time I spent either in the main administrative building, a Portacabin, or wandering around the site, observing, taking notes, occasionally taping, although most of my taping was done indoors for obvious reasons of audibility. As I am sure everyone knows, construction sites are often very noisy places. Sometimes I would wander around the site in my Donkey Jacket, jeans and hard-hat, looking much like any of the other workers on site. Other times I would sit in a corner in the site hut, at a table piled high with bits and pieces, including my tape recorder, in the room which the Project Manager occupied. Often I would have the tape recorder on as I read a book or a photocopied paper, breaking off every now and then to talk, to walk, to lunch and tea-break and so on.

By chance, I found that I was often present at a number of more or less impromptu site-meetings. Some were, in fact, quite formal ones. These became the major, although not the only, source of data for the study.

### 3 Previous research

The data that I collected from these meetings took the form of tape recordings of what actually was said as it happened during them. At first sight the findings that emerged from these data might not have seemed too dissimilar to some of the existing findings that had been produced by social researchers who had researched construction sites in the past. There had been at least one major piece of work done on British construction sites by social scientists. A project had been commissioned by the Building Industry Communications Research Project into 'communication' on sites. The work was undertaken by researchers from the Tavistock Institute and published as *Interdependence and Uncertainty: A Study of the Building Industry* (Higgin et al 1966). This was very much a piece of research into 'normal' sites: as the researchers said at the outset 'In selecting projects for study we concentrated on those which seemed likely to go well. No purpose was seen in criticizing projects which were obviously inefficient' (Higgin et al: 17). Despite this, they found that 'normally', 'none of the projects seemed to live up to expectations. . . misunderstandings,delays, stoppages and abortive work', resulting from 'failures in communications, and impressions of confusion, error, and conflict' were the norm.

While practitioners might find this troublesome for me it was reassuring. The data that I had collected did not seem to be aberrant: the misunderstandings, delays, stoppages and abortive work that the data demonstrated were normal, it was business as usual. When I encountered joiners on site, idly kicking a football about and then telling me 'there must be something wrong with this bloody job' (Clegg 1975: 'The Joiners' Tale': 87-91), which they went on to elaborate in terms of the lack of managerial control, I need not have worried; when I observed and taped the Project Manager, Office Manager, Measurement Engineer and General Foreman 'Cooking the Books' (Clegg 19075: 91-100) and constructing fictive figures to disguise the appalling weekly results, then I should not have been alarmed; when the Office Manager later insisted to me on the obdurateness of the figures which were being cooked (Clegg 1975: 'Them figures . . .are figures you can't argue with': 102-107), and claimed that they had a reality which was distinct from that which was being represented in the weekly accounting terms, I should have felt no surprise; when, later still, in terms which were almost Weberian in their lauding of efficient, formal and bureaucratic organization, he castigated the management of the present site for its alleged incompetence (Clegg 1975 'Al, the ideal typist': 107-119), I should have realized that this was just the everyday achievement of the British construction industry. No worries, no problems, nothing amiss: a typical site, the usual work, the characteristic organization.

## 4 Organization: formal, informal and faithful

How was this regular achievement possible? That was the sociological question which both I and the Tavistock researchers set out to answer. Before proceeding to my own interpretation, let us first consider that proffered by the Tavistock Researchers. At the outset they reject the terms which were often presented to them by industry figures. The record, they say, has 'commonly been seen in personal terms - incompetence, laziness, or financial greed of others for example, and although bitterness, and even hurt, can be given by accusations in such terms - these behaviours are seldom crucial' (Higgin et al, 1966: 52). Instead,they identify the real reason in a disjuncture between the 'formal' and the 'informal' system of organization.

The formal system of organization is identified as that which is laid down in formal tomes such as the RIBA *Handbook of Architectural Practice and Management*. In this formally sanctioned view of organizational practice are emphasised the independence and sequential application of tasks such as briefing, designing, design quantification, construction planning and control, manufacturing, sub-contracting, and so on. The formal model of organization assumes that these tasks have a 'sequential finality' which

> does not seem suited effectively to control a process characterised by the interdependence of its operation, fraught with uncertainty and requiring carefully phased decisions and continuous application of all control functions (Higgin et al, 1966:45).

Interdependence and uncertainty result, it is claimed, from the functional demands of the building process. Interdependence arises from the 'relevance of different streams of information to each other in particular contexts' (Higgin et al, 1966:45). The construction site is a complex inter-organizational world in which many different types of specialist knowledge are required at different stages in the construction process. Any decision taken at one time, with respect to a particular application of specific knowledge(s), may well have wider implications in space and time and for other forms of specialized knowledge in the site organization. It is this which introduces one potent source of uncertainty into decision-making. Yet, there are at least two further sources of uncertainty, they say:

First, there are the uncertainties engendered by the action of those not directly involved in the building process, such as government departments, planning authorities, public bodies, client organizations, and even the general public. . . Second, there are the uncertainties which stem from resources: labour, equipment and materials (Higgin et al, 1966: 34)

All these sources of uncertainty make the formal organizational model one which will not work well in practice. In consequence, when we actually observe what it is that professionals and other people do on construction sites we find that

the characteristics of the formal system are so much in conflict with the control functions required to achieve effectiveness in the system of operations that, in practice, the formal system cannot be closely followed. Rigid adherence to the procedures of the formal system would not be possible, under normal conditions,without unacceptable expenditure - particularly of time. In practice, reality forces a recognition of interdependence, uncertainty, phased decision-taking, and the continuous application of functions. It forces members of the building team to adapt themselves (Higgin et al, 1966: 46).

One can construct from their research an ironic juxtaposition of what formal organization recommends and the informal organization which actually occurs. I shall summarize the points. In theory, design is completed at an early stage. In practice it is not, to a far greater extent than is recognized by the provisional items in the Bill of Quantities. In particular this is the case with the design of services, related to the sequential manner in which the design process is usually handled. In practice, much of the detail of service design is worked out on the job, during tours of work after site meetings , for example. In theory, the quantity surveyor should quantify the job in detail prior to competitive tender. In practice, there is rarely sufficient information to do so. In theory, the full working drawings should precede the preparation of the Bills of Quantities. The RIBA *Handbook* is quite explicit about this: 'Final decisions on every matter related to design specification, construction and cost, and full design of every part and component of the building should be embodied in these drawings it insists. An ominous warning, in bold type, alerts one to the onerous consequences: 'any future changes in location, size, shape, or cost after this time will result in abortive work'. In practice, of course, this is a caution more often honoured in the breach than in obeisance to the formal model. Finally, the ironical juxtaposition between theory and practice reaches into the very heart of the social relations that occur on construction sites: their constitution by and as a *contract*:

The contract, in theory, is arrived at as a result of tendering procedure which is considered to be a legally and commercially rational bargain between the client and the builder - generally the builder who can undertake the work most cheaply. This view is based on an assumption that all details of the project have been finally decided and are specified in detail in the tender documents, and that the contractor can anticipate accurately at this time what all his costs will be. This is not so and it is not surprising, therefore that the builders pricing and the client's acceptance of any competitive tender must always be acts of faith (Higgin et al., 1966: 47-8).

At the core of the rationally binding contract is an act of faith! The act of faith lies in the tacit acceptance of a model governing construction site relations which is a formal fiction and of little practical moment in the mundane life of the site.

In many respects the observations that I made supported the general impression of the Tavistock researchers. The idea that the contractual documents are a series of instructions, or formally complete and binding rules for constructing a structure from its 'detail' cannot be sustained for long after one has observed a site in progress. Yet, at the point of *explaining* and *interpreting* why this should be so I would wish to proffer a somewhat different account to that of the Tavistock researchers.

## 5 Contracts, construction and conflicts

Construction sites are constituted by contractual relations to a greater extent than many other kind of organization. It is not only labour contracts with personnel which are central, but the contractualization of virtually everything: who can do what, where, when, in what sequence, with what materials,with what technologies, at what standard costs an so on. Virtually every contingency, according to the formal model, has been covered, yet still, in practice, it remains an act of faith. Why?

The Tavistock researchers argue that it is because in practice communication problems and uncertainties constantly undercut the formally secured organization which the contract seeks to concretize. This formal organization is a stable set of meanings, of interpretations of documents, which are supposed to govern the site. It is when these are challenged, as a result of contingent 'uncertainties', by some members of the site organization, that 'communication problems' arise. These exigencies then modify the formal organization into an adaptive informal organization.

I wish to proffer a contrary, but related, view. Uncertainty should not be seen simply in terms of anticipated if specifically unforeseeable informal modifications of the formal organization wrought by a reality recalcitrant to rationality plotted imperiously elsewhere and in advance. The 'misunderstandings, delays, stoppages, and abortive work . . . confusion, error and conflict' do not just represent 'communication problems'. These 'communication problems' are seen as resulting from a collision between an uncertain reality with the formal organization. The uncertain reality takes the shape of the actual, real, informal organization, while the formal organization ism, by contrast, somewhat unreal: it doesn't actually exist. The formal organization in this model has hardly any reality at all: it is just a set of symbolic signs, words and drawings, which have only a hazy and problematic relation to what people actually do on site. It is supposed to be 'uncertainty' which makes the formal organization unreal, unrealized, nothing. This is because actual practice on site, according to the Tavistock researchers, can not conform to the ideal model because the model does not recognize nor can it cope with the everyday 'uncertainties' which occasion the prolific number of 'communications problems'.

Uncertainty may be defined in a number of ways. I regard it, analytically, as a lack of knowledge about how to go on, an absence of rules for remedying surprise. Uncertainty, defined thus, does not characterize construction sites. People do carry on, buildings do get built, they can be recognizably related, in the future, to the future-perfect representations of them which are today's plans (Schutz 1967; Weick 1969). The formal organization is not a useless model, something which is simply a case for ironical treatment by social researchers. It is regularly and routinely invoked by as a part of the organization of social action on site. It is contained in something which my field-work demonstrated is referred to constantly in the normal course of the site work. It can be found in the detailed contractual documents comprising the 'bill of works' on which the contractor tenders. It is these which are constitutive of the specialist trades, professions and practices and their inter-relationships - the knowledges - which one can find on and around the site. The contract is only barely contained in these contractual documents. By this I mean that its meaning, its interpretation, is never self-evident. It is doubtful that in moderately complex organizations that can it ever be so.The contract functions as a potent symbol on site, waiting to be enacted, to be made meaningful, by the possessors of various knowledges, those who hold differential means of interpretation of the formally fictive unambiguity of these documents.

The contractual documents are never unproblematic, never unambiguous, because they can never be *unindexical.* Indexicality is a technical term. It refers to a situation where the meaning of something is always contingent upon someone interpreting it. Such an interpretation always 'indexes' the particulars of the occasion of its interpretation. It is dependent on who is making the interpretation, from what interests, from what knowledge, at what time in the unfolding drama of the site. The contract is some thing which is never, nor never can be, apparently matter-of-factual, that is, without need for interpretation. By the notion of an 'interested interpretation' one means to suggest that no interpretation is ever innocent of interest. Different knowledges, different positions in a hierarchy, different personnel in a network of inter-organizational relations, different times in the temporal flow of events or spaces in their spatial location, can always produce differentially interested interpretations of the matter-at-hand. Hence, indexicality is irremediable. It is, if you like, a part of the human condition. Thus, conflict is ever potential wherever there is indexicality. And, where there are attempts to frame matters unindexically, in complex organizational contexts extending across space, time and knowledge, through contractual documents, there will always be indexicality, thus the possibility of conflict. It is endemic.

The difference from the account of the Tavistock researchers should now be evident. There is not the (unreal) formal organization and the real (informal) organization. Actual organization on site, that which is real, is not something which uncertainty makes of the formal structure, modifying it into the informal organization. By contrast, I would maintain, it is something which the interested members of, in and around organizations make of uncertainty and the formal organization.

Uncertainty is not a naturally occurring state or an act of God. It is something made, produced, by the site organization members out of their grasp of the indexical nature of the documentation of the 'formal' organization contained in the contractual materials. These formal organizational contractual documents provide the constitutional and constitutive grounds and framework within which the meaning of the contract is negotiated, contested, and sometimes contained.

On site these processes take place through the medium of site-meetings, the meetings which I was able to tape in real time as a part of my study of organizational work as it happened, in *audio verite*. These are socially organized procedures for constituting, formulating, and discussing 'issues', what is issueable and what is not. With site-meetings I found that I had come analytically full circle. Construction sites, just as much as media organizations, afforded access to the creation of issues. Issues are instantly recognizable as such because they are addressed as something formulated as a problem. They provide a focus for practical reasoning about the issue of issues - what is to count as an issue and how it is to do so.

In a situation on sites where the contract formally covers every contingency the only remaining contingency is the contract itself. This is why contracts cause conflicts on construction sites. No contract can ever provide for its own interpretation because interpretation is not disinterested. In a complex inter-organizational reality such as a construction site the interests, embedded in different knowledges, different organizations, different hierarchies and different levels in the same hierarchies, are complex. Consequently, the interpretations are rarely uncontested. The contestation is not whimsical, not merely inter-personal (although that undoubtedly enters into it sometimes) but embedded in different and distinct knowledge-practices and associated interests which produce different ambiguities in, and different ways of seeing, the documents. Perhaps some examples may serve to make matters clearer.

## 6 Mocking up

Sometimes one would find techniques such as 'mock-ups' of models used to try and resolve an issue. It was during one such session that I began to grasp the way in which contracts function to create conflict on site. There was a discussion of the interpretation of an aspect of the contractual drawings, an interpretation which was being inhibited, the Project Manager said, by the inability of the Client's Architect to visualize in more than two dimensions. Hence the mock-up. The mock-up functioned as a device to make the drawings seem less indexical and to secure one interpretation over another. However, corollaries of the Project Manager's favoured interpretation rapidly emerged: 'V O's' (variation orders), 'star-rates', including a 'buggeration factor', all rapidly were made apparent (Clegg 1975: appendix 3; also pp.132-5) in the context of an exquisite appreciation of organizational time. The Project Manager introduced this 'discourse of temporality 'as follows: '. . . well, there's no skin off my nose really, in doing it, but I'm not doing it now, I will do it, if he gives me a V. O. to cover it, and thereby, it means he pays me extra . . . for doing it'. It turns out that 'now' indexes a time after the issue of 'normal clay'. There was time before and there was time after 'normal clay'.

## 7 Normal clay

In brief, the issue was simple. The Engineers drawings instructed excavation to a minimum of 600 mm. into 'sandy, stony clay'. It did not specify the depth at which this 'sandy, stony clay' occurred. Accompanying these drawings was a consultant's bore-hole report of a site-survey of ground conditions. This recommended that the contractor should excavate to two metres into clay. It did not specify that there was any clay other than 'clay', and it did not differentiate between 'puddle clay' and 'sandy, stony clay', a distinction raised by the Project Manager and elevated to some importance in the discussions that ensued. The Project Manager excavated the bases according to his interpretation of these details, which was to excavate to 'normal clay', which he defined as 'stony, sandy clay'. The resulting depth of his excavations, and the way in which he organized them, became the subject of an acrimonious letter from the Clients Architect to the Construction Company. The points at issue resulted from investigation of the claimed actual excavation levels', as the letter put it, by the Architects Department revealing 'little or no consistency'. The architect's letter advances as reasons, that first, additional excavations were made to the first two or three bases after they had been checked by the Clerk of Works as being at the 'specified levels', without his being explicitly informed. Second, the letter suggested 'that a general instruction' was given by Construction Co.'s site-staff to the excavator driver to go down as far as he considered necessary, irrespective of the consulting engineer's drawings' Third, the architects letter proposed that there was an inconsistency in levels because Construction Co.'s 'staff considered the basement-levels as shown (on the architect's drawings) to be incorrect' and had not informed the Clerk of Works of this. That the 'satisfactory bearing strata varied from bases to base' is not considered as likely a reason as those previously advanced, with the 'more likely' explanation being that the 'excavator driver carried out his work with wrong instructions and/or inadequate supervision'.

It was evident that the Project Manager felt strongly and personally about the whole issue: 'Yesterday when I got that letter, yesterday morning, I got the shakes, believe it, you know,I was so angry, and I'm still feeling a bit that way. Questions of competence and managerial control were at issue. Not just those of the Project Manager. He, in turn, was to accuse the Clerk of the Works of incompetence - 'your Clerk o' Works can't even read a level', he said. Moreover, in an interestingly unobtrusive measure, a part of the Project Manager's case for 'normal' clay develops around his estimate of a 'normal machine-driver':

| | | |
|---|---|---|
| 2.85 | Project Manager | But the other thing that absolutely proves that we did not turn a machine-driver loose, if you take the levels/ |
| 2.86 | Foreman | Yeah/ |
| 2.87 | Project Manager | off that sketch on our bases/ |
| 2.88 | Foreman | Yeas/ |
| 2.89 | Project Manager | they're pretty constant throughout the site, |
| 2.90 | Foreman | Yes, yes. |
| 2.91 | Project Manager | but . . . oh, sorry, they're not constant, some are up, some are down according to local conditions/ |
| 2.92 | Foreman | Yeah/ |
| 2.93 | Project Manager | it's the other way around/ |
| 2.94 | Foreman | Yeas/ |
| 2.95 | Project Manager | but if we'd turn a man loose, he'd have dug 'em all to the same bloody level, he wouldn't have, he's not going to go down an extra metre on one hole just 'cos it suits him. |

The numbering indicates the appendix and utterance numbering in the original transcription of the text in the appendices of Clegg (1975). Hence, 2.87 indicates the 87th utterance transcribed in appendix 2.

Of course, as quickly transpired in subsequent material, these matters of opinion or moral estimate of the 'normal' machine-operator, are of little or no moment. What is important is what can be settled by the 'documentary method' (Zimmerman 1971) of investigation. Features of the setting which are investigable matters that can be settled by documents need to be evoked. Had the Clerk of the Works, who, reputedly, can't even read a level, been available to check the levels to which Construction Co. excavated, there would have been documentary proof of the levels excavated and the Borough Corporation's agreement to these as the client. The Client Architect argues that the excavations were deliberately made when the Clerk of the Works was not available to check them.

The crux of the issue becomes the V. O. that Construction Co. claimed for the extra work. The extra work was claimed to be necessary because of the inconsistency between the bore-hole report and the consultant's drawings. While the latter specifies 'sandy, stony clay' the former only notes the existence of clay. The Project Manager claims to have followed the instruction on the drawing, rather than the recommendation on the bore-hole report, and to have excavated to 600 mm. into the 'sandy, stony clay'. The Project Manager's retort is that had the Clerk of Works checked the excavations he would have seen that it was necessary to have excavated to a greater depth than that recommended by the bore-hole report. He accuses the Clerk of Works of not checking the levels, because he does not know how to do so! It was now too late, after the receipt of the letter, to check some of the excavations, because they had concrete poured into them and had been 'backfilled' around the pour with a composition of pebbles and gravel.

The Corporation's formal accusation is that an unnecessarily large amount of earth had been excavated because Construction Co. had put in a claim for additional work involved in doing these excavations, over and above that which the bore-hole report recommended they would have to do. As the Consulting Engineer put it

> You'd think that, it all comes back to this, if you were asking for ten quid they'd give it to you rather than haggle with yer, (. . . ) but because you're asking for, probably three thousand to eight thousand, or whatever the figure is, that's where you're coming unstuck (Clegg 1975: appendix 4)

Everything becomes clear. 'Normal clay' has been raised as an issue by the Corporation's representatives, because, if conceded, the claim will cost them a considerable amount of money. And of course, if conceded, it will also earn Construction Co. an additional sum on a job in excess of the contracted-for expectation of contribution to profit. It was to prove to be the turning-point in the way in which the indexicality of the contractual documents was subsequently addressed. 'Normal clay' shattered that fragile faith of which the Tavistock researchers wrote. It was the Corporation who raised 'clay' to an issue, because they were not specifically asked to check and agree the levels and because they had found that some of the excavated bases which had not yet been poured and filled were deeper than the level recommended by the bore-hole report. They thus have one interpretation of the indexicality of the documents which proposes that two metres ought to have been the depth of the excavation. Using this 'documentary evidence' they are able to point to the actual excavations, and the claimed depths, and say that there was no reason for digging out to these depths. This is denied by the Project Manager in terms of his interpretation of the indexicality of the documents. He uses the engineer's drawing instruction to excavate to 'sandy, stony clay' as his 'documentary evidence' for the need to excavate more than the recommended two metres, because this 'sandy, stony clay' was not encountered until a greater depth than this two metres:

4.202 Project Manager We don't pour on any old kind of clay we dig up, it all comes down to what is clay, we couldn't have rooted out any higher on those bases, it was only puddle clay, we wanted that sandy, stony clay, whereas you say we can't differentiate between the two types of clay.

4.203 Architect Well, this, has been my whole point all along, what is clay, what is stony clay, and what will basically take three tons?

The Architect continued to maintain an interpretation in terms of the bore-hole report while the Project Manager continued to argue his interpretation in terms of the drawings, and the necessity, on occasion, of having to dig deeper than the two metres recommended, if he was to fulfil the instruction of excavating to 600 mm.. into sandy stony clay. He sought to support his claim further by invoking some investigable features of the dispute. It was necessary to dig deeper, he maintained, due to variations in the strata, variation in ground conditions - that some of the excavations were in old cellar bottoms, and because of the 'unreasonable' assumptions about the machine-drivers behaviour which would be necessary to sustain the fact of the different depths excavated.

I do not know to this day what the outcome of the dispute was nor am I interested particularly. My interest in 'normal clay' is and was purely analytical. What interests me is the way in which, throughout the dispute, it is the contractual documents, supposedly the source of unambiguous instruction in the formal RIBA model of the building process, which are the occasion and opportunity for the conflict. It is the necessarily and irremediably indexical nature of these contractual documents, documents which might well seek to formally stipulate almost every conceivable matter-at-hand on the site, but which can never stipulate nor provide for their own interpretation, which are crucial. It is the differing indexical interpretations of the documents which enables the matter to raise above the level of personal bickering.

'Normal clay' was not a one-off or isolated event. In the experience of the site, time became constituted as' time before' and 'time after' 'normal clay', in terms of the frames of reference of the protagonists. Subsequently, as became apparent, the frequency of what Tavistock would have called 'misunderstandings' were on the increase. It should be clear that I would see this not as a result of 'misunderstanding', 'communication problems' or 'personal troubles'. We are not dealing with just personality problems, a lack of understanding or communication problems. We are dealing with the predominant form or mode of rationality in a construction industry organized around the principle of the competitively tendered contract. 'Normal clay' exemplifies the normal mode of functioning of power/knowledge relations on construction sites. It represents the strategies of power of the participants in the site organization seeking to maintain control over costs and profits. 'Normal clay' threatens that control, because it threatens its measure.

After 'normal clay' the grounds of the Project Manager's 'understanding' of the documents became apparent. We can invoke his own words: 'Now' he will not do any 'little favours' like setting some 'ramps' for the architect, unless 'he pays me extra' (Clegg 1975: appendix three; also see p. 149). Now, unlike some time previously, he has both a grudge against the corporation, because of the letter, and he has to find some additional profit to cover the losses he is likely to sustain on 'normal clay'. We could, if one wished, concretize this as a rule: 'Always make (out) of the bill of works what you can'. This is not always made for direct payment for the' buggeration factor', and any other extras, through star-rates, variation orders, and the like. It may more readily be made by 'speeding up the programme' thus reducing some of the 'variable costs'. These would be costs such as the length of plant-hire time and the period over which a full wages bill has to be paid.

These various issues, and many more from the data that I have left unaccounted here (see Clegg 1975), are strategies employed by the Project Manager to try and reassert control. His 'understanding' of the bill of works that he is to build from varies with the state of play in the site organization. After 'normal clay' issues become far more transparent, because 'normal clay' makes them so. Issues are constructed by the Project Manager against the possibility of eroded profitability. 'Normal clay' presents the Project Manager with a view of himself which others had presented to me: as someone for whom control was slipping away. The job and the person are intertwined, so that a profitable job displays a worthy self. Where profit is attacked or eroded, its preservation becomes a personal matter.

The issues which have led to the characterization of construction sites as loci of 'misunderstandings', 'communications problems' and 'conflicts' share a common rationality. This rationality is not the result of a functional informal organization emerging and operating out of necessity because of uncertainties which undercut the formal organization. On the contrary, the formal organization, as it is instantiated in the contractual documents, the bill of works and so on, has a crucial role to play. It should not be consigned to the non-existant and negative role which the Tavistock researchers scripted for it. The issues of conflict which occur are constructed out of the constitutive basis of the organization as a rational occasion for its members to re-negotiate the formally-contracted-for building. This constitutive basis consists of the contract. Hence, as occasions for rational agents to exploit indexicality in their interest, contracts cause conflict.

## 8 What is to be done?

If conflict is normal and endemic to the form of contractual life which governs construction organizations what is to be done about it?

Perhaps the most important step is to admit a degree of realism in to the procedures. Little point attaches to seeking to remedy 'communications problems' through ever more attention to the communication process if these problems are not so much an inadvertent and unanticipated consequence of the building process, but endemic to it as it is currently organized. Nor does much point attach to endlessly re-writing the contractual documents to correspond more closely to the model of the formal organization. To repeat: no set of rules, however complete, can ever provide for their own interpretation. Where there are differentially embodied configurations of power/knowledge in organizations there will always be an interest in indexical exploitation.

This has a bearing on some of the recent discussions of construction conflict (Fenn 1991a; 1991b). Construction is contentious, it is plagued by disputes, and under the tighter economic conditions of recession, one would anticipate that there would be an increase in these figures in terms relative to levels of industry activity. Fenn (1991b: 47) asks the question: '*Why do a substantial percentage of construction contracts end in serious dispute?*' I think that the analysis provided in this paper goes a long way towards answering this question: because it is rational for them to do so.

Various strategies, including Alternate Dispute Resolution (ADR) have been mooted to deal with this state of affairs. The assumption behind ADR is that the parties have an interest in resolving conflict, while my analysis suggest that in the day-to-day management of construction sites there may well be normal circumstances in which this rule will not and does not operate. However, in so far as disputes appear to be irremediable, then perhaps the parties may be persuaded that, in some cases, the most effective and least expensive means of resolving them would be both appropriate and realistic. In other cases, of course, where the contracted-for contribution to profit on a job is making a less than adequate return, there may be very little interest in resolving matters economically and efficiently. In such circumstances realism might dictate the full majesty of the adversarial legal process in the hope of the return that a well briefed legal representative might deliver.

Other options exist, of course. Perhaps the most fruitful might be 'partnering' (NEDO 1991), but, if this is to develop fully, it will require a fundamental change in the cultures and rationalities of everyday life in the construction industry. Indexicality as an opportunity for the amplification of mutual understanding rather than particular advantage is a far cry from the more rugged forms of exploitative understanding which have been charted here. Research, and experience, suggest that such transformations in cultures and rationalities are not easily produced. It may be that I am being unduly sceptical, too much the sociologist? Perhaps, rather than the self-interested and eminently skilled knowledge-brokers of the everyday organizational construction-site life that I would see, you, as professionals in and around the industry, would see something quite different? If so, I look forward to hearing it and hoping that this conference will throw up some constructive suggestions to temper my sociological scepticism. Perhaps, then my contribution might only be to urge a focus on the empirical realities, rather than moral positions, grounded in the rhetoric of professional practice, which surround these at every turn. Such stories are not convincing; in the past, I doubt if they have convinced many people who felt drawn to this conference and, in future, they should not serve to make the issues any more confused than they need to be.

Let us be clear: normally, contracts cause conflicts. The reasons why these are particularly acute in the construction industry have to do with the way in which the contractual form has a specific relationship to the economic rationality that professionals, such as Project Managers, routinely use in this industry. It is by making indexicality out of the contract that a contribution to profit can often be enhanced, in a way which we do not find in manufacturing industry, for instance. Indexical sub-contracts which are turned into occasions of conflict do not enhance orders, profits or potential in manufacturing. There are plenty of other variables, in the labour process, the marketing, the technology, etc., which are open to manipulation. On construction sites the symbolic and material centrality of contractual relations which specify almost every conceivable variable contained in the bill of works precludes access to the range of remedies that other types of industry might have available.

In construction, where the one-off product is essentially a complex process, to be crafted for a previously contracted customer, bid for under a tight regime of competitive market discipline, there are no variables for remedying anticipated profit other than those which are specified in the contract itself. To vary those, reasons have to be found. Thus, the contract's indexicality. Where the agents own conceptions of what constitutes rationality lead them to exploit indexicality, where the profit margins are tight or even non-existent (the job being accepted to make a contribution rather than a profit per se), they would be foolish not to do so. Of course, not every job will fall under that judgement and so conflict will be variable.

There is independent confirmation of this insight from other research done on the design process in a firm called for the purposes of the research, Fraser, Railton and Springfield (Linstead and Grafton-Small 1990). In this work, looking at the site meetings from the design side, one encounters the 'fudge factor' (Linstead and Grafton-Small 1990: 409)

> whereby deliberately low tenders could be jacked up to profitability by dexterous use of the special effects clauses. The builders involved, in a rare uncharitable moment, called it 'pissing about.' It became clear as the meeting went on that many builders avoided this sort of work whenever possible. Others found it exciting and tendered for it with relish. Should one of the former have the contract, as was the case here, it fell to the designer to reassure the customer of the service being given and to spot any last minute alterations which might facilitate payment.

Later in their account we find out some more about how the contract got to be constituted in the first place, by Fraser, the senior partner in the design partnership.

> Fraser's own operation is an uncertain one, and he uses the environmental uncertainty to control his designers internally. Failure to work within tight budgets is 'screwing up.' These budgets are of course dictated by the 'objectivity' of completion, yet Fraser is remarkably possessive and secretive about the special way his deals are done. He inhabits the organizational boundary between two forms of ignorance - the designers who don't understand cost constraints and the customers who don't understand design. Ultimately, with no means to

assess in advance the competence of a highly technical profession, the customer has to rely on trust - the architect, in common with all professionals although it is rarely realised, has an important problem in developing this trust.

This can lead to boundary collusion between salesman and customer - 'a fast P.R. job over an expensive lunch' and reflect back on the internal tensions within each operation - 'we pick up the pieces of his bloody disastrous tendering'. The process of servicing the contract being problematic is backgrounded, the power dimension being explicit as the architects are treated in the same way as the jobbing builders - there's the job, there's the price, do it. The architects, however, cannot chose not to do it as the builder can, although builders who don't like the work must sometimes take it because they have to.

The case rests. Contracts cause conflicts because they are the rational occasions whereby indexicality can be exploited by self interested professionals in the design and construction process. The discrete charm of the RIBA draws a veil of professional rhetoric. In this paper I have sought to shift this veil in order to glimpse the reality of the construction and design process. What we see in these cases is just as problematic as the Tavistock researchers might expect, although the explanation that I have proffered as to why it should be so, is one which differs considerably.

As we have seen from the example of 'normal clay', one of the major sources of variability in the way in which indexical potential gets realized into actual conflict is the temporality of the site. Building does not just occur in space but also in time. Time, its passage and symbolization in the diverse rationalities of the site, embedded in diverse knowledges and bodies of people, is of the essence in determining the politics of indexicality. That is why the potential for conflict contained in indexicality does not always become realized. Future work will address the variable contractual conditions from partnering through ADR, as well as normal competitive tendering, in terms of comparative international experience, focusing in particular on the institution of contract as it is understood in East Asia, particularly amongst the Chinese. Experience suggests that these embedded, cultural factors are sufficiently important that the institution of contract varies greatly. However, it remains one of the ironies of the modern market system that, in fact, we have little comparative empirical analysis of one of its key institutions: that of contract.

Penultimately, one may note implications of a more political economic complexion. The 'contractualization of everything' as part of the economic rationalist thrust of the eighties may not be the efficient panacea that some of its more bright-eyed and bushy tailed protagonists have suggested. The centrality of contractual relations to models of 'postmodern management' also needs to be seen in this light.

In conclusion, castles made of sand, or at least their analytical equivalent, stand far less chance of being constituted in our interpretation of the relation of contracts and conflicts in the construction process if understanding starts from the bedrock which has been excavated here.

## 9 References


Clegg, S. R. (1975) **Power, Rule and Domination: A Critical and Empirical Understanding of Power in Sociological Theory and Organizational Life**. London, Routledge and Kegan Paul.

Fenn, P. (1991a) 'Alternative dispute resolution', **Construction Law**, 99-101.

Fenn, P. (1991b) 'Manage conflict, don't litigate . . .' **Professional Engineering**, November, 46-7

Higgin, G., Jessop, N., Bryant, D., Luckman, J., and Stringer, J. (1966) **Interdependence and Uncertainty: A Study of the Building Industry**. condensed and edited for publication by C. Crichton, London, Tavistock.

Linstead, S. A. and R. Grafton-Small (1990) 'Theory as Artefact: Artefact as Theory' in P. Gagliardi (ed.) **Symbols and Artifacts: Views of the Corporate Landscape**, Berlin, De Gruyter 387-419

NEDO (1991) **Partnering : Contract without Conflict**, National Economic Development Office (NEDO), London.

Schutz, A. (1967) **The Phenomenology of the Social World**, Evanston, Ill., Northwestern University Press.

Weick, K. E. (1969) **The Social Psychology of Organizing**, Mass., Addison Wesley.

Zimmerman, D. (1971) 'Record-Keeping and the Intake Process in a Public Welfare Organization', in Wheeler S. (ed.), **On Record: Files and Dossiers in American Life**. New York: Russell Sage Foundation.

# CONSTRUCTION MANAGEMENT INTEGRATION:AN ANALYSIS OF THE DEGREE OF INTEGRATION BETWEEN CONSTRUCTION PROFESSIONALS AND PROJECT PERFORMANCE

PHILIP TURNER-WRIGHT,
Leadbitter, Oxford, England

Abstract
This paper presents an analysis of the effect that the degree of integration of the site management team has upon project performance. It examines, commencing with an evaluation of the integration model, how conflict can influence the willingness of construction professionals to work in unison. In addition its consequential effects upon project performance. Proposed is that in order to identify the cause of, and productively control conflict, the implementor of such tasks must be conversant and sympathetic to the negative and positive values of conflict which disrupt an harmonious and integrated management unit.
Keywords: Integration, Conflict, Perception, Physco-Productive Environment, Project Performance

## 1 Introduction

In the first instance 'Managing Conflict' may be assessed as being instigated by one of two sources: (a) managing from within a team, (b) the managing of a team's conflict by a party external to that team. This analysis considers the relationship between integration and conflict.

Once the integrative qualities of a team has decreased or become non-existent, the task of re-establishing unity will be more difficult to implement than controlling a herd of wild horses in an open corral. Furthermore, the need to manage conflict either by developing or reducing it will have been established from managerial deficiencies within the team and or a breakdown of it's structure.

Conflict is borne when a team member, or several team members, attempt to obtain their own goals relative to, or as a priority to the goals of another; or indeed when a team member interferes in the attempt of another to reach his goals. Moreover, unless all team members are equally informed of the project goals, differences between individual goals and situational perceptions will arise. These may well be at a variance to the projects overall goals and objectives.

## 2 Integration Model

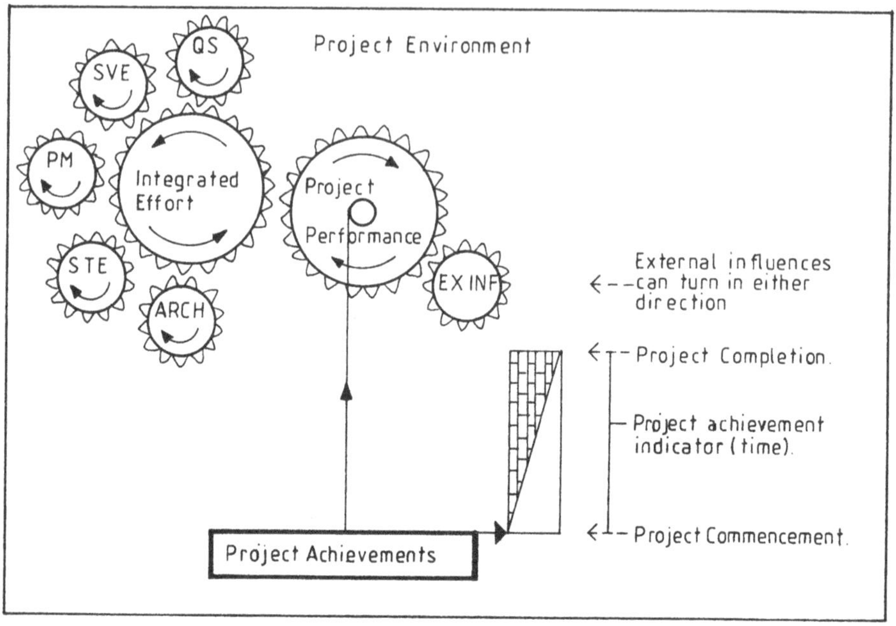


KEY

QS – QUANTITY SURVEYOR
SVE – SERVICE ENGINEER
PM – PROJECT MANAGER
STE – STRUCTURAL ENGINEER
ARCH – ARCHITECT
EX INF – EXTERNAL INFLUENCES

### 2.1 Introduction

The above model is presented to demonstrate the effects of a management team's integrated effort: When all the team members are fully locked into integrated effort they collectively pull the weight of project achievements towards project completion. There may not be clear indications as to any variance in the actual or perceived effort required from individual members, thus one or more team member may have a greater influence upon the teams degree of integration and unity.

### 2.2 The effect of variance in team members effort

It is apparent that once a team member reduces his individual effort, be it during the period of establishing a pathway for own goals or from non–aligned perceptions of individual importance, the result is to increase the collective effort required from the remaining team members; that is, if the project's productive momentum is to be sustained.

The velocity of a project's performance is further retarded when a non-integrating team member attempts to operate in an opposite direction. In such circumstances the team member, or members, attempting to work in a non-aligned direction may find the collective force of the remaining team members disengage their efforts or contributions; the result would be to leave the management team inchoate, hence reducing project performance opportunities.

Should the occasion arise when the majority of team members are effectively working in the opposite direction - from that required for direct attainment of the project's goals, such as late delivery of design or construction information, or the degree of conflict delays decision making and procedural agreements - the cog of project performance will reverse and lower the level of project achievements, thus widening the gap to it's completion.

### 2.3 External Influences

Full evaluation of the integration model requires consideration of 'External Influences'. These are categorised as influences imposed upon the team or individual from outside the management team itself, usually instigated by the client, hierarchy of independent departments or practices, or essential resources.

## 3 Managing conflict to enhance project performance

### 3.1 Introduction

Conflicts are an integral part of a physco-productive environment. They can be attributed in part to the differences and variations of the construction industry's professional members attributes. Also, to some extent, in the levels of 'interprofessional ignorance' that exist in evaluation of and between individual abilities, training and professional perceptions.

Conflicts arising from an ineptness of interpersonal skills contain personal emotions such as, anger, envy, pride and frustration. Primary importance can be given to frustration. Sooner or later man becomes aware that his knowledge, innate intelligence and experience fail to achieve his individual goals.

When a person attempts to achieve a goal and something interferes with it, which he is unable to positively respond to or control, he becomes frustrated. The reaction to the degree of frustration depends upon his tolerance, the tolerance of others, the teams' degree of integration and to what extent it effects the project's goals. The guidance for control in such circumstances is contained in achieving an approach towards satisfaction, the avoidance of obstacles to satisfaction and achieving a mediation which will provide an opportunity to develop an environment of productive contentment.

### 3.2 Regulating conflict by increasing integration

While it is unwise to consider conflict solely as a destructive element of a management structure, levels of conflict induced by interaction and non-interaction must be contained and controlled at a level which provides stabilisation of unity, collaboration and the co-ordination of efforts and attributes of all team members.

To achieve acceptable working levels of conflict we would have to acknowledge, evaluate and establish a dependency network for the management team. An individual team member's effect upon another suffers from relative importance; that is to say, there is little value or group influences caused by the effect that one team member may have upon another. However, each member has a relative importance upon the 'management quality' of the team for three reasons:

A member is important for contributing to the resources in a network to the extent that other members use his contributions to add merit to their own, and towards achieving the projects goals.
A team member is an important contributor to the network to the extent that he contributes to other 'important' members of the network, and the degree to which this effects the attainment of the project's goals.
A team member is an important resource to the team's efforts in attaining the project's goals, irrespective of the level of his contributions, because he has an intrinsic value to the project and other team members.

**March & Simon (1958)** stated: 'Greater interdependency meant a greater urgency to come to joint decision making.' While this does not imply there is no longer any chance of conflict it clearly leaves an opportunity (through integration) for individuals to consider the effect of decision making upon social relations and future negoiations.

Perception of power and dependency is reliant upon one member regarding the activities of another as important and that it is difficult for him to find suitable alternatives. There will remain a constant desire from some members to achieve an equal or higher regard from others, this may be achieved in maintaining their own importance within the dependency network. The direct path for this achievement is by increasing the degree of ones contributions and integration within the decision making framework. Furthermore conflict will arise during the presentation of ideas by such motivation. During evaluation one must reconsider the intrinsic importance of all team members and be aware that their useful contributions may be from serendipity and not extensive experience.

### 3.3 A higher degree of integration to attain project goals

All interactions, whether written, verbal or even gesture are sensitive to the reactions of other team members. Thus to achieve a unity in understanding the benefits of communication one refers to dependencies and interdependencies: How will proposals affect future interactions? How will the level of conflict be expressed, and to what extent will it be instrumental to achieving the goals of the implementor, other team members and that of the project?

Relations and dependency are influenced and regulated, to some degree, by people attempting to demonstrate their competence and in strengthening their respect with others. Moreover, effectively linking outputs and developing mutually accepted decisions about production, design, and technical improvements are often a source of friction.

Conflict is generally seen as winning or losing, we or they, and competitively, between individuals, groups or sub-groups. Thus I propose that controlling conflict be viewed as a three phased programme which is aimed at achieving an integrated decision-making format and one which includes contributions from all team members.

Firstly, co-operation is required from all team members in presenting their own goals, to all other team members as accurately as possible.

Secondly, negotiations are set into operation where all of the team members personal goals are openly discussed. This procedure, while demanding strength of presentation, will identify those values which are relative to both individuals and the project's goals..

Thirdly, an integrated and tactical use of the outcome of the previous phases is used to influence the attainment of the project's goals.

## 4 The physco-productive environment

### 4.1 Introduction

The quality of the management teams output will be enhanced through pre-evaluation of the attributes and knowledge levels required from its members; that they may significantly contribute to the teams degree of integration.

The fundamental requirements of the management team is that it is able to achieve its members' goals and productively emit influential management and technical criteria to achieve the project's goals.

The physco-productive environment is reliant upon dependency relationships. These will be borne from within the teams social relations and technical abilities. In such an environment conflict can be contained at productive levels through 'management engineering'.

### 4.2 Management Engineering

Management engineering is a pre-requisite to achieving a physco-productive environment and is attained by selection of team members by their individual social, technical and managerial abilities. These requisites are aligned to an holistic view of interrelated criteria, that have been identified as necessary for the effective achievement for the project's goals. When building or compiling a management team the absence of specific technical or managerial abilities will erode integration, breed conflict and create delays in production. This will be due to the team being unable to provide the project with necessary assistance to maintain constant production.

Also members may look to others for solutions whilst knowing they are unable to provide them.

## 5 Conclusion

The objective of this paper is to present a perspective which demonstrates the need for a higher degree of integration within a construction site management team. There can be little progress towards greater control of a projects outcome while we fail to acknowledge diminishing project performance levels induced by non-interaction, frustration and non-aligned perceptions of each others and the project's goals. We cannot progress alone. We develop our perceptions and individual requirements as required for the attainment of a project's goals and by presenting them to our fellow team members knowing they will be open to criticism. The conflict of such procedure will generate, enhance and establish social and dependency relationships. And improve interaction levels while providing a clearer pathway towards the attainment of project's goals.

In acknowledgment of the attributes and perceptions of each others profession team members are provided with the best opportunity for enhancing project opportunities and for 'Continuing their Professional Development'.

## 6 References


Bisno, H. (1988) **Managing Conflict**. Sage Publications, California, USA.

Brown, D. L. (1983) **Managing Conflict at Organisational Interfaces.** Addison-Wesley, Massachusetts, USA.

March J.G. & Simon H.A. (1958) **Organisations.** Wiley, New York, USA.

Mastenbroek W. F. G. (1987) **Conflict Management and Organization Development.** John Wiley & Sons, Chichester, England. – Translation of: Conflicthantering en organisatieontwikkeling.

# THE FRENCH APPROACH TO HANDLING CONFLICTS AND TO NEGOTIATING: CERTAIN NOTABLE FEATURES

Sir C.A. LEEDS, Bt
University of Nancy II, Nancy, France

Abstract
This paper outlines some of the main styles for handling conflicts and for negotiating as a background for a discussion of certain features relating to the French approach. The respective headings of 'impersonal' and 'personal' provides the context for an analysis of the two style negotiating model dominating-integrating. One conclusion is that French specialists in this area including mediators have a similar approach to the problem-solving or integrating strategy but prefer the concept of concertation to describe the processes involved.
Keywords: Individualism, Affective, Systematic, Organic, Integrating, Avoiding, Concertation, Dominating.

## 1 Introduction

This paper discusses certain ideas and methods which have influenced the French approach both to negotiating in general, and to handling intra-firm interpersonal conflicts. In many respects diplomatic negotiations between companies from different countries. In addition the areas of both conflict management and negotiations overlap, as reflected in the development by scholars of similar styles, strategies or techniques for use in these two fields. Consequently many of the points referred to will apply not only to French diplomatic or business negotiations, but to how interpersonal conflicts are handled in businesses.

### 1.1 4 style model

Francois de Callieres, a French diplomat wrote the first book for negotiators in 1716. Felice, another Frenchman, wrote a treatise on negotiation in 1778.

The conclusions of Willem Masternbrook, a Dutch specialist on negotiating history, is that both these writers discussed the basic dilemmas of negotiators, which originate in basic or primary impulses or instincts. Should a person adopt a fighting or aggressive style, a cooperative or lenient one, or a possible third option of avoidance, restraint, evasion or postponement, when the negotiator refuses to reveal his position?

Mastenbrook (1991) believes in a four style approach to negotiation based on the respective dimensions of fighting versus cooperating, and avoiding versus exploring. He associates cooperating rather negatively as yielding or accommodating behaviour. To escape from the triangular dilemma of cooperating, fighting or avoiding, Mastenbrook advocates exploring. Both sides are frank and open, and try through flexible problem-solving methods to find common interests and a solution of virtually equal benefit, as in a positive sum game of win-win.

**1.2 5 style model**

Various American scholars since the 1940's have also favoured a 5 style model in relation to handling negotiations or interpersonal conflicts. Afzal Rahim (1985) has recommended that the respective styles be labelled as follows:

1. Integrating (called Exploring by Mastenbrook)
2. Dominating (called Fighting by Mastenbrook)
3. Avoiding
4. Obliging (associated with cooperating and yielding by Mastenbrook)
5. Compromising (associating by Mastenbrook with either avoiding or yielding)

**1.3 French 2 style approach**

The French are noted for possessing a contradictory strain in their mentality and behaviour, and for a capacity to live in paradoxical situations. This is partly explained by their diverse origins, and the early and continued influence of contrasting cultures from neighbouring areas, notably the Latin and Mediterranean on the one hand, and the Atlantic and Germanic on the other. Consequently, the French tend to maintain an equilibrium, a balance between contrasting tendencies, such as acceptance of centralised systematic, formal practices, control and order on the one hade and a rejection of this tradition expressed in a preference for individualistic, organic, informal practices, autonomy and even disarray on the other.

This tradition is traceable to a pre-1789 French administrative maxim of a 'strict rule but a lenient practice'. These two facets of French behaviour I have called, for purposes of simplicity and clarity, the 'impersonal' and 'personal' approaches, developed respectively in section 2 and 3 below, in relation to conflict and negotiating and regarding 'competing' and cooperative styles.

Bearing in mind the above remarks, it is not surprising that the French style in negotiating has been described respectively as competitive or distinctive, akin to a zero sum game (win-lose), and also as a cooperative or problem-solving style (win-win). In reality the French approach tends to be a combination of both these styles as noted by Weiss (1983) which reflects their appreciation for lively confrontational debates and for reconciliation, harmony and consensus.

The two style approach to negotiating, competing and cooperating (in the sense of integrating rather than yielding) has already been developed in American literature on negotiations.

A metaphor which best captures the idea of the 'impersonal' approach is systematic whether applied to human behaviour or to organisations. In systematic cultures a close correlation exists between official procedures and practices and what actually happens in practice. In many respects the French are highly systematic. This is a reflection of first the strong centralising policies of successive governments. For many centuries, .... the all-pervasive influence of the state in the life of the people, particularly through the impact of an elaborate range of laws, and third the French attachments to certainty and order. Security and dependence not unsurprisingly many large organisations in the private sector including businesses, tend to be hierarchical and bureaucratic like the civil service organisations. Problems and mistakes are regulated by reference to the normal reporting channels.

The influence of 'systematic thinking' is shown in the French respect for detailed legibly binding written agreements when interfirm negotiations are finished. More importance is attached to the letter of the contract than the spirit whereas the British, for example, give importance to 'oral understanding' and the spirit in which an agreement is concluded and applied.

At various stages during long negotiation the French sometimes desire various points agreed to be formally written down and initialised by the parties involved.

**2.1 Dominating**

Most comparisons tended to have traditional or directive management styles so not surprising a forcing or competitive strategy used to be the main option of superiors in handling conflicts with subordinates up until the end of the 1960s.

Sometimes managers or supervisors devised special methods for winning. One tactic was to give the subordinate a difficult task and when he failed to do it satisfactorily he would lose self-confidence and become more dependent and submissive. Another technique was to divert a conflict a superior had with two subordinates by getting them to compete with each other on identical assignments.

Since 1968 employees have been less willing to be subjected by purely dominating strategies by superiors. Managers have also been encouraged to adopt more flexible strategies so as to help improve management-employee relations in companies.

Both in intra and in inter-firm negotiations the French have been and sometimes still are reputed for their rigid confrontational strategies more or less in the dominating or competing mode. Obviously in many situations such tactics will not work, particularly when power, skills or resources are equally shared. However, the French may gain by a strategy of repeatedly saying "No" which will eventually lead to the other party yielding. Alternatively the French delegations, particularly in diplomatic negotiations, have been known to abruptly terminate meeting when they have not got their own way. Various reasons have been suggested for French inflexibility such as their approach to negotiations in general.

In the past the French tended to regard the process of negotiating rather sceptically as a valueless exercise or as an attempt to reconcile divergent aims of two opposed groups. Neither side favoured creative solutions for resolving a conflict which was seen as a power struggle. The result inevitably was a fruitless confrontational stalemate, as in the case of disputes between management and workers which invariably involved ideological dimensions. Often the government intervened to impose a solution. Harrison (1987) has observed that negotiation is "quite far the list of preferred French methods for dealing with problems and conflicts either domestically or internationally".

The French, encouraged in their education to develop a logical mind, sometimes see negotiations as an activity counter to logic. The truth is unequivocally to be found by reasoning. Therefore, if one side is right, the other must be wrong.

This attitude has influenced French behaviour during business or diplomatic relations and earned them the reputation for sometimes being inflexible. A French negotiating team customarily prepares its position meticulously in advance, which they may feel convinced is in the interests of both sides.

At the actual negotiating table they adhere their carefully prepared position without making an effort to understand the other side's needs and motivations. They seem reluctant to abandon attachment to a particular concept, idea or principle unless the other party finds a flaw in their reasoning.

Another reason for French inflexibility has been attributed to the hierarchical centralised nature of their decision-making. Too often French negotiations have little authority, autonomy or leeway to vary pre-arranged positions. Delays have occurred when new proposals have had to be referred back to superiors at headquarters.

### 2.2 Compromising

Another reason for the French view towards negotiating, contributing towards inflexibility, is the dislike for bargaining, making concessions or compromises. To Anglo-Saxons the idea of "splitting the difference" seems a way of breaking a stalemate and a possible or natural outcome of negotiating. Many French do not share the English notion of 'Fair play'. To them the compromise strategy suggests a nil contest, half measures, a lose-lose situation which satisfies neither side.

Often the practice of compromising suggests making a dishonest opportunistic or shady deal ("une compromission") or as something faulty ("un compromis boiteaux") as in the case where a manager asks two people to make concessions which leaves them irritated and does not solve the problem. A compromise has been regarded as an accommodation, more apparent than real, made between two equally strong sides. The disagreement may continue in a latent form.

However compromise may be viewed positively, particularly if words are used such as consensus and entente. Individuals in conflict sometimes recognise the need to transcend their differences to reach a consensual views, which enables them to continue working together while "agreeing to disagree".

## 2.3 Conflict avoidance

French business is reputed for conservative safe management, and efforts to achieve perfect or ideal solutions.

Managers are adept at discussing options or alternatives to a proposals, but not necessarily in making the final decision. As people dislike making errors, a problem may not be treated directly or immediately, based on the grounds that insufficient information exists, or on the belief that things will turn out for the better without the need for intervening at all.

Managers in business often develop various devices for avoiding or suppressing inter personal conflicts. Examples are appointing a friendly in-group or committee of 'yes-men' (les hommes liges), stress in communications on loyalty good spirit and cooperations in suggesting ambiguous solutions which allows each of two people in dispute to believe he is the winner, passing the responsibility to others, or breaking off relations entirely for a certain period.

To an important extent the notion of "organic" captures the idea of the personal approach. In contrast to systematic organisations a considerable variation exists in organic organisations between the official or formal system and what happens in practice. There are usually less formal rules and structure, and greater reliance on mutual trust and confidence between the people involved based on tacit understanding and unwritten customs.

French organisations may in practice be quite flexible. Even if a clear systematic structure does exist and decision making centralised, normal hierarchical or reporting channels may not be much respected. In addition certain jobs may not be too closely defined or the staff too rigorously supervised, employees appreciating independence and autonomy in how decisions are implemented.

The desire to escape the burden of the official system, rules and laws imposed by the State has contributed, according to Vachette (1984), to the dynamism of French individualism. Often this takes the form of nonconformity, or non-compliance with impractical or incomprehensible rules or their modification, behaviour known as system D (self-reliance or le debrouillargise).

Many French managers adopt a personalist or social motion of authority according to Laurent (1986), power being an attribute of the person exercising it, rather than stemming from one's role or the company's rules. This is in contrast to American and British managers, for example, who tend to interpret authority functionally as stemming from a person's role.

In sensitive matters, when tension is high or if subordinates are asked to perform lengthy monotonous tasks - which require no skill or creativity, then superiors should treat them with particular care and respect, (wains d'Iribarne). A Frenchman usually is an individualist with a strong sense of personal identity. Whereas in most companies people usually feel they work for someone else, in a French company employees like to imagine they work for themselves, answerable only to their conscience or "sense of honour".

How tasks are executed may depend less on a sense of company loyalty than pride in one's profession, and as personal relationships with a superior.

Instructions need to be conveyed in ways most likely to elicit cooperation of subordinates. In a study carried out in ten European countries, described by Barsoux and Lawrence (1990), French employees came last, only 25% replying 'Yes' to the question about whether they automatically followed orders. However the French came first, with 57% saying 'Yes' in reply to the next question: "I will only follow the instructions of my superior when my reason is convinced".

Organic structures frequently facilitate the handling of conflicts. As regular vertical communication between managers, foreman and workers may not occur, informal horizontal alliances develop, groups cooperate, creating their own rules and settling problems.

However serious problems and conflicts may arise because of too much slack or looseness within organisations. Too much variation in management or leadership style may exist between departments which create their own rules and traditions. Managers, to reinforce their power position, may only pass on certain information to associates within their personal private network. Supervisors avoid contacts with subordinates, denying them information, but expect them to be informed. Subordinates may not report regularly to superiors. Losers in these struggles may be isolated, and not invited to attend meetings.

Rivalries develop between competing groups, as in the case of struggles between top management and separate manager cliques or between rival worker clans. Often groups rivalries overlap with different departments, each developing its own mentality, rules and work methods, while some departments consider their function or role more 'noble' than others.

**3.1 The affective dimension**

A minority of the countries in the world (mostly confined to Northern Europe, Northern America, Australia and New Zealand) tend to be 'task' oriented.

Neutral relationships are the norm, and work is not usually mixed with social fraternisation during negotiations with clients. If the technical aspects are right, negotiations from such countries will normally conclude as agreement, irrespective for the state of their personal relationships with each other. People tend to be monochroic, doing one thing at a time, and attach importance particularly to information from written or printed sources.

Most other countries in the world, including France, tend in various degrees to be 'people' orientated. Affective relationships are customary, and polychromic behaviour, where people may do many things at one time. They do not mind being interrupted at work, and attach importance to information received verbally from private networks or relatives, friends and associates.

It is common to try and establish good personal relations with others in business so that problems can be discussed from the human angle. If group A succeed in achieving good mental rapport with group B, then formal appointments, meetings, schedules and agendas are less important as means of solving difficulties. At the end of a inter-firm negotiation, people from such cultures are likely to conclude as agreement if they have got on well together which might be the basis of a long term business relationship even if the technical conditions for one or both of the parties were not perfect at this particular time. For example the price may be less important than the question of volume and prospect of future orders which will enable a firm to increase its market share.

Numerous writers on French negotiating characteristics have emphasized the importance of the affective element. It has been said that for Latins interpersonal relationships intervene in a negotiate earlier and more intensively than they do for Anglo-Saxons and that, while Americans attach importance to a persons professional accomplishments, a Frenchman also looks for qualities within a person and for personality. The French feel that if they share some of the interests, outlook or assumptions of their clients this will help build up a long term relationship extending beyond the initial negotiation. Rodgers (1989) has commented that the French are subjective, liking someone to know and like the same things as themselves.

**3.2 Individualism**

Although individualism is referred to in different parts of this chapter, the topic warrants a short section of its own. The French respect highly independence of thinking and the creative mind.

In many respects they are proud in not being like others, and are not overly interested in ideas developed elsewhere.

The considerable influence in conflict and negotiating of an important number of people in business is reflected in a quite sizeable published literature on these subjects.

However the author's aim is often, as in the case of some academic talks on the subject, not to develop along the lines of existing research trends, but to make a quite independent and personal contribution to the subject.

The French appreciate a person who develops his own theoretical model, if it is explained in a logical, clear manner, even if no explicit mention is made to past studies on the subject, to reference works or to concrete evidence to substantiate the observations. The French have a fascination, as Harrison (1987) notes, "with grandiose, elegant schemes rather than feasible projects". The past research on different conflict handling styles, referred to in Section One, has been dismissed by some French specialists on conflict as 'classical' or 'old hat'. One French writer on conflict made the same remark in relation to a standard French book on negotiation by Dupont (1990), largely because it summarised existing work on the subject rather than creating any new ideas or models.

What is particularly notable about much of the published texts in recent years is that, broadly speaking the French are developing various aspects of the exploring problem solving or integrating approach to handling conflicts and negotiations.

**3.3 Integrating**

This method of dealing with conflicts or negotiations 'integrating' is virtually a synoymn for ideal best practices. It involves collaboration between partners, openness of approach and full exchange of information and study of differences so that a mutually acceptable solution is reached. The real conflict needs to be uncovered first and this style has been said to involve a first element of confrontation. Each side is direct and candid with the other. Then the problem solving or creativity stage follows.

Some negotiations may conform to that 2 style format, following a distinct competing (distributive WIN LOSE) phase first, both sides testing the strengths of the other, and if neither yields then, they enter a cooperative (integrating WIN WIN) stage.

Many negotiations may involve a variety of styles being used, and sometimes a 'mixed approach' by which negotiations may try to be both tough and friendly, firm and flexible, at the same time.

Negotiations may start with competing, and then try another strategy, possibly yielding or compromising. The French, being individualistic, do not usually appreciate being categorised in terms of styles, no doubt because of a desire to develop their own methods as circumstances dictate, which may not fit any textbook model.

However, since the French are reputed for being both inflexible at times, and then adopting an integrating position rather late in the proceedings, the 2 style model frequently fits how the French actually negotiate.

Both in diplomacy and business the French are reputed for being flexible, particularly at later stages of negotiations, or when informal small group discussions are held as opposed to full official meetings or when determined to strike a bargain. The French, sometimes unpredictable and impatient, may suddenly alter their positions or style according to circumstances and the degree of personal relations established between the teams. Integrating is frequently called cooperating, although as mentioned earlier cooperating is also used in the sense of yielding, which exposes one side to possible exploitative behaviour by the other. Consequently Mastenbrook (1991) and others prefer terms such as exploring or problem-solving.

Ironically some French scholars have also criticised integrating or WIN WIN solutions as sometimes being a false or artificial situation, if one side has been able to cleverly manipulate the other. The integrating outcome has also been viewed as an improbable situation of unanimously or uniformity. Instead concertation has been suggested as the preferred style; perhaps surprising since in practice the two concepts 'integrating' and 'concertation' are close in meaning.

### 3.4 Concertation

This has become a fashionable word in common parlence since the Paris disturbances of 1968. It has been applied both to conflict handling, negotiating and as a special form of consultative management. Very close in meaning to 'integrating' the word concertation denotes working together, parties regarding each other as allies or friendly colleagues, as in the case of musicians in the same orchestra or two government departments working on a joint project.

In 1978 the government tried in vain by Law to introduce concertation into business as a modus vivendi for operations. Then in 1981 the new socialist government through the law on the collective right of expression of employees attempted to develop the spirit of concertation, in the sense of greater worker-manager cooperation.

Basically concertation involves the various repair strategies for breaking deadlocks and handling conflictual situations and getting parties to negotiate in a problem solving or exploring manner, which have been developed by practitioners, mediators and scholars in the field of conflict, peace-building and social work, whether in France or elsewhere since the 1960s.

Obviously concertation also has particular significance in the French context. Until 1968 many French top managers or owners still adopted very authoritarian or paternalist attitudes towards workers. When conflicts arose at work rarely did workers or managers resort to available conciliation or arbitration machinery. Instead management tended one, to adopt avoiding strategies, thereby enabling conflicts to continue in a latest form, or two, to attempt to stifle or crush conflicts, or third to allow a situation of stalemate to arise, which in serious cases tended to be resolved by government intervention. After 1968 governments and consultants have encouraged managers to adopt consultative styles of management, and an 'integrating' negotiating approach to handling internal problems. Greater emphasis has, since the early 1970s, been placed by management on identifying and coping with problems before they become serious on an informal basis. The aim has been to prevent interpersonal conflicts escalating. Many of the larger and more dynamic companies have encouraged managers and supervisors to develop 'people' skills appropriate for implementing the principles of concertation. d'Iribarne (1989) has highlighted the delicate balancing act a factory foreman may have to perform to get the best results, both knowing his men and what is going on, yet remaining invisible so as not to threaten individual autonomy. Frequently problems are arranged amicably and informally.

Broadly speaking, skills and techniques recommended by French specialists so that managers or supervisors can successfully intervene in conflicts involving others, or handle their own, whether involving superiors, peers or subordinates, include the following aspects below:

1 <u>Mutual understanding</u> Open-mindedness, objectivity and listening skills need to be developed so that ideas can be exchanged in a spirit of 'give and take'. Instead of seeing the issue as 'black and white', each discovers that his viewpoint may only be part of the truth.

2 <u>Conflict features</u> Some analysis is required of some of the common processes involved in conflict situations, so that the parties can appreciate that their situation is not unique. Soft strategies for breaking the conflict cycle might be proposed such as those applied in the martial arts.

3 Causes Many levels, elements or dimensions may be involved in any particular conflict. Clarification is vital as to the grievances/needs of each party and how each can help the other.

4 Solution The aim is to develop mutual empathy, or at least a situation where both sides make concessions, and work towards constructive solutions, superordinate goals, or as evolution of their different perspectives into something new by a process of synergising.

Nothing is particularly novel or new in this summary which only serves to indicate that in the field of conflict handling and negotiating there is much common ground in the thinking of specialists from France, Britain and the United States.

## 4 References


Barsoux, J.L and Lawrence, L (1990) Management in France. Cassell, London.
d'Iribarne, P (1989) La logique de l'honneur. Sevil, Paris.

Dupont, C (1990) La negociation: conduite, theorie, applications Dalloz, Paris. 3rd Ed.

Gruere, J.P and Moreb, P (1991) Cadres Francais et Communications Interculturellas Eyrolles, Paris.

Harrison, M M (1987) France - The Diplomacy of a Self-Assured Middle Power, in National Negotiation Styles (ed. H Binnendijk) Centre for the Study of Foreign Affairs. Foreign Service Institute: US Department of State, pp 75-103.

Hofstede, G. (1991) Culture and Organisations McGraw-Hill, Maidenhead.

Laurent, A (1986) The Cross-Cultural Puzzle of International Human Resource Management. Human Resources Management, 25, No 1, 181-199.

Lebel, P (1984) L'Art de negociation L'Edition D'Organisation, Paris.

Mastenbrook, W (1991) Development of Negotiating Skills, in International Negotiation (ed V A Kremenyuk) Jossev-Bass Publishers, Calif, 379-401.

Mole, J (1990) Mind Your Manners, The Industrial Society, London.

Rahim, M A (1985) A Strategy for Managing Conflict in Complex Organisations Human Relations, 38, No 1, 81-89.

Rochefordiere, Y. de la (1990) Du Conflit au dialogue - tous ganasts Les Editions D'Organisation.

Rodgers, I (1989) French-British Negotiations - A Cross Cultural Comparison, Inter-Cultural Management Associates, Paris.

Salzer, J and Romano, C (1990), Enseigner C'est aussi savoir communiquer Les Editions D'Organisation, Paris.

Vachette, J.L (1984) Le Modele Francais du Changement, Revue Francais de gestion No 47 September/October, pp 119-122.

Weiss, S E and Stripp, W (1983) Negotiating with Foreign Business Persons Graduate School of Business and Administration, New York.

# SUBSTANTIVE TECHNIQUES FOR CONFLICT RESOLUTION: AGGREGATE EXTRACTION IN SOUTHERN ONTARIO

D.C. BAKER and A.G. McLELLAN, Faculty of Environmental Studies, University of Waterloo, Ontario, Canada

Abstract
This paper examines creative means to reduce conflict among disputing parties by viewing aggregate mining as a landscape opportunity. The creation of a new landscape that meets the goals and needs of all the parties involved provides a substantive tradeoff that allows both mining and the rebuilding of valued, ecologically sensitive areas. Within this context, the opportunity exists to demonstrate that there can be community enhancement, landscape improvement, increased diversity, and profit realized from an approach that recognizes that landscape is a valuable resource beyond the extraction of mineral aggregates and traditional reclamation practises. A substantive approach to resolving locational conflict incorporates bilateral participation by affected parties in the design of compensation strategies to mitigate the negative effects of mining. Compensation strategies for aggregate mining provide a unique opportunity to reduce local conflict through recompension of both environmental and social impacts.
Keywords: Compensation Strategies, Development Opportunity, Landscape Design, Aggregate Mining

## 1 Introduction

Land use changes often generate conflict. A large class of locally unwanted land uses (LULUs), such as airports, landfill sites, or prisons tend to intensify the hostility of conflict when placed in a local community. Although these land uses provide a benefit for the greater interests of society, they are often vehemently opposed at a local level because of their undesirable impacts. This type of land use conflict is often characterized by the localised cost to the community versus the dispersed benefit for society.

The extraction of construction aggregate resources in southern Ontario provides a good example of a LULU, where the negative effects of mining have caused disruption to local landowners and communities. Aggregate extraction has generated intense conflict among municipalities, provincial agencies, affected citizens, the aggregate industry, and concerned lobby groups over the past three decades. At the heart of much of the conflict has been the environmental and social disruption caused by the mining of sand, gravel, and bedrock.

The proximity of aggregate sources near market demand is a crucial factor in the location of available extraction sites. With over 50 percent of the delivered cost of the product resulting from transportation costs, aggregate sources are often sited as near to major urban demand areas as possible. In many cases, within the urban fringe, suitable sites for subdivision, building, and recreation are also ideal for aggregate extraction (Bryant et. al., 1982). Local property owners adjacent to aggregate operations are usually affected in an adverse manner. As well, there are negative effects that extend beyond the site, referred to as "shadow effects"(Marshall, 1982). Dust, traffic, noise or vibration all contribute to the shadow effect beyond the mine site.

Consequently, the public response and perception of sand and gravel sites tends to be imbued with negative connotations. Different interest groups, such as local naturalist clubs or adjacent property owners, have the potential to be affected in a negative way by the mining of aggregate sources. Concerned members of the public may want to restrict aggregate extraction as a result of:

a) environmental disruption to flora and fauna,
b) adverse visual impact,
c) noise and dust disturbance from the mining site,
d) truck traffic hazards and road damage on haul routes,
e) disturbance to the water table and local wetlands,
f) fear of lowered property values, and
g) lack of past pit and quarry reclamation.

As a result, the approval and licensing process in Ontario is becoming increasingly restrictive and adversarial.

The present appeal process for planning decisions in Ontario that deals with land use conflict is based on a semi-judicial hearing that places disputing parties in a win-lose situation. There is very little opportunity nor encouragement for different sides to investigate areas of common interest and search for potential solutions to resolve the conflict generated by mining activities. Hence, the opposing positions taken at hearings are often entrenched and bitter.

The focus of this paper is to examine substantive methodologies to help resolve conflict in the mining of aggregate resources. Aggregate extraction is an interim land use, and provides a unique opportunity for landscape change and reclamation as the product is mined. Compensation strategies can be formulated between producers, adjacent property owners, municipalities, and environmental concerns to reduce conflict. Substantive strategies to reduce conflict between disputing parties focus on interest-based bargaining, where tradeoffs can be made between different values in a changing landscape.

## 2 Conflict within the present planning process

Planning control of aggregate extraction has evolved with municipal and provincial jurisdiction over a period of two decades. At the provincial level, the Pits and Quarries Control Act (1971) provided the first comprehensive legislation in Canada to attempt uniform regulation of the aggregate industry. The Act, for the first time demanded site plan preparation, licensing procedures, rehabilitation, and other fiscal regulatory requirements. The provincial controls focused on an attempt to regulate the most evident negative environmental effects of gravel pits. This legislation has been modified in the ensuing years and was replaced by the Aggregate Resources Act (1989), but environmental conflict over aggregate mining is as intense as ever!

Municipalities have attempted to control the environmental problems associated with aggregate development through the Official Plan where the Province has failed to regulate the negative impacts of pits and quarries with the Pits and Quarries Control Act and environmental legislation. The Official Plan is a planning document that sets out land use restrictions and zoning regulations within a township. Municipalities have the option to control hours of operation, set-back requirements, noise and dust levels, and other restrictions such as limitations on truck traffic.

There has been a traditional conflict between municipalities and the Province for the control and planning of resources. Aggregate resources provide a good example of provincial expertise and "provincial interest" attempting to enforce legislation and policy on municipal governments who are often reluctant to follow the guidance of provincial mandates. Municipal governments tend to be more responsive to local citizen concerns, and often, elected municipal counsellors support anti-aggregate sentiments within the townships. Local responses to the environmental problems caused by aggregate mining are frequently enacted in municipal politics and then formulated within the Official Plans. Pressure has been frequently placed on local politicians to ban aggregate mining, with many municipal elections based on stands "for or against" aggregate extraction. A failure of planning authorities to coordinate and manage the extraction of aggregate mining can be attributed, in many cases, to different agendas of the various local and provincial agencies. The present planning system and legislation alienates residents who are affected by aggregate mining in terms of benefits derived from a local resource. Aggregate resources, as a local product, are being exported with little compensation being returned to the community from which they are derived. Aggregate production involves a considerable environmental and social impact resulting from both mining and transportation of material. Residents and municipal coffers are faced with infrastructure costs and environmental uncertainty as a result of mining. However, there is minimal compensation paid to the people affected by these externalities.

The present means to resolve conflict for the development of aggregate sources are based in both substantive efforts to regulate the resource with the

Official Plan and legislation, and procedural attempts through the Ontario Municipal Board (OMB) Hearings. The OMB is an appointed board that holds regulatory tribunals to resolve land use conflicts and planning issues within the Province of Ontario. The hearings are conducted on a formal basis and involve testimony from expert witnesses, cross examination by legal counsel, and rules for the admission for evidence. The process is based on an adversarial system.

The conflict that has characterized aggregate development appears to be a recurring theme. Over 130 OMB hearings have convened since 1971 to settle disputes involved with objections to aggregate mining. These disputes are costly (for example, in 1990-1991 the 16 month OMB hearings at Puslinch Township are estimated to have cost the disputing parties and the public approximately seven million dollars) and often embitter opponents with distinct winners and losers. "Aggregate wars" are now a tradition in southern Ontario. With this tradition have come entrenched positions for parties that have rallied for or against the development of aggregate resources. For example, the Foundation of Aggregate Studies operated an information and resource centre based in Toronto to oppose aggregate developments; while, the Aggregate Producers Association has lobbied in favour of aggregate extraction.

The polarization of groups "for and against" aggregate mining has served to impede the resolution of conflict at a local level. The entrenchment of positions is a frequent consequence of on-going or long term disputes and forms a difficult obstacle to possible resolution. The OMB, in many cases, may not resolve conflict between disputing parties. Although the hearings may rule on a particular dispute at a specific site, the process may not address the basis for the resolution of conflict within a given area or township (Cullingworth, 1987). As a result, often if residents are not satisfied with an OMB ruling for one site, the resentment and dissatisfaction continues within the area or township and quickly resurfaces at the next opportunity. The OMB hearings may exacerbate conflict in many cases because:

a) of the adversarial nature of the hearings and its emphasis on winning rather than examining effective compromise;
b) the inequity of resources in the form of scientific information and legal counsel between citizens and aggregate companies at the hearings; and
c) of the extremely high costs of the hearings in terms of time and money for both producers and opposition.

The planning process may attempt to resolve conflict on the basis of legislation or the "set rules", but this does not mean that the conflict is resolved at the "interest" level for all parties concerned.

The land use conflict that has been generated by aggregate extraction has been an expensive exercise in planning and decision-making. It is continuing to demand time and resources from citizens, municipalities, proponents, and the province. In summary, the dimensions of this problem consist of: a long tradition of conflict that has polarised sides and public opinion; a struggle for control of aggregate resources between several actor groups; the OMB hearing system, which may not serve to mitigate conflict; and the issue of compensation to the individuals and communities affected by aggregate mining. The substantive and procedural basis for planning is inadequate to address the present diversity of interests, and lacks the flexibility to accommodate an interest-based solution to resolve differences.

## 3 Windows of opportunity

The development of construction aggregate sources requires an "interest-based" approach in order to resolve much of the conflict that is generated through mining activity. Ury, Brett, and Goldberg (1988) advocate the need to build an effective conflict resolution strategy on the basis of party interests rather than the defined "rights" of parties, or a solution that is contingent in "who is most powerful". The present system of planning in Ontario uses both a rights and power based method to resolve conflict. The continuation of the three decade old "aggregate wars" indicates the ineffectiveness of this conflict resolution strategy.

The authors suggest that there are "windows of opportunity" that can enhance an interests-based approach to resolve some of the difficulties of mining aggregate resources. As aggregate mining is an interim use of land, parties can search for things to trade which they may value differently (Susskind and Cruikshank, 1987). Thus, in many cases there is an opportunity to remove mineral aggregates from a site, and restore a wildlife corridor or wetland habitat according to the area's biotic needs. This approach focuses more on substantive means of settling disputes between parties by creating opportunities of mutual gain for "most" of the interests affected by mining. We emphasize "most" because not everyone can be appeased or compensated, some people will always be opposed because of their moral or personal stands.

A substantive approach to resolving conflict among disputing parties goes beyond singular, unilateral compensation strategies. Rather, this method of dispute resolution incorporates bilateral participation in the compensation process, where the proponent and affected parties determine the context of aggregate extraction in the community. As Lake (1987) has observed, locational conflict often arises when decisions are made separately from the process of facility siting.

Inherent to the resolution of localized conflict are a series of steps that provide a base to ensure that development and community interests can be compatible: site specific inventory, community "net gain", local expertise, and agreement on what cannot be mitigated.

### 3.1 Site specific inventory

The assessment of the potential site must be comprehensive. An inventory of the geological, ecological, and cultural characteristics will enhance sensitive design. The proponent must accept responsibility for an exhaustive study and possible restrictions on mining operations. The site inventory provides a base-line data source from which to evaluate potential environmental and social impacts of the operation and determine the potential landscape changes.

### 3.2 Community "net gain"

The mining of aggregate resources needs to be placed in the context of the community, and should no longer be tolerated as a degrading practise. With this approach in mind, a physical and/or social enhancement of the community can be incorporated with the extraction of sand, gravel, or bedrock resources. Aggregate extraction is an interim use of the landscape. Part of the profits realised from aggregate production should also benefit the community in various ways, such as the creation of desired habitat and recreation sites, or the design of new landscapes for future intended land uses.

### 3.3 Local expertise

An important part of the site design process is to incorporate the local resident's expertise. This is not always easy. Often, hostile groups take an adversarial stand, and oppose the project. It is essential to realise that the public who is against the project are not to be ignored and must be incorporated into the design process. The local residents must be given a chance to inform the proponents about their local environment and what is important to their lives. Local opposition should be encouraged to bring constructive amendments to the process and the proponent's plans.

### 3.4 Agreement on what cannot be mitigated

Even after considerable consultation to ensure a sensitive site design, not all the impacts of aggregate mining can be mitigated. For example, with truck haul routes, there will still be disruption to the residents along roads with the additional heavy traffic. It must be recognized that in the public's interest, inevitable intrusion on private lives will occur. In cases where people are affected by the negative impacts of pit or quarry development and mitigation strategies are not possible, individuals should be compensated for the inconvenience. In these situations, the proponents may have to provide at their own cost: site specific landscaping, noise scaping, or appropriate personal services (such as air-conditioning for houses).

## 4 Case studies

Two case studies will serve to illustrate potential conflict situations where mining activity was carried out in sensitive sites that normally would not be considered for aggregate extraction. Each case study is unique in its location and opportunity to work with different interest groups.

### 4.1 Snyder Flats - Bloomingdale, Ontario

In Ontario, Conservation Authorities have a statutory mandate for conservation and water management within watersheds. A primary objective of Conservation Authorities has been to control flooding, and in the past this has involved the construction of numerous dams and reservoirs. Their flood control policies have decreased flooding, but in doing so, the floodplains and lower terraces have no longer been inundated by flood waters. This has the unfortunate effect of removing the traditional spawning habitat of many species of fish, and has decreased the ecological diversity and unique aquatic and terrestrial environments.

Conservation Authorities have been traditionally opposed to extraction of aggregate resources in areas within their jurisdiction. The floodplains and terraces of major rivers in southern Ontario, now contain much of the remaining aggregate resource base of the province. However, they are also considered sensitive, valuable ecological areas for habitat protection and recreation. Until recently, the Conservation Authorities have been reluctant to allow access to the valuable aggregate resources they contain, and often have vehemently opposed aggregate mining in proposals near to their properties.

A proposal by an aggregate company to the Grand River Conservation Authority offered an opportunity for a pit to expand its operations into a river meander bend, below the water table. Normally this type of operation is restricted by provincial licensing regulations, floodplain controls, and the municipal Official Plan. However, after detailed site examinations, it was discovered that the 250 acre site at Bloomingdale, which is owned by the Grand River Conservation Authority, could actually be enhanced in terms of its ecological diversity by a judicious plan of aggregate removal. The plan involved aggregate extraction above and below the water table, and in so doing, created ecological niches no longer found on the site. For example, the following new habitats are incorporated into the mining plan: cold water ponds fed only by groundwater for cold water fish species; warm water habitat which is seasonally flushed by river waters; and prairie-savanna grasslands.

Some problems with the site design were encountered due to complete removal of aggregate materials and vegetation. Leaving a portion of the available aggregate resources with the landscape and vegetation intact could have cost less than rebuilding a new landscape and provided for more rapid habitat rehabilitation. Additional site problems consisted of: difficulties with diverting river waters through the system without creating flood erosion and bank instability; reforestation during drought conditions; and the reconstruction of promontories into several lakes for species habitat. Despite these problems, the project has been remarkably successful and already the number and diversity of seasonal migratory waterfowl has increased. For example, in April,1992, Common Mergansers, Buffleheads, and Canada Geese were observed nesting on the created habitats. The fish habitat that has been carefully constructed in below water-level areas contains old tree stumps and sheltered areas to ensure maximum fish spawning opportunities. From a habitat perspective, the completed design is considered a success by the Grand River Conservation Authority.

The Grand River Conservation Authority is now using the Snyder Flats experiment as a "show piece" to demonstrate the enhanced river habitat that has been realised by the extraction of several million tonnes of sand and gravel. In addition to the creation of a valued and enhanced ecological landscape, the Conservation Authority gained over a million dollars in mining royalties from the project over a ten year period, which in times of budget duress are a welcomed addition.

In order to achieve both the economic and ecological benefits from the Snyder Flats site, the traditional battle-lines between the Conservation Authority, provincial licensing agency, and aggregate producer had to be altered. Common ground was found over a considerable time horizon in which the site changed from barren excavation to a restored habitat. Input from all parties, with continuous monitoring, ensured interests were fulfilled.

### 4.2 Urban aggregate rescue - Kitchener, Ontario

Commonly in Canada, many aggregate resources have been sterilized by the expansion of urban development, and yet, it is urban expansion that has generated the demand for construction aggregate materials. It would seem more sensible to ensure ready access to these proximal resources rather than the continued relocation of new aggregate mining sites to more distant locations where increased social, environmental, and long distance haulage costs are the unattractive consequences.

The development of aggregate sources within an urban setting is termed "urban aggregate rescue". Sometimes property becomes available within city limits that contains suitable deposits to mine; thus, the opportunity is created to develop a source within the urban demand. This type of operation usually involves a variety of conflicting interests because of the proximity of mining to other established urban land uses. In order to develop a source within this context it is important to have a fast, short-term mining operation, and ensure that the reclamation plan facilitates local interests.

In Kitchener, Ontario, the authors have been involved in an urban aggregate resource rescue operation within the city limits. The project involved connecting two potential sources with private haul routes to a producing plant. Old aggregate producing plants, originally located on the urban fringe are often now found inside the urban area, adjacent to industrial parks and customers (redi-mix concrete plants, block plants). In order to extend the life of an aggregate processing plant within the city limits of Kitchener, a private haul route (thus avoiding the use of public highways) was arranged to connect with a nearby 120 acre industrial subdivision site and an abandoned 80 acre farm site. Each site contained large volumes of sand and gravel resources, but were originally scheduled to be used for subdivision purposes without considering the aggregate resources.

By lowering the proposed industrial subdivision by 15-25 feet, 2.5 million tonnes of commercial aggregate were "rescued". The owners of the site realised an additional financial reward from the aggregate royalties, had their landscaping completed at no cost, and lost only two industrial lots as a result of the revised multi-purpose plan. At the other near-by farm site, located adjacent to the main highway into Kitchener, a haul route was designed beneath the highway to reduce traffic conflict. The project yielded over two million tonnes of aggregate, while the landscape was changed in a pre-planned way to provide for several different land uses:

a) The upper elevation yielded approximately 30 feet of high quality sand and gravel which was removed and transported by off-road routes to the processing plant. In this situation, because of the site's proximity to the urban area, the local generation of building refuse (and the concomitant dumping fees of $82 per tonne) allowed the infilling of the areas vacated by aggregate resources. The land was subsequently infilled to a similar elevation to the original landscape and now is used for light industrial and commercial use.

b) The area slightly below this in elevation, but above the regional floodline (therefore capable of being developed) has now been converted into 25 commercial lots after the removal of 20 feet of sand and gravel.

c) The lower area, below the regional floodline (thereby excluded from urban development) yielded over a million tonnes of sand and gravel. Within this area, a recreational site has been developed with the creation of two landscaped lakes, and picnic sites for the residents of the local community.

Opposition to the aggregate rescue plan was initially encountered from both government agencies and the public. The local Conservation Authority was concerned with extraction below the floodline and the possible siltation of the adjacent river. The Urban Parks and Recreation Department and city planners were apprehensive with mining occurring in the city limits, near a recreation corridor. Public opposition to the proposal was primarily due to the transportation dangers and noise of large aggregate trucks on the local highway and roads.

Consultation with members of the public, and the different agencies required an examination of different site scenarios and potential compensation strategies for the affected parties. An alternative haul route was constructed to remove the potential traffic problem, settling ponds were installed to handle any siltation problems during mining, and a recreational site was designed as part of the reclamation plan.

A multi-purpose plan was created which has avoided the sterilization of valued aggregate resources in locations close to an old existing processing plant, which might have otherwise been closed due to exhaustion of available reserves. Agencies that have traditionally been placed in an adversarial stand to the mining of aggregate resources are now realising the benefits of cooperation. Indeed, from this experience the city of Kitchener has induced its planners to pass a by-law to utilise, where it is possible, all aggregate resources on development sites in an urban rescue attempt to save resources.

Problems remain, however, such as creating the opportunity for the rapid mining of urban aggregate sources to reduce conflict among neighbouring land uses. Situations within the urban environment that allow rapid excavation and reclamation of potential sites, still must be placed through a slow approval process similar to full-scale mining operations. This is a problem because urban land values are too high for land to remain idle for a long period of time. The regulating ministry, despite its mandate to provide aggregates at reasonable social and environmental costs to the Ontario public, is very reluctant to expedite this "window of opportunity".

## 5 Conclusion

Hostile reactions to LULUs are still an obstacle to overcome. The history of land use conflict generated by the mining of aggregate resources in southern Ontario provides an insight into how people can be entrenched into positions with very little room for compromise. Many groups and individuals remain loath to constructively participate in doing anything that will increase the likelihood of success for aggregate operations because of this bitter history. This is unfortunate, for it is clear that opposition groups have much to offer in rendering the operations more sensitive to their concerns. It is significant to note that the attempts to initiate and develop strategies for conflict resolution have come from consultants acting on behalf of proponents, not from the public nor regulating government agencies.

In order to overcome the traditional hostility between "pro" and "anti" aggregate stands, the authors have been using an enhanced view of the current demand for "no net loss" in Ontario developments. Mining is seen as an opportunity rather than a constraint. We have, as consultants, been promoting in our designs and with our clients the position that their proposal should examine in detail all opportunities for net environmental and social gains to the community. The vision that the landscape change concomitant with aggregate removal will lead to community benefits and landscape enhancement could be seen as a positive gain in changing current social attitudes in southern Ontario. Certainly, the success rate at obtaining aggregate mining licenses with this approach (20 licenses out of 22 applications) indicates the value of substantive conflict resolution methodologies.

In both case studies outlined, it has been demonstrated that it is possible to coordinate land uses to reduce locational conflict. Creative and imaginative landscape change can often ensure that perceived incompatibilities are readily overcome. Mining, instead of being a degradational activity can create new landscapes that better fit future intended land uses, and can yield substantial financial benefits whilst creating new ecological and land use opportunities.

## 6 References


Bryant, C. R., Russwurm, L. H. and McLellan, A. G. (1982) **The City's Countryside: Land use Management in the Urban Fringe.** Longman Group Ltd., London, England.

Cullingworth, J. B. (1987) **Urban and Regional Planning in Canada.** Transaction Books, New Brunswick, New Jersey.

Lake, R. W. (1987) **Resolving Locational Conflict.** Centre for Urban Policy Research, Rutgers, New Jersey.

Marshall, I. B. (1982) **Mining, Land Use, and the Environment.** Lands Directorate, Environment Canada,Ottawa.

McLellan, A. G. (1990) Massive Landscape Change in Ontario: Integrating Geomorphology, Landscape Changes, Resources Management, and Human Activity, in **Ontario: Geographical Perspectives on Economy and Environment** (ed. B. Mitchell),. Department of Geography Publication Series No. 34, University of Waterloo, Waterloo, pp. 205-235.

Susskind, L. and Cruikshank, J. (1987) **Breaking the Impasse: Consensual Approaches to Resolving Public Disputes.** Basic Books Inc., New York.

Ury, W., Brett, J., and Goldberg, S. (1988) **Getting Disputes Resolved: Designing Systems to Cut the Costs of Conflict.** Jossey-Bass Publishers, San Francisco.

# "DO-IT-YOURSELF" HOMES - MORE OR LESS CONFLICT PROBLEMS?

CRISTINA COSMA
Polytechnical Institute of Iasi, Romania

Abstract
Between the parties signing a contract, the client and the builder, there will always be disputes, in the first place because of human nature. Such conflicts appeared in Romania in the past too, when the state was the only entrepreneur whom you could contract with. We will certainly find more conflict in the future, while the user's requirements are growing. The purpose of this paper is to suggest a way of settling these disputes in that the user can be in the same time the builder of his own private home. The relationship between client-builder turns into a buyer-seller one, which is much simpler. However, with this change other types of conflict may appear, and the paper has emphasized them.
Keywords: Transition, Requirements, Methods, Risk, DIY Concept.

## 1 Introduction

In Romania, the transition from a central-planned economy to a market economy, based on competition, requires a lot of changes in actual legislation. With regard to the construction field, a first step was made with the Government Decision No. 291/1991, providing a competitive bidding system, with tenders for contracts being regarded as a main instrument for creating market relations along with the abolition of some previously adopted documents.

So far centralized investment and projects were ordered by the state, on the basis of a five-year plan, irrespective of the client's wishes and needs or the builder's capacities. This is the reason why although the growing rhythm of building fund accelerated after 1960, the dwelling deficit could not be eliminated. On the other hand, the number of dwelling units increased with the diminution of their quality and this has led to conflicts between the client and the builder. But neither of them were free agents, so they could not freely determine the terms and conditions of the contract which made implicit the settlement, on a legal basis, of the disputes that arose. This does not mean that a legal basis did not exist, but an individual could not put the state on trial - it was a sort of unwritten rule, and people knew they had no chance of winning.

Moreover, the client was very different from investors in market economies. He could not invest money in a project and demand performance from contractors. He also could not employ an agent to represent his interests. In Romania, between the future owner and the building company there existed an intermediary state company called, "The Personal Propriety House Building Office" (OCLPP), that cashed an errand from the future owners and provided them with technical assistance concerning the project's quality, cost and terms. OCLPP played the owner's representative role and was the only company to do this. But both the builder and the "owner's representative", were state enterprises under the local Town Hall's subordination, so that all the risks remained with the client.

## 2 Factors causing dispute appearance

### 2.1 Owner's requirements

A common opinion is that the client/owner does not get the best product for his money. However, what does he really want?

As a human being, the future owner will have "human requirements" that can be arranged under the various facets of man:

Physiological requirements;
Psychological requirements;
Sociological requirements;
Economical requirements.

These are generally first thought of in qualitative terms, when they are perhaps best described as goals or objectives for the building to fulfil. At this level of detail, the concepts are very much those that would be included in an explicit, functional brief for a building, discussed and agreed between the client and the designer. Correspondingly, these will also be the terms in which the degree of success of a completed building will be judged by its owners, and if one or more of these requirements are not respected it could be a source of conflict.

### 2.2 Levels of dispute appearance

There are four main levels where disputes between the future owner and the builder may appear:

(a) When signing the contract;
(b) During the performance of the building;
(c) At the reception of the building;
(d) During the guarantee period.

The last three levels depend upon the first one. A suitable standard form of contract could ensure an appropriate allocation of risk between the two parties, but the client can be a real keystone in this process. The client always asks himself if he has sufficient leverage to change the terms of business in case he does not like them anymore. In Romania this thing was not possible - it was a matter of take it or leave it.

### 2.3 The C-T-Q factors

In dispute appearance the real "Bermuda Triangle" remains the COST-TIME-QUALITY factors.

Cost overruns, extensive delays in the planned schedule and very serious problems in quality lead to an increasing number of claims and litigation. Some authors think that these failures are caused by the current bidding system used in the public sector because the selection process in based on only one element; namely cost.

## 3 Methods of settling disputes

In the first place, because of the pacifist nature in Rumania, alternative methods of settling disputes are preferred such as negotiation, conciliation, mediation and adjudication. If after going through all these stages the two parties are still in dispute, arbitration and endless litigation follow.

### 3.1 Negotiation

For negotiation to be efficient the rights of the two parties involved must be very clearly defined. As the interests of the two parties can be very different, they will often have to negotiate cost exchanges that will set them both in a better position than before. Generally, when they are not being asked to negotiate people adopt positions which may prove to be very expensive for the others involved.

### 3.2 Conciliation, mediation

A third party that is interposing between the two parties in conflict will clarify the issues between them, or a solution acceptable to both parties may be proposed. Usually, in Romania the third party is the designer or the architect as he is the state's employee and not the client's agent like in other countries. Design institutes only recently became autonomous companies that can provide project design, consulting and other services.

### 3.3 Adjudication

Adjudication clarifies the real rights that the two parties have. A third party will decides these rights. The problem is solved by investigation and not by choice.

Finally cooperation between these two parties can be realized by conviction or by compulsion. When negotiating, supplementary options are offered in order to settle the problem. If options are reduced the settlement is based on compulsion.

## 4 DIY Concept

The initials DIY come from "Do-It-Yourself" - a method of advice that is now often heard in Romania.

In the building industry a lot of changes take place: the separation line between the producer and the user is gradually diminishing. People are beginning to take over certain services that until now were performed by different companies. The production of do-it-yourself prefabrication kits, where the owner buys packages and builds the home himself, or acts as his own contractor is increasing day by day. In the U.S.A. 20% of single-family residences are being built by the homeowners themselves and there is a rapidly growing market for goods and services well suited to this type of housing construction and home improvement.

It is important to involve users, and in particular specially knowledgeable ones, in obtaining products. Many authors think that the role of owners is crucial, the importance of real estate owners and the creation of groups of owners with long-term perspectives have not been emphasized enough in the past.

This concept requires a whole new way of thinking and interacting with the entire decision-making team.

The factors that determined the appearance of the DIY Concept were:

(a) Inflation;
(b) Lack of Quality;
(c) Difficulty of finding skilled workers;
(d) Increase of free-time.

In Romania, accelerating inflation has had a staggering effect on the housing field. Most of the people find their "middle" incomes insufficient to purchase a new home. To make housing affordable we must first get inflation under control.

## 5 Relationship between buyer - seller

The previous separation between the two parties - producer and consumer - created a double personality. The same man who, as producer, was taught to be well-disciplined, as a consumer he will search for an immediate reward/profit. He is an entirely different man.

In the existing procurement systems, the risk remains more with the client, whereas in a Do-it-yourself system a balance of responsibilities is intended.

Risks can be proportionally shared between client/buyer and seller, who can differ from the producer and will take over some of the producer's risks.

In a Do-it-yourself system the relationship between owner-builder turns into a buyer-seller one. Let us see, in this case, what happens with the factors that could cause the appearance of disputes.

The buyer's demands can now be promptly satisfied. The buyer can see with his own eyes and even touch the materials, elements or parts of his future home. He is free to select whatever type of house he wants and if he is not a very knowledgeable buyer, manuals, reviews and a lot of other explanation relating to materials, revealing the performance these elements can reach and how to use them, stay at his disposal, along with the seller's explanations. For this to happen, construction materials and elements must become well-known, like any other consumer commodities. Publicity and the mass-media have an important role to play here.

The cost problem remains with the seller this time with payments being made in advance or by installments. The seller is free to deliver the goods before or after the payments. The ball is in his court. But both the buyer and the seller must be careful to avoid dealing with an insolvent company or person.

Time is not a problem any more. The buyer comes, sees and buys. Delays in delivery may happen but are not comparable with those that occur on site.

Quality remains a factor which may continue to cause conflicts. For the buyer, quality is a measure to determine if money is well spent. Since the buyer is not always ready to pay more for better quality, he could prefer to choose those elements having a quality suited to his budget ("cheap and good" is often said in Romania").

The length and nature of warranties given by the seller must also be clearly defined, as he will certainly be seeking to limit his liability.

## 6 Conclusions

The total or partial transfer of the building work to the future owner - called the externalization of the cost of working force - is not a totally new phenomenon. Factors like inflation, lack of quality, difficulty in finding skilled workers and even the existing construction industry plagued by disputes, have led to the appearance of a new concept, "Do-it-yourself", that turns the relationship between client-builder into a buyer-seller one, removing at the same time some of the factors that previously caused disputes. In this new system the seller takes over some of the client's risks, such as cost and duration, and shares them with the producer. Even the significance in changes of these factors, for example duration, is not anymore the duration of the construction but the duration of delivery.

The Do-it-yourself system is still not very well known and relatively undocumented, but it is a way of avoiding and settling some of the disputes that can arise in this field. The problem remains open to a great variety of specialists in building materials, technology, strength, economy, marketing, management and of course specialists in law, who must be put together, not necessarily geographically, but in their thoughts and actions.

## 7 References


Blachere, G. (1988) 'Building principles'. **Office for Official Publications of the European Communities**, Luxembourg, 9.

Fenn, P. (1990) 'Settling disputes on construction projects'. **Proceedings of the International Symposium on Procurement Systems**, Zagreb, Yugoslavia.

Herbsmnan, Z. and Ellis R. (1991) **A multi-parameter bidding system - an innovation in contract administration**

Heyne, P. (1991) **The Economic Way of Thinking** (in Rumanian), E.D.P., Bucharest, Romania, 265-267 & 293.

Parker, J. and Cover, M. (1991) 'I.T. Systems'. **Building Research and Information**, vol. 19, No. 1.

Toffler, A. (1983) **The Third Wave** (in rumanian). E.P. Bucharest, 364-373.

Westling, H. (1991) **Technology Procurement**. Swedish Council for Building Research, Stockholm.

# TRANSITION AND MANAGEMENT OF UNCERTAIN RESOLUTION

RADU LUPASTEANU and EDUARD ANTOHIE
Polytechnical Institute of Iasi, Romania

Abstract
The activity in the Rumanian construction field was marked by the characteristics of the communist society. The works were distributed according to a plan and not as a result of a competition between the construction companies. The management of those was characterized by a dictatorial style so that the conflicts between the enterprises and the beneficiaries or the enterprises and the surveyors were solved by the party's militants, putting pressure upon one of the sides in conflict. The arbitrary methods were also used in the relations between the superintendents of work and their men, the supreme words being the directives of the party. Those ones came off victorious over the technical or economic criteria. This paper is meant to present these aspects, as well as some suggestions for improving the new relationship between partners in the construction field in Romania. The authors consider this fact absolutely necessary for this period of transition to a market economy - taking into account that many managers in the construction field had an old fashioned management style. This paper tries to grasp the differences between a leadership by oneself and a group leadership so intending to improve the management in the construction field for this period in Romania.
Keywords: Construction, Management in the Transition Period, Uncertain Decisions, Romania.

## 1 Introduction

The activity in the Rumanian construction field was marked by the characteristics of the communist society. The result of centralized system and the communist mixture between politics and economy was the party militants' intervention in case of conflicts. This "Damocles sword" above the construction managers influenced management style and its effectiveness.

Through this transition period, crossed by Romania, new leadership methods are attempted for coping with the construction firms' specific nature framework. It is hard to apply fixed economic or management models, even models verified in other countries because the specific economic and social aspects can falsify the efficiency of the models. The specific aspects of the transitional period require also more and more risk and uncertainty assessments.

This paper gives a brief review of the aspects, causes and implications of the centralized and dictatorial communist management style because if the system was changed, the managers and employee are the same and they have to function on the new background. This inheritance being spotlighted it will be easier to change what must be changed in construction managers' style to increase their effectiveness.

## 2 The uncertain and management styles

### 2.1 Models of decision

The act of decision is borne because of the reality impact and it is the most important moment of the leadership thinking process. The decision links thinking with action.

The literature in matter emphasizes a lot of classifications of the decisional models:

Decision in terms of certitude
Decision in terms of risk
Decision in terms of uncertainty
Cybernetic model

The classic models (certitude and risk) have a rigorous mathematical foundation so it's relatively easy to deal with them. They are appropriate especially in cases with high level technics and/or high level of knowledge. The model of decision in terms of uncertainty has to be considered when we deal with low levels of knowledge and/or with high influence of the human factor. The cybernetic model is not a decisional model, not based on theoretical knowledge but through the intermediate feedback. The need for decision is not formulated but it's felt. For supporting this model one may mention the American philosophical trend founded by Sanders Peirce "the pragmatism". The pragmatism appeared as a response against the "absolute truths" and considered as obviously only the practical checking of ideas' effects. The truth of an idea is not a result of value of an idea in itself or a result of its source. Only practise may confirm the ideas' truths.

As we mentioned, one of the main factor which directly influences the decision under uncertainty is the leader's level and his quality of knowledge. We may point out the relation between the knowledge's level and the degree in certainty/uncertainty through four zones:

Zone IA: Low level of knowledge, hazy image of facts;
Zone IB: More knowledge but yet insufficient , may lead to a rapid increase of the subjective certainty. The stability of image is

| | |
|---|---|
| | precarious. |
| Zone IIA: | The increase of knowledge leads to a sharp critical spirit, to increase the doubts. The alternative solutions are formulated. |
| Zone IIB: | The alternative solutions are obvious and well differentiated. The image is sound. |

Another aspect is the reducible/irreducible nature of uncertainty. This specific feature becomes manifest through opportunities so that by increasing the level of knowledge the uncertainty diminishes or not. The perception of this aspect determines the behavior of one who has to decide through all stages of the resolution process (the formulation of the problem, finding alternative solutions, the differentiation of alternatives, decision, decision applied).

The decisional factor has to realize the uncertain zone in which he lies, he has to estimate if the cognition effort is rewarding and he has to adopt the proper management style, also according to his personality. The different degrees in uncertainty lead the decisional factor to different management styles. At minimum levels of knowledge, the analytic decisional procedures become efficientless.

So we may characterize "the first arrival" procedure. The first identified solution is adopted. The uncertainty is artificially absorbed by investing the decision stage and the solution with surplus value (zone IB).

**2.2 Aspects and consequences of the centralized leadership**

Often in the relationships between the party militants and construction managers the production plan and the taboo communist party's directions were involved in order to support a pressure attitude from the first one towards the latter. With such a frame, very often the economic effectiveness and normal human relations were neglected.

The lack of time, insufficient equipment and often insufficient project preparation and last but not least the party militants direct implication (the communist state being the unique owner) led to a "first arrival" decisional procedure. For the communist militants the decision, or rather the directive, was not founded on the cognition procedure but on the communist dogmatic set of rules.

The decision of the party activist, as coming from someone on higher position, came off victorious over that of the people from the building firms. So an uncertain situation which leads to a decision full of surplus value, not ending artificially the incertitude, doesn't settle on the basis of proficiency, but on the basis of social status.

This one is the basis of a certainty from IA zone or IB and not one from IIB zone as it would be proper. But accepting the incertitude would contradict the farseeing policy of the Communist Party so the activist's decisions were controversial and arbitrary.

So decision wasn't seen as an iterative process and was to be applied strictly without taking into account those who had to put it into practice. We have here a typical situation of overrating of the determinator and of underrating of the group.

We shall try to analyze the effects of this managerial style on the level of construction firms' managers.

As we showed before the relationship between the party leaders and those in the construction firms generate anxiety for the latter. This situation full of stress and threatening was caused by a permanent latent conflict. This situation to which we might add the hard specific character of the work brought about some consequences for the building managers:

The group of builders consolidate their cohesion and strengthen their fellowship, sometime in the way of overrating their position. To a certain extend there was a motivation for the increase of competence and the improvement of performances. A negative process was that of the increase of frustration which leads to conflict managers-subordinates. A neglect of quality in favor of the quantity.

The researches of J.P. Dany (1966) and F. Schermer (1987) showed that once we passed over that step (which motivates the increase of competence) the efficiency of management diminishes so then for the reduction of anxiety if we don't eliminate the cause then we resort to defensive devices. In this way the individual expresses his obedience to a person or to the rules, his conservatorism, his stereotyped thinking.

We must also take into account the fact that in that period they built a lot in Romania and in spite of the aspects mentioned before, the rumanian engineers didn't neglect the safety devices and the resistance of the buildings. The proof is the resistance of these buildings to the last three earthquakes (1977, 1986, 1990).

**2.3 Decision and styles in management in the transitional period**

One might imagine that the fall of communism in Romania was followed immediately by a period of complete annihilation of the causes for an authoritative style in management. As we have already seen this style was founded to a high level of incertitude and gave birth to anxiety.

If we analyze the situation now we may point out that:

The state is still the main owner. The level of technical knowledge and technical equipment are approximately the same. There isn't any longer the leadership of the party but there is a lack of funds for building. Prices rose.

As a result investment has considerably diminished. The construction managers and employees are almost the same, having a crystallized style in work and management. The result is inertia in acting through nchanges. There is a wide tendency of contesting authority at any level.

A quality project management requires a great deal of skill on the part of the management at the technical level of the project he is managing; at the interpersonal level of the human relations he has to handle and at the level of knowledge of good project management practices. The manager must cope with many differences in values and attitudes, must anticipate problems that may arise in connection with this project and must know how to act in order to solve them. In order to achieve these aims we must eliminate the causes which might disturb the management efficiency, or at least minimize their effects. We do this by:

Passing to a mate complete private property in construction, small firms having the advantage of accepting the leader, a larger mobility, smaller capitals and terms. Coopting new funds opening new construction sites by external investments. General straightening of economy. New education of the managers according to the new economical structure. Harmonization of the human relationship by increasing the living and cultural standard.

All these changes are meant to replace the authoritative style of management with a democratical one. This one is more stable but necessitates some conditions as: the possibility of finding and confronting alternative solutions, stable social relationships, the transition from the decision in incertitude top decision in terms of risk, which make possible the evaluation and the anticipation of the facts in terms of probabilities.

## 3 Conclusions

The paper tried to show that it is difficult to give categorical solutions which do not apply in Romania today, even though they were successful elsewhere. A comparative analysis made by R.T. Pascale and A.G. Athos (1981) among the American and Japanese managers showed that the social-cultural differences among the two generate significant differences in their style of decision.

The American manager choose a determined conduct, wishes to give the impression of a determined person, so acquiring a high social status. The American style annihilates quickly the ambiguities.

In Japan, the decisional behavior is not a prime source of power and of social status. So the decisional process is an open, iterative one.

We may take into account two styles in management in Romania. The authoritative style and the democratic one. The first one, we speak of an authority based on competence is supposed to offer new structures for coherent actions in the frame of dynamic environment full of incertitude.

The democratic management is supposed to support the group to create himself such a structure. In order to choose one of this variants the managers will take into account the level and the type of incertitude (reducible or irreducible), the psychological structure of the employee, the external world of the firm. They will have to anticipate facts for a more flexible attitude based on probable evaluations. As we have already said our analysis is not supposed to give solutions but to point out a number of aspects which have to be solved by the construction managers in Romania.

## 4 References


Berkeley, D., Humphreys, P.C. and Thomas, R.D. (1991) Project risk action management. **Construction Management and Economics**, volume IX, February 1991, 3-17.

Boboc, A. (1980) **Filosofia contemporana. Orientari si tendinte in filosofia nemarxista din secolul XX.**, Editura Didactica si Pedagogica Bucuresti.

Le Bon, Gustave (1990). **Psihologia Multimilor**. Editura Anima

Pascale, R.T. and Athos, A. (1981). **The Art of Japanese Management**. Penguin Books.

Zamfir, C. (1990). **Incertitudinea, o perspectiva psihosociologica**. Editura Stiintifica Bucuresti.

## Part Three

Claims, Litigation and Arbitration

Claims procedures are in a state of flux; and it has been suggested that adjudicative procedures have failed to serve the parties involved in construction projects. The papers in this section discuss the techniques available.

'Adjudication procedures: a temporary diversion' (Bentley) describes the development and implementation of adjudication in UK construction contracts.

'Can construction claims be avoided?' (Revay) argues that the ambiguity of language in construction contracts leads to disputes, identifies major reasons for claims and discusses ways in which they can be avoided.

'Review of Australian building disputes settled by litigation' (Watts and Scrivener) investigates the sources of disputes on building contracts in Australia, court records in New South Wales between 1989 and 1990 are examined.

'Costs in arbitration proceedings' (Quick) outlines the present law of costs as it operates in arbitration proceedings and questions the likelihood of change.

'Construction contracts: towards a new relationship' (Colledge) explores the nature of commercial relationships in the construction industry and the framework for drafting and selection of construction contracts.

'Construction contractors liability in Saudi Arabia' (Assaf and Al-Hammad) highlights liability issues in Saudi Arabia, focusing on the Saudi Standard Public Works Contract which governs all public projects.

'The role of integrated cost and time models in conflict resolution' (McGowan, Horner, Zakieh, Jones and Thompson) describes the need for models which reflect the interaction between cost and time. The development of new models is outlined.

'The position of materials re payment and ownership in construction projects in the UK' (Bowles and Gow) investigates the position of materials on and off-site with reference to advance payments, ownership and transfer of title to the employer. The differing legal structures in Scotland and England are highlighted.

'Statistical Modelling of Claims Procedures and Construction Conflicts' (Dalton and Shehadeh) considers post contract conflict on construction projects.

# ADJUDICATION PROCEDURES : A TEMPORARY DIVERSION?

BRUCE BENTLEY
Dibb Lupton Broomhead, Sheffield, England

Abstract
This paper considers the development and implementation of adjudication procedures in United Kingdom (UK) standard forms of construction contracts. A background to the reasons behind adjudication is provided with reference to case law. The adjudication procedures in UK standard forms are examined; their position in relation to other disputes procedures is considered. The extension of adjudication to other areas currently precluded is discussed.
Keywords: Adjudication, UK standard forms of construction contract, litigation, arbitration, mediation.

## 1 What is adjudication?

Adjudication is a procedure where power is given by the Contract to an independent third party to make interim decisions on disputes between the parties arising under the Contract. Its essential characteristics are:

It is a contractual disputes procedure;
It is intended to operate during the course of a construction project, and not after it;
The Adjudicator can seek information to make a decision rather than simply respond to information or representations provided by the parties involved;
It provides a short time scale from reference to announcement of the decision.
Decisions of the Adjudicator are usually binding on the parties until subsequent agreement between them, or until overturned by litigation or arbitration;

## 2 Why has it developed?

The ground conditions in the Construction Industry giving rise to the need for an adjudication process are not difficult to establish, and can be summarised as follows.

The Construction Industry is labour intensive and highly competitive. Profit margins are historically low. Contractors (or Sub-Contractors) have not expected to have to fund construction work from start to finish.
Its economic structure has therefore been based on a "pay as you go" short and repetitive valuation and payment cycle. Contract payment clauses have usually been intended to operate so that interim payments will be made to the Contractor or Sub-Contractor of the approximate value of work done, with an accurate post-Contract evaluation and adjustment of the Contract Sum.
Performance of construction Contracts involves the co-ordination and sequencing of work by numerous parties, who have usually promised to carry out their part of the project for a price that assumes there will be little or no interruption delay or difficulties in their performance.
There is therefore potential for numerous inter-related legal and factual disputes between parties whose size and bargaining power differ.
The efficient completion of projects necessitates that disputes can be managed or contained in the short term so that the project can proceed without the parties being diverted into wrangling about their respective rights, rather than performing the Contract.
John Maynard Keynes was not speaking of Building Contracts and Sub-Contracts when he said (or reputedly said) "In the long run we are all dead"but the comment is apposite in relation to Building Contractors and Sub-Contractors if the resolution of contractual disputes is left as a matter of course to the lengthy dispute cycle that characterises litigation or arbitration.
The need has become apparent for a method of speedy and flexible interim resolution of disputes, pending (if necessary) their detailed legal consideration by arbitration or litigation. Adjudication procedures are intended to provide just that.

## 3 Development in contract forms

Interim payment provisions have always been the basis of Building Contracts, and any interference with or restriction on the flow of cash on a regular basis to Contractors or Sub-Contractors has invariably brought about commercial difficulties for the Contractors or Sub-Contractors concerned.

Historically, interim payment clauses did not of themselves provide that the person carrying out the work was entitled to receive the value of the work done at the date of assessment without deduction. On the contrary, Building Contracts have invariably provided that the paying party was entitled to make deductions or set offs of proper claims available against the Contractor or Sub-Contractor concerned. Even if they had not contained such set off provisions, the Common Law had always recognised that a contracting party was able to "set off" or set up a cross claim as a defence in proceedings brought by the other contracting party if that cross claim either arose out of or was connected with the Contract the subject of the initial claim. The construction industry was not regarded as being any exception to that rule.

In the 1960's, concern therefore grew, particularly amongst Sub-Contractors and their professional organisations, as to the ease with which Main Contractors could interfere with or restrict cash flow under the Sub-Contracts concerned, by the setting up of optimistic, excessive or even spurious set off claims to avoid payment under Sub-Contract interim certificates.

In 1971, temporary relief appeared to be granted to Sub-Contractors in particular, by the Court's decision in the case of Dawnays v Minter [1971] 1 WLR 1205. In that case, the Court of Appeal decided that a certificate issued to the Sub-Contractor under a Green Form of Nominated Sub-Contract (used with a JCT 63 Standard Form) was "equivalent to cash" and the normal rules of set off did not apply to it. Certified monies in these circumstances therefore had to be paid, save only for liquidated and ascertained cross claims established and admitted as payable. Unliquidated and disputed cross claims could not be used as a set off or deducted by the Main Contractors. Dawnays case was followed in a number of subsequent cases. See for example GKN Foundations Ltd. v. Wandsworth [1972] 1 Lloyds Rep 528 Token Construction Ltd v. Naviewland Properties Ltd (CA 11 May 1972) Carter Horseley (Engineers) Ltd. v. Dawnays Ltd The Times July 5 1972.

This period of joy and freedom from oppression for Sub-Contractors was brought to an end by the House of Lords in its decision in Gilbert Ash (Northern) Limited v Modern Engineering (Bristol) Limited [1974] AC 689.

The House of Lords expressly disapproved Dawnay's case and decided that

(i) Building Contracts were no different from other commercial Contracts to the extent that the employer under the Contract (including a Contractor so far as a Sub-Contractor was concerned) was entitled to exercise the rights of set off available at Common Law against certified payments, unless the Contract specifically excluded those rights of set off.

(ii) There was no general presumption in contracts entered into between Contractors and Sub-Contractors that it was intended that the Contractor's right of set off should be excluded.

The consequence was that if an arguable set off claim was raised, summary judgment on an interim certificate could not be obtained to the value of the set off claim. Payment of the deducted sum would either have to wait until the outcome of litigation or arbitration as to its correctness or otherwise, or the parties reached some commercial agreement to resolve the dispute.

Too often, the result was that hard pressed Sub-Contractors were forced to compromise claims for payment by settling set off claims on disadvantageous terms. The primary problem in Contract disputes was therefore perceived to be control of the abuse of set off claims by Main Contractors. Bona fide Sub-Contract claims for payment could be denied or postponed by the Main Contractor for long periods by inflated or colourable set off claims.

What was needed was a speedy process for independent review of the set off claims, as and when they were put forward as a basis for withholding payment of interim certificates.

The solution adopted in the Standard forms was the introduction of an Adjudicator whose sole concern was to consider any set off claim put forward by the Main Contractor under the Sub-Contract, and decide upon its merit.

The clauses introduced thereafter were as follows:

Set off - Adjudication Clauses were introduced as follows:-

| | |
|---|---|
| 1976 | JCT 63 Sub Contract forms (Green and Blue Forms). |
| 1980 (DOM1) | NSC4/4A forms for use with JCT 80 Standard Form of Main Contract. |
| 1981 | DOM2 Sub-Contract for use with JCT 81 With Contractors Design Main Contract |
| 1984 | NAM/SC and IN/SC named and domestic Sub-Contract forms for use with Intermediate form of Contract (IFC 84) |
| 1987 | JCT Works Contract for use with JCT Management Contract |

Extension of application to other issues and other contract forms

| | |
|---|---|
| 1983 | BPF form of Building Contract - Adjudication provided in respect of a wide range of issues beyond set off claims. Optional adjudication was available in its precursor form The ACA form of Building Contract 1982 |
| 1985 | GW/S Sub-Contract for use with GCWorks 1 Edition 2 1985 (set off/adjudication clause similar to NSC4/DOM1 Sub-Contract) |
| 1988 | JCT 81 With Contractors Design Main Contract - Adjudication is available in relation to a wide range of issues |
| 1989 | GC Works 1 Edition 3 (not limited to set off) |
| 1991 | ICE Draft New Engineering Contract adjudication of all disputes as an interim procedure |

## 4 Method of operation of common clauses JCT contracts

### 4.1 NSC4 forms used with JCT80 standard form or main contract

Relief Available

By Clause 24.3, Adjudication is limited to the consideration of Main Contractors set off claims. The Adjudicator is empowered to decide in his absolute discretion whether the amount of the set-off claim shall be:-

i) Retained by the Contractor or

ii) Pending arbitration, deposited by the Contractor with the named Trustee Stakeholder or

iii) Paid to the Sub-Contractor or

iv) Dealt with in some combination of i), ii) and iii).

Note

(a) Although the Adjudicator has "absolute discretion", clause 24.3.2 requires that his decision should be such "as he considers to be fair, reasonable and necessary in all the circumstances of the dispute as set out in the statements."

(b) By clauses 24.4.1 and 24.4.2. the maximum amount payable to the Trustee Stakeholder or the Sub-Contractor is the amount otherwise due from the Contractor as the interim payment under clause 21.3 (in respect of which the Contractor sought to exercise the clause 23 right of set-off)

Procedure

The procedure can be invoked when the Main Contractor has given notice under Clause 23.2.3 of a set off intended to be made against an interim payment. The adjudication procedure is implemented as follows (Clause 24.1):-

The Sub-Contractor disagreeing with the Contractor's notice of set-off must at the same time:-

a) send a written statement to the Contractor setting out the reasons for disagreeing with the claim to set-off, and particulars of any counterclaim (quantified in detail and with reasonable accuracy) arising out of the sub-contract;

b) request (in writing) action by the named Adjudicator (clause 24.1.1.2) and provide him with a copy of the statement provided to the Contractor and the notice of set-off to which it relates.

c) give notice of arbitration to the Contractor in respect of the dispute (clause 24.1.1).

The time limit for taking these steps is within 14 days of receipt of a clause 23 notice from the Contractor.

No submission is required under clause 24 by the Contractor in response to the Contractor's notice, unless a Sub-Contractor has made a counterclaim, or the Adjudicator actually requires the Contractor to make a statement. The Contractor can however make a submission, and the Adjudicator will consider it as long as it is available in the timescale provided by the procedure.

The Adjudicator also has a discretion to ask for such further written statements as may appear to him to be necessary to clarify or explain the ambiguity of the written statement of either the Contractor or the Sub-Contractor (clause 24.3.1).

The Adjudicator's decision is given without a hearing. A reasoned decision is not required (clause 24.3.1).

The decision should be made, at the latest, within 21 days of the Sub-Contractor's request for action. There is no express provision to this effect but it seems to be the applicable period by virtue of the operation of clauses 24.2 and 24.3.

Effect of Decision

The Adjudicator's decision is binding on the Contractor and the Sub-Contractor until the matters upon which the Adjudicator has given his decision have been settled by agreement or determined by an arbitrator or the Court (clause 24.3.1).

The Arbitrator appointed under the contract is free to vary or cancel the decision of the Adjudicator at any time before his final award on the application of either party (clause 24.6) if it appears just and reasonable to him to do so. The Arbitration of the set off claim and the Adjudicators decision need not be postponed until after practical completion.

1987 Amendments

In 1987, amendments were made to the set-off rights in clause 23 in the NSC 4/4A forms. These confirm the Contractor's rights to set-off and can include loss and expense suffered or incurred.

The 1987 amendments also amended clause 24 itself to:-

a) require the Sub-Contractor to provide brief particulars of the sub-contract to the Adjudicator when requesting action under clause 24.1.1.2;

b) provide for the appointment of an Adjudicator by the Sub-Contractor from a list maintained by the BEC if the parties fail to name one in the sub-contract and the appointment by the Adjudicator of the deposit taking bank as Trustee Stakeholder if no Trustee Stakeholder is named in the sub-contract;

c) clarify the wording dealing with the time limit for the Adjudicator's decision (although the amendment raises as many problems as the wording it replaces).

An almost identical adjudication procedure is contained in NSC/C issued in 1991.

**4.2 DOM1**

This is the domestic Sub-Contract form for use with the JCT 80 Main Contract form. The DOM1 form contains adjudication procedure which is almost identical to the NSC4 forms. Differences to note are:-

a) the Adjudicator's consideration of set-off claims by the Contractor

attributable to delaying completion by the Sub-Contractor are not dependent on the issue of a certificate of non-completion by the architect (the architect does not issue certificates in relation to domestic Sub-Contractor delays);
b) the time limit for the Contractor to notify the Sub-Contractor of his intention to exercise a right of set-off is three days before the payment becomes due.

**4.3 JCT87 works contract**

The Works Contract contains adjudication provisions intended to operate with the Main Contractor set-off provisions. The relevant clauses are 4.33 to 4.44 of the Works Contract. They are essentially the same as the NSC4/4A provisions as amended in 1987.

Differences of note are that the set-off claims covered are claims for loss and/or expense and/or damage. No qualifying wording is included (presumably because the Management Contractor does not actually suffer the losses but he is recovering them for his employer).

Deduction of set-off arising from a delay in completion is not dependent on the architect's certificate of non-completion (see clause 3.34.1.) Arbitration of the set off is not postponed until after Practical Completion.

**4.4 DOM2 (JCT81 design and build)**

The DOM2 Sub-Contract Conditions are the DOM1 conditions save only for amendments necessary to deal with the involvement of the Sub-Contractor in design and the use of the form with the JCT 81 Form instead of JCT 80. The set-off and adjudication provisions in the DOM2 form are therefore virtually identical to the DOM1 provisions.

The 1987 amendments to DOM1 were included in the DOM2 form.

**4.5 IN/SC Sub-contract used with IFC84 intermediate form of contract (IFC84)**

This sub-contract form, when issued, contained set-off and adjudication provisions in Clauses 21 and 22 substantially the same as those in the original DOM1 form dealt with above.

Amendments to these provisions were made in 1989. The amendments are the same as those in NSC4 and DOM1 issued in 1987.

**4.6 NAM/SC Sub-contract used with ISC84**

The set-off and adjudication provisions in Clauses 21 and 22 are identical to those in IN/SC.

**4.7 JCT 81 with contractors design**

This form has an adjudication procedure introduced in February 1988 by the issue by the JCT of Amendment 3 to the Form issued in 1981.

It represented two further stages of development in the use of adjudication procedures.

i) It introduced the adjudication procedure to a Main Contract form (although the procedure is optional by virtue of the provisions in Appendix 1).

ii) It extended the range of issues that could be referred to adjudication beyond the set off procedure in the Sub-Contract forms.

Relief Available

The matters that can be referred to adjudication are set out in Clause S1.2 and are:

1. Any adjustment or alteration to the Contract Sum.
2. Whether the Works are being executed in accordance with the Conditions and the Supplementary Provisions.
3. Whether or not the issue of an instruction is empowered by the Conditions.
4. Whether either party has withheld or delayed a consent or statement or agreement where such matters are not to be unreasonably withheld or delayed.
5. Extensions of time and loss and expense to the limited extent provided for in Supplementary Provisions S6 and S7.
6. Various specific matters identified in Clause S1.2.5 in relation to the Supplementary Provisions themselves.

The procedure is set out in the Supplementary Provisions.

Note that the termination provisions of the Form in sections 27 and 28 are not capable of adjudication.

Procedure

If the adjudication provisions are operative, and assuming the dispute refers to an "Adjudication Matter" as defined in Clause S1.2 then:

i) either party gives notice to the other that an adjudication dispute has arisen;

ii) by the expiry of 14 days after the date of that notice both parties are to submit statements to the Adjudicator setting out the matters in dispute on which the decision of the Adjudicator is required (S1.3.1).

iii) Within fourteen days of issue of the statements (or such other time as the parties may agree) the Adjudicator is to notify the parties when he expects his decisions will be given. He may request further information or documents from either party as he reasonably requires to enable him to reach his decision

(Note: a failure by either party to comply with any requirements does not invalidate the decision).

iv) The Adjudicator gives his decision, acting as expert and not arbitrator (S1.3.2).

Effect of Decision

A novel idea has been put in place for ensuring the validity of the decision. It shall "be deemed to be a provision of this Contract (an Adjudicated Provision) and such adjudication provisions shall be final and binding on the parties unless referred to arbitration as provided in S1.3.4 or S1.4" (S1.3.3).

Conflicts are dealt with by the Adjudicated Provision prevailing over "any other provision of this Contract" where there is a conflict.

If either party is dissatisfied with the Adjudicator's decision he can notify the other party within 14 days of the receipt of the decision and the dispute then is dealt with under the arbitration provisions in article 5 and clause 39, subject to the time constraints imposed in Clause S1.3.4 that arbitration cannot proceed until after Practical Completion. The Adjudicator's decision remains in force until dealt with by the arbitrator.

Each party so bears its own costs incurred in the adjudication (Clause S1.6). The Adjudicator's fee is shared equally between the parties (clause S1.7).

Other Practical Points

The Adjudicator is not disqualified from being a witness in any arbitration by virtue of acting(S1.3.5).

Disputes or differences arising between the parties in respect of an adjudicated provision are to be dealt with under Clause S.1 and Article 5 and clause 39 in the same way as a dispute or difference under any other provision of the Contract.

**4.8 Other forms**

ICE New Engineering Contract.

The Guidance Notes issued relating to the NEC state:-

"In NEC an intermediate stage of dispute resolution has been introduced in the form of adjudication. It is the intention that all disputes should be resolved by the Adjudicator. However if either party is dissatisfied with the Adjudicator's decision he may refer the dispute to arbitration."

Clause 90 contains the adjudication provisions.

Relief available

The party complaining can proceed to adjudication on disagreement with an action of the Project Manager or Supervisor (or where it is considered such action is outside their authority). There is a time limit of four weeks from the action complained of for exercise of the right to adjudication (Clause 90.1).

Procedure

i) Adjudication is by the Adjudicator named in the document;

ii) Notice of the dispute existing is given to the Adjudicator and the other party;

iii) The Project Manager is to submit to the Adjudicator the information upon which the disputed action was based (Clause 90.2);
iv) The Adjudicator makes a decision as to the correctness of the action and if he considers it was not correct, what action should be taken and the time and cost consequences thereof (Clause 90.2).
v) The decision is due within 4 weeks of the matter being referred to adjudication (Clause 90.3).
vi) A party dissatisfied with the Adjudicator's decision has 8 weeks in which to refer the matter to arbitration (Clause 91.1). Clause 90 gives no indication of the effect or Adjudicator's decision pending arbitration, but if the time limit of 8 weeks expires it must be assumed the decision has binding effect.

Effect of decision

The Adjudicator does not normally alter decisions made by the Project Manager. If he disagrees with a decision made by the Project Manager, he will say so and award the Contractor compensation for the fact that the decision was wrongly taken. He will not however overturn the Project Manager's decision, except in respect of amounts certified for payment, or other financial decisions and in certain circumstances, extensions of time (see below).

If the Adjudicator disagrees with the amount certified, the Project Manager will be required to make the necessary correction in the next Certificate.

If the Adjudicator disagrees with the Project Manager's decision about extension of time, the Adjudicator will overrule the Project Manager's decision and the completion time will be set in accordance with that which the Adjudicator decides. If however it is too late to allow the Contractor to re-programme, the decision will only be a financial one.

Notes.

i) The Adjudicator's immunity from liability (Clause 90.5).
ii) The Adjudicator's fees are shared equally.
iii) The new Engineering Sub-Contract contains similar adjudication provisions. If there are simultaneous disputes under the Contract and Sub-Contract the Contractor can refer the Sub-Contract dispute to the Adjudicator and the two disputes are treated as one (Clause 90.4).
iv) The ICE Engineering Contract 6th Edition does <u>not</u> have an adjudication procedure. A conciliation procedure is provided (see clause 66(5) followed by arbitration under clause 66(6).

GC Works 1 Edition 3 1989

A procedure described as adjudication was introduced in the 3rd Edition in Clause 59.

Relief Available
Adjudication is available in respect of any dispute difference or question arising out of or relating to the Contract other than one as to which a decision is expressed to be final and conclusive. (Interim payments cannot therefore be adjudicated.)

The dispute etc must have been outstanding for at least 3 months.

The Adjudicator is selected by the person named in the Contract and is to be an officer of the Authority (or a person acting for it) who has not been associated with the letting or management of the Contract.

BPF Building Contract

This Contract form is part of the BPF System for Designing and Constructing Buildings issued in 1983.

The Building Contract itself contains an adjudication procedure which refers all disputes to adjudication.

Relief Available
Clause 25 provides that adjudication may take place in relation to:-

(i) Adjustment or alteration of the Contract Sum;
(ii) Extensions of Time;
(iii) Whether the Works are being properly executed;
(iv) Availability of the Contractual rights of termination;
(v) The Clients representatives rights of access to the Works and workshops to test and inspect;
(vi) The reasonableness of any objection by the Contractor to a replacement clients representative and to execution of work or installation of things by others.

Procedure

Adjudication is dealt with by the named Adjudicator. The procedure is commenced by either party requesting adjudication. There are no time limits as such for commencement (although note clauses 7 and 17). A decision is to be given within 5 working days of a request for it.
The Adjudicator can request either party to provide oral or written statements, documents or information to assist him.

If the Adjudication dispute relates to a matter upon which either party relies to terminate the Contract, that right of termination is suspended until after the Adjudicator's decision.

If either party is dissatisfied with the decision, or a decision is not given within the stipulated time, either party may give notice of arbitration within 20 working days of the receipt of the Notice of Decision or expiry of the time within which it should have been given.

The Adjudicator acts as an expert.

Reference of a dispute to the Adjudicator does not relieve either party of its obligations under the Contract.

Effect of Decision

Clause 25 provides that the Adjudicator's decision "shall forthwith be given effect to" by the parties. It is final and binding upon the parties until Taking Over of the Works. Arbitration is available, but if either party fails to give notice of arbitration in relation to the adjudication matter within 20 working days of receiving notice of the Adjudicator's decision, the decision is final and binding (clause 25.5).

## 5 Interaction with other disputes procedures

Is adjudication intended to work with these procedures, or to replace them?

It has been recognised for many years that arbitration and litigation have considerable shortcomings as methods of dispute resolution. Adjudication is one approach that has been developed to meet those shortcomings, but there are others such as conciliation mediation and ADR.

### 5.1 Arbitration

As stated at the outset of this paper, adjudication is a contractual dispute procedure. It should be noted that all the Contract forms I have referred to as having adjudication procedures, also have arbitration clauses intended to be available for the resolution of disputes and differences under the Contract concerned.

Moreover the adjudication clauses described have not (until the New Engineering Contract) been drafted so as to deal with all disputes that might arise under the Contract. As a matter of policy therefore, adjudication has been developed to complement arbitration rather than to replace it. Its purpose has been to deal primarily with disputes that might affect the day to day cash flow of the Contract.

There are obvious differences between the two procedures namely:

(i) In arbitration, the Arbitrator acts in a judicial or quasi-judicial capacity. The Adjudicator acts as an expert.

(ii) the Arbitrator must decide the reference on the representations and evidence provided to him by the parties. An Adjudicator is not so limited, and can carry out his own investigation into the circumstances of the dispute.

(iii) the arbitration procedure provides for statutory rights of appeal or determination of preliminary points of law (see the Arbitration Act 1979). The adjudication process is not subject to legislation. Its procedure is set out exhaustively in the Contract itself.

(iv) an arbitration award is enforceable under the Arbitration Acts as a judgment of the Court, following registration. Adjudication decisions are not enforceable in that way. If the decision is incorporated into the Contract as a term, it will be enforceable by action on the Contract, and by summary judgment.

The differences in methodology are consequences of the different purposes of the two procedures

**5.2 Litigation**

The same comments made in relation to arbitration are equally applicable to litigation as a dispute resolution procedure. Two points in particular are worth noting in this respect:

i) In set off cases in particular, the Court is likely to grant summary judgment in a case where an Adjudicator has decided a claimed set off is not proper, as it will rely on the Adjudicator's decision as to whether any arguable set off or defence exists to the interim payment claim. The adjudication procedure therefore assists the use of the separate Court procedures for enforcement of contractual rights.

ii) Adjudication decisions, if ignored, are likely to depend on the Courts and litigation for their implementation, by enforcement through the Courts as contract terms.

Adjudication therefore provides a dispute procedure which can be invoked by a party that otherwise has no remedy in the short term, or would have to rely on the Courts unassisted by the Contract machinery. Adjudication decisions are interim and reviewable as a matter of course through arbitration or litigation. The history of adjudication therefore shows that its development has been to support litigation or arbitration rather than substitute for it.

Its virtues of speed and economy (and the accepted interim nature of decisions made in it) should allow it to meet the perceived shortcomings of delay and expense in litigation and arbitration.

**5.3 Mediation**

Mediation is always available to the parties in dispute, irrespective of the existence of Contract procedures. It has similarity to adjudication in is overall purpose and procedure, but it is usually intended to be a means of obtaining final and conclusive disposal of disputes, rather than an interim resolution of them. Like adjudication, it can be a speedy and economic way of dealing with disputes.

Its drawback is however that it relies for its existence and effectiveness on the co-operation of the parties in dispute. That is not always guaranteed. Its development as a dispute resolution procedure therefore depends to some extent on a change of attitude in the construction industry away from the games playing and position taking that contributed to the development of adjudication procedures. The existing adjudication provisions do not make any attempts to accommodate or work in parallel with mediation procedures. For the moment mediation must therefore be assumed to be a "competitor" of adjudication in the dispute resolution process.

## 6 Should adjudication clauses be optional or mandatory in standard forms?

One answer to this is that the choice is already there, in the sense that as a contract procedure, adjudication provisions can be deleted by the parties from any standard form to be used as the Contracts between them.

Its deletion does not otherwise upset the operation of the Contract. It may however be that an express provision in the Contract to the effect that the procedure is optional may result in one party at the outset insisting that the provisions be excluded.

This may be for no other reason than existence of the procedure becomes apparent because of the need to consider and make an amendment to the form. It is not always the case that parties using standard forms acquaint themselves with every item in them.

There is no evidence available to support the point one way or the other, but the ability to exclude the set off provisions in any event suggest that it is better to have the provisions in the Contract form as a matter of course, rather than have provisions that adjudication will/will not apply, and then leave the parties to make a positive choice at the time of completion of the Contract.

## 7 Future development

There seems little doubt that adjudication has merit as a dispute resolution procedure, offering benefits not otherwise available in the traditional procedures. To that extent, the current procedures should be used and supported. The question arises as to what developments of the procedure could usefully be undertaken. Some prospective developments are as follows:

(i) Jurisdiction. Most of the adjudication procedures currently used are limited in extent. It seems sensible that they should be extended to cover more areas of dispute. They are particularly useful in disputes where the delays associated with arbitration or litigation can cause unfair prejudice to either of the parties. It would seem therefore sensible to have the procedures available for all interim payment related issues, loss and expense and extensions of time.

(ii) Nature of decision. The procedure is intended to be a rough and ready solution without the benefit of a full presentation by the parties to the Adjudicator. That right is sacrificed in the interests of speed. The question therefore arises as to whether or not the interim nature of the decision should be preserved. The proper course seems to me to be to ensure that they remain of an interim nature, but they are made effective until such time as the award of an Arbitrator or the Court is made finally resolving the dispute. This should enhance their use as a complement to litigation and arbitration.

(iii) Enforceability. It is unfortunate that they are not enforceable as the decisions of an Arbitrator. They should be drafted so as to ensure they can be quickly enforced, if ignored, through the Courts by the summary judgment procedure. Alternatively the arbitration clauses in the forms should recognise that interim adjudication decisions are to be implemented by the parties until finally reviewed, and if necessary altered by an Arbitrator's award.

Overall, adjudication is perceived as a useful tool for Contract dispute management, as part of the contractual framework implemented by the parties at the outset of the project. Whether that theory is borne out in practice remains to be seen. Much will depend on the attitude of the Courts, and the parties themselves.

# CAN CONSTRUCTION CLAIMS BE AVOIDED?

S.G. REVAY
Revay and Associates Limited, Montreal, Canada

Abstract
The likelihood of disputes (and/or claims) developing on most construction projects today is great, if for no other reason but because the language of the usual construction contracts is seldom so clear as not to leave room for disagreement. More importantly, the owners' ill-advised attitude of trying to save money by shifting more and more risk and responsibility onto the contractor (or supplier) and a similar approach by contractors towards their sub-contractors are clear invitations to disputes. But the acknowledgement of the likelihood of disputes occurring on most construction projects need not force us to accept that they are inevitable. Conflict management does not start when the dispute first raises its ugly head. Rather it begins with the selection of the philosophy of contracting that could eliminate (or at least reduce) potential areas of disputes. This paper will identify frequently re-occurring reasons for claims and will investigate ways how those claims can be avoided and with what financial consequences.
Keywords: Disputes, Claim Avoidance, Changes in Design and/or in Job Conditions, Partnering

## 1 Introduction

The dictionary definition of the word "claim" is "an assertion of a right to something" or "the demanding of something rightfully due to one". Both of these definitions invoke a "right" that the claimant allegedly has. This "right" on a construction project does, of course, depend on the interpretation of the governing terms of the contract; and the interpretation frequently depends on the interpreter, e.g., the contractor, the engineer, the owner, or perhaps even an independent tribunal. Construction contracts are seldom written in such clear and precise language as not to leave room for differing interpretations, particularly when the interpreter's financial interest could be jeopardized by the meaning of a given clause. These potential disagreements concerning the rights and/or obligations of the parties to a construction contract have long been identified as a root cause of claims. It has also been recognized that he who authors the contract can tilt the final outcome in his favour.

The owners, or the engineers acting on their behalf, are often accused of shifting more and more responsibilities onto the contractors and in fact trying to write totally one-sided contracts where the owner has all the rights and the contractor all the responsibilities, if for no other reason than because such an allocation, in their opinion, would eliminate all the rights the contractor may have to additional compensation, therefore, it would preclude all claims. This, of course, is an unrealizable dream. One cannot very well write a contract which would hold the owner harmless of any and every eventuality and still expect to receive bids. More importantly, however, if disclaimer clauses are introduced with a view to safeguarding the owner's financial interests, then perhaps the cost effectiveness of such an approach ought to be first analyzed. It is not always in the owner's interest to pass on all potential risks to the contractor. This is particularly true with respect to government bodies or institutional owners. Even in situations where most of the risks have apparently been successfully shed by an owner, his exposure is not necessarily eliminated. The misunderstanding by owners and at times by their engineers concerning the true protection they enjoy under a contract on the one hand and their obligation towards the contractor on the other is a very frequent reason for successful claims. Added to this is the owner's misguided desire to save money at the wrong end of the project. Together, they probably represent the causes for most claims. It is not unusual today to hear of projects where the owner paid more to his legal counsel to defend against claims asserted by contractors than to the engineer for the design of the project. Claims cannot be eliminated by trying to make contracts more watertight. Moreover, this tendency could conceivably increase the bid prices, particularly when the market conditions allow it.

## 2 Reasons for claims

Perhaps the most frequent causes for claims today, which can be traced back to the owner's misguided desire to save money at the wrong place, are the following:

1. inadequate site and/or soil investigation prior to starting the design;
2. starting design efforts too late and/or unduly limiting the cost of engineering/design;
3. calling for bids with an incomplete set of drawings;
4. endeavouring to complete the design through shop drawing review;
5. introducing untimely design revisions without allowing commensurate time extension for the completion of the project or without recognizing the contractor's right to impact costs;
6. interfering both with the sequence and the timing of construction (e.g. to compensate for the delay in the delivery of owner-supplied equipment /

material);

7. continuing to introduce changes under the disguise of correcting deficiencies.

The larger and more complex the project, the greater will be the likelihood of several major claims. There have been attempts to cope with some of these problems while maintaining an ambitious construction schedule through phased construction, usually by introducing a construction manager between the owner and the trade contractors. Undoubtedly, this management method of contracting has its advantages, mostly in the duration of the construction, but it is totally useless from the point of view of eliminating claims. History tells us that projects built under the construction management method end up generating more claims than a similar project built by a general contractor, simply because with construction management there is great danger for the project to be carried out on "fast track". Further reason for the increased number of claims is the construction manager's attempt to escape the responsibility for the coordination of the various prime (e.g. trade) contractors. A general contractor would never dream of delegating project-coordination to one of its subcontractors for fear of losing control of the job and henceforth of the cost of construction; but a cost-plus construction manager is usually not concerned about the potential cost overruns of the various trades that often result from nonexistent or poor site coordination. Going the construction management route, therefore, is not the answer. This method of contracting, in fact, can increase, not reduce, the opportunities for disputes, particularly concerning risk allocation.

How much hardship results out of construction claims? The question is asked more and more frequently, if for no other reason but because the answer could alter the way construction projects will be purchased in the future. There is no simple answer to this question, nevertheless even cursory examination of past claims points to severe financial consequences particularly on the part of the claimant; claims have been also known to destroy the viability of the entire undertaking for the owner.

Engineers working for our company have analyzed a total of 175 projects in three independent studies. The governing criteria for the selection of these projects, out of the nearly two thousand on which we were consulted in the past, was the availability of suitable information. These projects were carried out between 1975 and 1991, mostly, but not exclusively, on the North American continent. The combined total of their original contract price was in the order of $1 billion. This amount, if adjusted for 1992 dollars, would equal five percent of the yearly contract-construction in Canada. As such, it may not be considered by some fully representative of the entire industry, nevertheless, it is sufficiently significant to deserve attention. These projects, which embraced industrial, commercial, institutional and heavy engineering construction, generated in excess of $300 million worth of change orders and/or claims, the value of some exceeding the original contract price. The reasons for the claims were manifold, most projects were affected by more than one problem, some by as many as four or five. The common feature of all these projects was the hurriedly and incompletely prepared bid documents, giving rise to design changes, extra work, quantity fluctuation, etc., during the currency of the project. The total direct cost of these changes represented 11.4 percent of the combined total of the original contract prices, that amount by itself may not be so alarming to some. After all, the owners probably would have had to pay that much even if the work covered by the changes had been part of the original contract. This, unfortunately, is a fallacious argument: firstly, because contractors seldom offer the same competitive prices for changes as included in their bid, secondly, and more importantly, because changes interjected during the currency of a project may, and usually do, give rise to severe cost and time impacts. Contractors, particularly if eager to get the contract, will not analyze the completeness of the design and limit their bid to the narrow, obvious scope covered by the documents issued for that purpose. Simply stated, their planning, work sequencing, the resource allocation and henceforth their cost estimates provide little or no room to accommodate additional requirements introduced once the work commenced. Such unforseen (at least by the contractor) requirements resulting from the ongoing design and/or procurement efforts could and often do interfere with the orderly and efficient performance of the project. Interference with the original planning, sequencing and resource utilization do give rise to extended duration and additional costs.

The average cost of an extended duration sustained by the sample projects represented 8.9 percent of their original contract price. On some of these projects the contractor was instructed or forced to accelerate the rate of progress by either working overtime or mobilizing additional resources with a view to buying back the time lost while coping with the changes. The average cost of such acceleration represented 1.7 percent of the original contract price of the sample projects. This figure, however, does not include the cost of lost productivity resulting from overtime, overmanning or congestion, that is the usual consequence of acceleration. Acceleration is not, however, the only cause that gives rise to loss of productivity. It may result from working during inclement weather, such as when an activity is delayed from summer into winter. Additionally, interference with the orderly sequence of the work can also give rise to loss of productivity, such as when new "late-in-the-day" requirements give rise to stop-and-go operation. The average cost of lost productivity experienced during the sample projects represented 11.5 percent of the original contract price. All in all the real cost of the variations resulting from the incorrect or incomplete bid documents represented an average of 33.5 percent of the original contract price. Simply stated, the real cost of the variations injected during the currency of the project was triple that of the direct.

Some of these changes would have been unavoidable even in the best of circumstances, but those represent a small percentage of all the ones that could have been prevented with better prepared and more complete bid documents.

There will be some who would argue that even such a high premium can be justified by the owner as long as the facility starts earning revenue on the scheduled date. Unfortunately, experience does not sustain such an argument. The analysis of 145 projects out of the total 175 (i.e. the projects for which appropriate information was available) revealed that nearly all of them suffered significant delays, notwithstanding the attempted acceleration. The average delay was 5.69 months per project, representing nearly fifty percent overrun in the planned duration. Thus, the projects examined took on average one and a half times as long to complete as intended and cost one-third more than estimated. Had the owners spent four to five months more in investigating, planning and designing these projects than they actually did, they would have saved at least twenty percent of the actual cost (sustained either by them or by the contractor), even had they paid approximately fifty percent more to their designers.

Obviously not every project is put out to bid without proper site investigation or prior to completion of the design, therefore, paying an extra fifty percent for the design would not automatically save twenty percent in the cost of every project. It is safe to say, nevertheless, that an owner would save significantly more in the cost of the project than the extra fee paid to the designers, particularly if one considers the cost of potential claims and the associated legal fees.

In 1980 our company was retained by an owners' association to study and report on the prevailing productivity in a Canadian province in comparison with some generally recognized standard (e.g. an estimating manual). As part of the information gathering, we interviewed 190 companies (i.e. contractors, subcontractors and owners) and obtained 120 offers to cooperate. Although not all of them ended up actually participating, we received extensive cost and scope information on 150 projects (constructed during the 1978-1980 period) and were able to analyze, in depth, the actually achieved progress and productivity of 1,200 operations. The results were categorized both with respect to the type, (e.g. industrial, commercial, etc.), the size (e.g. less than $15 million, $15-50 million, over $50 million, etc.) and the location (i.e. urban or rural) of the projects. Additionally, we differentiated the results, with respect to all categories, between those obtained on firm price contracts and on cost reimbursable ones. The findings, in the report, were summarized as follows:

> "Comparable tasks on firm price contracts consistently show better productivity varying between 5% to 91% on an individual basis and 30% to 40% on a global basis."

Further examination revealed that most of the cost-reimbursable projects were "fast-tracked" apparently because their design was insufficiently advanced at the time the construction started.

It is contended that the results of the above survey further support the earlier assertion that jobs which were put out to bid (or where construction started) prior to completing the design (at least with respect to a given package out for bid), take longer and cost more than estimated. The extra cost may not necessarily accrue to the owner. In certain circumstances, primarily as a result of the governing terms of a particular contract, some of the extra costs are left to the contractor to support, often, after a lengthy and costly dispute (e.g. litigation).

## 3 Conclusion

Can claims be avoided at all? The answer is perhaps, but not always. With respect to the owners the answer is simple, although at times difficult to put into practice: spend more money at the front end of the project and give more time to your engineers to plan, engineer and design the project in a more meaningful manner before calling for bids. With respect to the engineers the answer is even more straightforward, but equally difficult to accept: be prepared to shoulder your classical responsibilities and do not try to get out from under your liabilities by shifting them onto the contractors. Admittedly these ideas may sound utopian if not outright insulting to some; nevertheless, with a little effort both owners and engineers could go a long way toward a claim-free construction climate.

Even a cursory examination of trade journals and seminar brochures proves that today we are more interested in finding a cure for the symptom than for the disease. Most of these periodicals and seminar programs start with the assumption that construction disputes are inevitable, therefore all the attention centres around more expeditious and less painful resolution. Even those procedures which are advertised as means of claim avoidance (and there are a few), in reality are nothing more than proposals for third party intervention at an early stage of the dispute. It would be a major mistake not to commend the efforts and recognize the results that have been achieved in the field of alternate dispute resolutions. The construction industry is slowly emerging from the dark ages where the owner had all of the authority and the contractor carried all the risks. The introduction of expeditious ADR methods deserve a lot of credit for the already accomplished transformation. Unfortunately, the success that can be achieved by curing the symptom only is limited. It is true that the very fact that expeditious dispute resolution procedure is available prompts face-to-face negotiation by the parties and therefore tends to reduce the bitterness and financial hardship associated with lengthy litigation. It is also true that an early resolution of a dispute helps to limit the ultimate cost overrun, but it does not eliminate it in its entirety. Delays still occur, there still are increased costs due to the resulting extended project duration and the argument concerning the responsibility for the low productivity remains.

Claims (or disputes for those who dislike the word claim) and the cost overruns will not be entirely eliminated until the buyers of construction services (i.e. the owners) and the sellers of those services (e.g. the designers, contractors, suppliers, etc.) treat each other as partners and not as adversaries. What the owners must realize is that disputes, in the final analysis, are always about unanticipated extra costs. Or more precisely, the essence of a construction dispute is always the responsibility for the potential cost overruns. Therefore, the only way to avoid disputes and/or claims is either to eliminate or at least to reduce significantly the opportunities for cost overruns, such can be achieved only with better prepared and more comprehensive bid documents. It goes without saying that the elimination or drastic reduction of opportunities for disagreement is in the interest of both parties, which is the reason for the suggested partnering.

The ultimate answer may not be found unless the construction industry is restructured in a way that the selection of the successful bidder is based on the competitiveness of the resulting product (e.g. the area of rented space or the manufactured product, etc.) and not on the competitiveness of bid price. The extent and the details of such a partnering would, of course, depend on the character of the project and the nature of the ownership. Example: the partnering agreement where government is the buyer of the construction service would have to be totally different from the one where the buyer is a private corporation. Similarly, the measure of competitiveness would be different for an office building, a highway, etc. The principle governing those agreements ought, nevertheless, to remain constant: the goal must be the elimination of cost overruns to the extent possible and where overruns nevertheless do develop, to assure that the burden is divided in a cost-effective and equitable manner.

# REVIEW OF AUSTRALIAN BUILDING DISPUTES SETTLED BY LITIGATION

VICKY M. WATTS
Tutor, University of Melbourne, Australia
JOHN C. SCRIVENER
Professor of Building, University of Melbourne, Australia

Abstract
Judgments of building disputes from the Supreme Courts of New South Wales and Victoria, Australia over the 1989-1990 period are reviewed. 59 different categories of dispute are recognised within a total of 117 sources of dispute. The most frequent sources of dipute were : determination of the agreement caused by the claimed failure of the builder or sub-contractor, and claims for variations
Keywords : Arbitrations, Building : Law, Building : Defects, Contracts : Law of, Law Courts

## 1 Introduction

The paper represents the first stage of a research study, the aim of which is to improve documentation and administration processes used in the building industry so that the number of disputes and their cost may be lessened.

Knowledge of the frequency of occurrence of disputes within the building industry and the manner in which they are settled is an essential basis for this study. As dispute resolution is so often a private matter between the parties, court judgements of building disputes provide an obvious and useful source of data. Thus building disputes resolved by litigation are reviewed in order to identify the sources of dispute in each case.

Data has been taken from cases which finally reached the Supreme Court of New South Wales or Victoria or the Court of Appeal of Australia, in 1989 and 1990. By review of the claims discussed in the judgements for these building cases the types and frequency of the sources of dispute were able to be identified.

## 2 The sample

Data for the study came from 22 Victorian (Vic) and 46 New South Wales (NSW) judgements of the state Supreme Courts, plus one Victorian and three New South Wales judgements of the Court of Appeal of Australia. In the sample period (January 1, 1989 to December 31, 1990) the Building Cases List indicated 80 judgements of building disputes but only 72 of these were available. The majority (90 %) of these judgements were unreported. Whether a particular case identified a legal precedent or not was of no concern.

Information was extracted from the judgements regardless of whether they were final or interlocutory judgements. In identifying the sources of dispute from the judgements the data includes all claims discussed by the Judge.

As a number of the cases had more than one judgement, 56 cases generated the 72 judgements.

59 different categories of dispute were recognised within the total of 117 sources of dispute.

In Table 1 the cases are arranged into four subgroups according to the resolution method adopted.

Table 1. Resolution methods

| | NSW<br>% | Vic<br>% | Total<br>% |
|---|---|---|---|
| Arbitration and Court | 41 | 58 | 47 |
| Court only | 30 | 37 | 32 |
| Reference Out by Court | 24 | 0 | 16 |
| Separate Question to Court | 5 | 5 | 5 |

In 47 % of the cases, arbitration proceedings were involved. In some instances the dispute reached the court through an appeal to the award made by the unsuccessful party. In other instances, while the arbitration was on foot, the court was approached in an attempt to halt the arbitration proceedings. In others the court was approached with an application for an interlocutory injunction. Regardless of why the dispute reached the court, if details of the dispute were given in the judgement then this data was included in the study.

32 % of the cases were the subject of normal Supreme Court trials. Adequate details of the nature of the dispute were usually given in these judgements and so the sources of dispute could be identified.

The provision, governed by Part 72 of the NSW Supreme Court Rules, to make references out of court to a referee was used in 24 % of the NSW cases. In most of these cases details of the dispute were available from the judgements as the matters were discussed at some length by the judges when considering whether or not to adopt the referee's report. While a similar provision, Order 50, exists in Victoria there were no Victorian cases in this category.

Under Supreme Court Rules Pt 31 NSW, and r47.04 Vic, separate questions can be tried. Five percent of cases fell into this category.

## 3 Sources of dispute

### 3.1 General

As each case could contain more than one source of dispute the 56 cases generated 117 sources. Table 2 groups similar sources of dispute under six headings and gives the frequency of occurrence.

Table 2. Sources of dispute

| Sources of dispute subgroup | Frequency |
|---|---|
| Determination of the agreement | 31 |
| Payment | 30 |
| The site and execution of work | 19 |
| Time | 19 |
| Negligence and nuisance | 15 |
| Final certificate/payment | 3 |
| Total | 117 |

In 80 % of the cases, the parties to the disputes were the proprietor and the builder, or the builder and a sub-contractor. The other 20 % of the cases involved disputes between the proprietor and the architect or engineer. Sometimes the local Council, the builder, or an adjoining owner were also involved as parties and generally these claims were founded in tort.

### 3.2 Determination of the agreement as a source of dispute

The Determination of the agreement subgroup of sources are listed in Table 3, where they are grouped into various categories and the frequency of each source is also shown.

The category 'failure by the builder or subcontractor' at 15 % of the 117 sources of dispute was the largest category of all the sources.

Disputes involving questions of determination of the agreement between the parties were slightly more likely to go to arbitration (7 cases), than to the Supreme Court (5 cases). One case was the subject of a Pt 72 Reference Out of Court.

Table 3. Determination of the agreement subgroup

| Source of dispute | Frequency |
|---|---|
| Failure by the builder or sub-contractor | |
| Builder fails to proceed in a competent manner | 2 |
| Builder fails to proceed diligently and at a satisfactory rate | 4 |
| Notice of default served on builder | 2 |
| Notice of default served on sub-contractor | 1 |
| Proprietor determines contract with builder | 7 |
| Builder determines contract with sub-contractor | 2 |
| Work and the site | |
| Builder ceases work on the site | 1 |
| Sub-contractor ceases work on the site | 1 |
| Proprietor takes over the site and denies access to builder | 4 |
| Builder denies access to site for the sub-contractor | 1 |
| Builder accepts proprietors repudiation | |
| Proprietor repudiates by denying site access | 1 |
| Proprietor repudiates by not paying progress claim | 1 |
| Proprietor repudiates by being non-financial | 1 |
| Proprietor repudiates by claiming to determine agreement | 3 |
| Total | 31 |

### 3.3 Payment as a source of dispute

The Payment subgroup of sources of dispute are listed in Table 4, where they are grouped into various categories and the frequency of each source is also shown.

Variations, as a source of dispute, had a frequency of occurrence of 12 % and was the second most frequent source.

The variation claims ranged through : claims due to design changes necessitated by unsuitable foundations; claims due to design changes required by local authorities; claims due to footing design changes; claims due to additional excavation being required; claims due to a fall in the Australian Dollar causing associated rise in material costs.

The claims for variations were generally made by the builder to the proprietor, and sometimes by a subcontractor to the builder.

In some cases the question of whether variations needed to be in writing under the contract was an issue.

The bulk of the variation based disputes were in the domestic (50%) and engineering (33%) fields and only one case was on a commercial building. It maybe that standard forms of agreement used on most commercial projects have adequate clauses to facilitate the claiming, approval and payment of variations, to avoid these claims triggering a dispute. Further study on this may be fruitful.

Table 4. Payment subgroup

| Source of dispute | Frequency |
|---|---|
| Variations | |
| Claimed by builder | 7 |
| Claimed by sub-contractor | 2 |
| Non-payment of approved variation claims | 3 |
| Requirement for variations to be in writing | 2 |
| Progress claims | |
| Non-payment to builder | 2 |
| Non-payment to sub-contractor | 2 |
| Payment for work done on "do & charge" basis | |
| Non-payment to builder | 1 |
| Delay costs | |
| Rectification of rate in appendix for delay costs | 1 |
| Time extension costs claimed by builder | 2 |
| Time extensions costs claimed by sub-contractor | 1 |
| Liquidated and ascertained damages | |
| Rectification of rate in appendix | 1 |
| Liquidated damages charged against the builder | 3 |
| Retention monies | |
| Not held in separate bank account by builder | 1 |
| Builder acting as trustee | 1 |
| Rise and Fall | |
| Builder claims sum for Rise and Fall | 1 |
| Total | 30 |

The source of dispute in one case concerned retention monies deducted by the builder from payments due to subcontractors. It was held by the Judge that the builder was acting as trustee of the money and as such he must keep the money in a separate bank account. This judgement is likely to have an impact on the manner in which many builders in Australia operate their finances.

### 3.4 The site and execution of work as a source of dispute

The site and execution of work subgroup sources are listed in Table 5, where they are grouped into various categories and the frequency of each source is also shown.

Table 5. The site and execution of work subgroup

| Source of dispute | Frequency |
|---|---|
| Foundations | |
| Footing design change due to site conditions | 3 |
| Lack of temporary support during excavations | 1 |
| Quality of materials | |
| Claims of negligent manufacture and supply | 1 |
| Defective materials | 1 |
| Quality of workmanship | |
| Defects in brickwork | 1 |
| Defects in tiling | 2 |
| Defects in footings | 1 |
| Defective wall construction | 2 |
| Standard of design | |
| Insufficient footing design | 3 |
| Notice to remedy defects | |
| Non-compliance by sub-contractor | 1 |
| Non-compliance by builder | 1 |
| Materials : non-payment by builder | 1 |
| Scope of work : differs greatly from tender documents | 1 |
| Total | 19 |

### 3.5 Time as a source of dispute

The Time subgroup sources are listed in the Table 6, where they are grouped into various categories and the frequency of each source is also shown.

Table 6. Time subgroup

| Source of dispute | Frequency |
|---|---|
| Extension of time | |
| Time extensions claimed by the builder | 6 |
| Time extensions claimed by the sub-contractor | 2 |
| Delays | |
| Builder delayed due to the local Council | 2 |
| Sub-contractor delayed due to builder | 3 |
| Damages for delays claimed by builder | 2 |
| Damages for delay claimed by sub-contractor | 3 |
| Validity of notice of practical completion | 1 |
| Total | 19 |

Claims for time extensions were generally made by the builder to the proprietor, with some claims being made by the subcontractor to the builder.

### 3.6 Tort as a source of dispute

Table 7. Tort subgroup

| Source of dispute | Frequency |
|---|---|
| Negligence | |
| By the local Council | 1 |
| By the architect | 5 |
| By the engineer | 4 |
| By the builder | 2 |
| By the supplier | 1 |
| Nuisance | |
| By the engineer | 1 |
| By the builder | 1 |
| Total | 15 |

In 12 of the 56 cases the claim was in tort. The tort subgroup sources are listed in the Table 7, where they are grouped into two categories and the frequency of each source occurring is also shown.

Claims made against the professional designers were often due to structural faults in the building and were generally founded in negligence based on a breach of duty of care.

It was not uncommon for claims of negligence to be made against the architect or the engineer, or both, and sometimes even another party, such as the local Council.

Most of the cases involving the engineer in claims for tortious liability went to the Supreme Court, whereas in cases involving the architect there was a slight tendency for the cases to be the subject of a Pt 72 Reference out of Court. A reference out of court was also the path taken in cases involving breach of duty of care by the builder.

Disputes with claims against engineers were all on commercial projects. However in the claims against architects the projects ranged from domestic, to commercial and engineering developments.

Some of the claims made against the engineer were:

(a) Failure to inspect, to survey site, to do underground exploration, to predict future settlement, to supervise;
(b) Faulty design, drawings, computations;
(c) Negligence in allowing adjoining building to collapse;
(d) Nuisance in allowing adjoining building to collapse.

Some of the claims made against the architect were:

(a) Defective design ;
(b) Failure to supervise structural design and construction, to comply with building standards, to appoint a competent builder, to employ competent consultants, to inspect;
(c) Errors in drawings, and in certificates;
(d) Breach of duty to warn of special conditions

### 3.7 Final certificate and final payment

Table 8. Final certificate and final payment subgroup

| Source of dispute | Frequency |
|---|---|
| Final certificate | |
| Validity | 1 |
| Non-payment of amount certified | 1 |
| Validity of final account | 1 |
| Total | 3 |

The Final certificate and final payment subgroup sources are listed in Table 8, where they are grouped into various categories and the frequency of each source is also shown.

## 4 Triggers to court

### 4.1 General

'Triggers to court' are claims which brought the dispute to the court and they do not relate to the nature of the dispute itself. 41 of the 46 triggers to court arose from arbitrations. The other five cases were from miscellaneous triggers.

Some cases allowed the identification of both sources of dispute and triggers to court. For example, a case may have begun due to a disagreement between the parties over the valuation of a variations claim. An award may have been published following an arbitration hearing. However the successful party may have resorted to applying to the court to enforce the award. In such a case a claim for variations would be identified as the source of dispute and would be classified in Section 3.3. The application to enforce the award as a judgement would be identified as a trigger to court and would be classified in section 4.2.

### 4.2 Triggers from arbitrations

In Table 9 the types of application which allowed a case to get to court are shown, although the main dispute was subject to arbitration proceedings. The frequency of each trigger is also shown.

Table 9. Applications to the court from arbitrations

| Reason for application to court | Frequency |
|---|---|
| Consolidation of arbitration proceedings | 1 |
| Restraint on arbitration continuing | 4 |
| Restraint removed by Court of Appeal | 1 |
| Stay of Court proceeding to allow arbitration | 3 |
| Leave to appeal on award | 12 |
| Appeal on award | 7 |
| Removal of arbitrator for misconduct | 3 |
| Enforce award as judgment of the Court | 3 |
| Interlocutory injunction | 9 |
| Total | 43 |

In six out of the 26 cases involving arbitration proceedings the dispute between the parties, which was the subject of the arbitration proceeding, was not commented on by the judge so the sources of that dispute could not be identified. However, when the judgement did allow identification of the sources of dispute these have been included in the foregoing Section 3.

### 4.3 Miscellaneous triggers

The Miscellaneous claims, which triggered court proceedings, were : application for injunction to restrain sub-contractor from filing to wind-up builder; application to remove referee for misconduct; application for injunction to assist in return of unfabricated materials to the builder; application for injunction to restrain the proprietor from calling up insurance bonds; application for leave to discontinue proceedings.

## 5 Conclusion

The most frequent sources of dispute found in this review of New South Wales and Victorian building cases which went to court over a two year period were :

(a) claimed failure by builder or sub-contractor resulting in the attempted determination of the agreement (15%),
(b) claims arising from variations (12%),
(c) claims of negligence in tort (11%),
(d) claims of delay including damages (9%),
(e) claims claims for extensions of time (7%),

It is proposed that these frequent sources (except tort) will become the focus during the second stage of the research project, when documentation weaknesses and failures in administration techniques are identified.

Acknowledgements

The advice of Mr Ron Fitch, arbitrator, the financial assistance of the Australian Institute of Quantity Surveyors with the Postgraduate Fellowship for 1991 to Vicky Watts and the financial assistance of the Department of Architecture & Building, University of Melbourne for travel of Vicky Watts to Sydney are gratefully acknowledged.

# COSTS IN ARBITRATION PROCEEDINGS

R.W. Quick
Partner, Morris Fletcher & Cross, Solicitors, Brisbane, Australia

Abstract
This paper outlines the present law of costs as it operates in arbitration proceedings and considers whether the law is likely to change.

The term arbitration proceedings is usually used here as meaning references under consensual submissions. Statutory arbitrations are not considered although references under an order of the court are considered briefly.

In England the legislation regulating references out of court under consensual submissions is the Arbitration Act, 1950 (U.K.). It consolidates the Arbitration Act, 1889 (U.K.) and the Arbitration Act, 1934 (U.K.) and is supplemented by the Arbitration Act, 1979 (U.K.) and the Arbitration Act, 1975 (U.K.), the 1975 Act applying only to arbitration agreements which are not domestic arbitration agreements.

## 1. Introduction

### 1.1 The rules or policies regulating the liability for costs

In the common law legal systems there are three principal rules or policies regulating the liability for the costs of legal proceedings:

(i) a policy giving a party costs not because of the role they play in the legal proceedings but because of his or her success in them. This is the English Rule: (Sometimes called the costs indemnity convention or costs indemnity rule.) Who loses, pays;

(ii) a policy requiring a party, win or lose, to bear his or her own costs. In Australia s.117(1) of the Family Law Act, 1975, No. 53 (Cth.) furnishes an important example of this policy. This is the American Rule, so called because of its prevalence in the United States of America: (In criminal proceedings the principle that the Crown neither pays nor receives costs effects a result similar to that brought about by the American Rule);

(iii) a policy awarding a party costs only if he or she plays a particular part in the proceedings e.g. if he or she is a plaintiff or defendant. The policy is discernible in some American civil rights legislation and less obviously in the rules of court of some Australian courts dealing with the costs consequences of offers of compromise: (See for example O.26 r.9 of the Rules of the Supreme Court of Queensland, Rule 41.05 of the Rules of the Supreme Court of South Australia and O.26 r.8(2) of the Rules of the Supreme Court of Victoria.)

(This classification is adapted from T.D. Rowe, "Predicting the Effects of Attorney Fee Shifting" (1984) 47 Law & Contemporary Problems 139 at 140.)

## 1.2 Some definitions

### 1.2.1 Costs of the award

These are the arbitrator's or umpire's fees and expenses; his or her remuneration.

The costs of the award will include expenses which can include legal expenses such as counsel's opinion on a point of law: (Mason v. Lovatt (1907) 23 T.L.R. 486) or a solicitor's costs for drawing an award: (Re Becker Shillan & Co. & Barry Bros [1921] 1 K.B. 391; Re Collyer-Bristow & Co. [1901] 2 K.B. 839; Threlfall v. Fanshawe (1850) 19 L.J.Q.B. 334).

### 1.2.2 Costs of the reference

These are the costs of the arbitration proceedings as a whole. The costs of the reference include the costs of the award: (Re Walker & Brown (1882) 9 Q.B.D. 434; Re The Autothreptic Steam Boiler Co. Ltd v. Townsend Hook & Co. (1888) 21 Q.B.D. 182) and all costs incidental to the reference: (Minister for Home & Territories v. Smith (1924) 35 C.L.R. 120).

The costs of the reference will include the costs of interlocutory applications such as an application for directions or an adjournment: (Re The Portland & Western District Freezing Company Limited v.

Austral Otis Company Limited (1897) 23 V.L.R. 462). The effect of making no separate order as to the interlocutory costs is to include them in the costs of the reference: (Evmar Shipping v. Japan Lines [1984] 2 Lloyd's Rep. 581 at 585) and unintended consequences can follow from this: (See for example The "Angeliki" [1982] 2 Lloyd's Rep. 594)

## 2. References under an order of the court

### 2.1 The jurisdiction

#### 2.1.1 In general

A reference to arbitration under an order of the court can arise either because:

(i) the parties wish it; or

(ii) because the court exercises a statutory jurisdiction to refer to arbitration the proceedings or some part of them.

A third category of case, where the reference is the outcome of an order of the court staying an action that one of the parties to a consensual submission has commenced, is properly a reference under a consensual submission.

Where the statutory jurisdiction is exercised there is an action and a reference. That makes it necessary to consider whether any distinction needs to be drawn between the costs of the action and the costs of the reference. Where the reference is one to determine the action or particular questions in it, it will result in an award and then other questions may need to be answered according to the statutory provisions regulating the jurisdiction, in particular:

(i) whether the court has power to review the award; and

(ii) what is the proper order for the costs of the reference and of the action when:

- (a) no provision is made for costs in the order of reference or the award or both;
- (b) the amount awarded in the reference could have been recovered by proceedings in an inferior court.

## 3. References under a consensual submission

### 3.1 The jurisdiction to award costs

Section 18(1) of the Arbitration Act, 1950 (E.) provides:

> "Unless a contrary intention is expressed therein, every agreement to arbitrate shall be deemed to include a provision that the costs of the reference and award shall be in the discretion of the arbitrator or umpire who may direct to and by whom and in what manner those

costs or any part thereof shall be paid and may tax or settle the amount of costs to be so paid or any part thereof and may award costs to be paid as between solicitor and client."

3.2 **The duty to deal with costs**

Where they are within the reference, an arbitrator must deal with the costs of the award and the costs of the reference because the award must deal with all matters referred.[1] Commonly the jurisdiction to award costs is exercised in a final award. That will usually exhaust the arbitrator's power to award costs, but the power is not exhausted when:

(i) he or she makes an interim award not dealing with costs;
(ii) when he or she exercises power to complete an award by providing for the amount payable or for payment of that amount;[2]
(iii) when the court remits an award to an arbitrator.
(iv) when he or she exercises power to correct the award under the "slip rule".

3.3 **The general rule in exercising the jurisdiction**

After initial doubts[3] it was decided by Wright J. in LLoyd Del Pacifico v. Board of Trade [1930] 46 T.L.R. 476 that an arbitrator must exercise his or her discretion as to costs as a judge would. In the absence

---

1 Re Becker Shillan & Co. & Barry Brothers [1921] 1 K.B. 391; A.V. Jennings Industries (Australia) Ltd v. Roman Catholic Bishop of Perth [1967] W.A.R. 3. In this and other respects a statutory arbitration may be different, see for example Sutherland Shire Council v. Kirby (1961) 78 W.N. (N.S.W.) 989.

2 Montrose Canned Foods v. Eric Wells (Merchants) Ltd [1965] 1 Lloyd's Rep. 590 at 601; See Hall, "The Taxation of Costs: an unnecessary inhibition" Arbitration, August 1985, p.434. The author suggests that in a final award the arbitrator can reserve to himself a power to tax costs if these are not agreed; ibid at p.436.

3 See for example Gray v. Ashburton [1917] A.C. 26 at 34 and 37 where the arbitrator was acting under the Agricultural Holdings Act 1908 (E.). The Act was considered to impose a comprehensive requirement as to how the discretion as to costs should be exercised.

of special circumstances he or she should award costs to the successful party and if he or she does not the court will review the way in which he or she has exercised the jurisdiction. There is no distinction here between the costs of the reference and the costs of the award: (Smeaton Hanscombe Co Ltd v. Sassoon I. Setty Son & Co (No. 2) [1953] I.W.L.R. 1481 at 1483-1484). The English Rule applies to both.

An arbitrator does not exercise his or her discretion if he or she applies an inflexible rule he or she has as to the granting or withholding of costs.[4]

The English Rule is also summed up in the further statement 'costs follow the event'. Both statements require further questions to be answered namely who is a successful party and what is the event: (See The "Aghios Nicolaos" [1980] 1 Lloyd's Rep. 17).

The event is to be determined distributively if there is more than one claim by the claimant or if there is a claim by the claimant and a counterclaim by the respondent: (See again The "Aghios Nicolaos" [1980] 1 Lloyd's Rep. 17).

The practice until comparatively recently was to make special orders as to the costs of particular issues according to a party's success or failure on those issues although in the arbitration of building or engineering disputes, whatever the issues in pleading terms, there is only one issue as the parties see it namely who should pay the balance finally determined to be due: (Hudson, Hudson's Building & Engineering Contracts, (10th Ed., 1970) pp. 870-871), and in court proceedings there is increasing recognition that orders as to costs of issues often result in complicated, expensive and unpredictable taxations. For this reason where there are separate issues on which a successful party has failed it is increasingly common to make a "fractional award" of costs giving the successful party a proportion of his costs of the proceedings.

### 3.4 Particular problems with the exercise of the jurisdiction

#### 3.4.1 Efforts to settle the proceedings

There seems no reason why the submission should not require the arbitrator to consider matters such as a party's efforts to settle the proceedings: (See the form

---

4 James Allen (Liverpool) Ltd v. London Export Corporation Ltd [1981] 2 Lloyd's Rep. 632. The case is an illustration in arbitration proceedings of the Scherer principle called after Scherer v. Counting Instruments Ltd [1986] 2 All E.R. 529

of arbitration clause considered in Messers Ltd v. Heidner & Co. [1960] 1 Lloyd's Rep. 500 at 502). However in the absence of such a requirement in statute or the submission, an award of costs based on the presence or absence of efforts to settle the proceedings rather than the parties' success or failure will not be a valid exercise of the discretion to award costs: (Lewis v. Haverford West R.D.C. [1953] 1 W.L.R. 1486).

### 3.4.2 Restrictions imposed by the submission

The terms of the agreement to arbitrate can attempt to regulate the exercise of the jurisdiction to award costs, but account has to be taken of statute. Section 18(3) Arbitration Act, 1950 (E.) states that the parties cannot provide in an agreement to arbitrate future disputes that either or both of them should pay their costs of the reference or award in any event.

### 3.4.3 Multiple arbitrations[5]

Under a consensual reference the parties make the arbitrator the master of the procedure to be followed in the arbitration. However, it is an implied term of their agreement to arbitrate that strangers to the agreement should not be allowed to take part in the hearing and conduct of any arbitration under the agreement.

It also follows from the private nature of arbitration that what an arbitrator or umpire has jurisdiction to award, is the costs of the arbitration referred under the agreement to arbitrate. The arbitrator does not, in the absence of express agreement have power to deal with the costs of any other arbitration even if by agreement this is held at the same time: Maritime Transport Overseas G.m.b.H. v. Unitramp (The "Antaios") [1981] 2 Lloyd's Rep. 284).

Where there are multiple arbitrations, a rigid application of the general rule that an unsuccessful party should pay a successful party's costs in each of the arbitrations can produce unfair results if one looks at the series of arbitrations as a whole: (See for example The "Catherine L" [1982] 1 Lloyd's Rep 484).

### 3.4.4 Offers in settlement

It is often necessary to consider whether a party who

---

5 This term and the analysis which follows owes much to M.J. Mustill & S.C. Boyd The Law & Practice of Commercial Arbitration in England (2nd ed. 1989) p.148.

has made an offer to dispose of an arbitration on terms is to be considered the successful party.

In court proceedings where the rules of court allow a payment into court, the defendant who has paid into court an amount exceeding the amount awarded to the Plaintiff at trial is considered a successful party entitled to his costs from the date of the payment into court: (Wagman v. Vare Motors Ltd [1959] 1 W.L.R. 853; Lauchlan v. Hartley [1979] Qd.R. 305) unless there is a good ground for exercising the discretion as to costs against him.

In arbitration proceedings the equivalent of a payment into court is an offer to dispose of the proceedings on terms. Here the matter is regulated by contract, not rules of court, so that, for example, the offer can lapse and become incapable of acceptance: (The Elbe Ore [1986] 1 Lloyd's Rep. 176 at 180) or it can become incapable of acceptance because of a counterclaim: (Huron Liberian Co v. Rheinel G.m.b.H. [1985] 2 Lloyd's Rep. 58). The offer can be a 'without prejudice offer', an 'open offer' or a 'sealed offer': (For an explanation of these terms see Tramountana v. Atlantic Shipping [1978] 2 All E.R. 870 at 876).

Whilst a 'without prejudice offer' or 'without prejudice' negotiations: (Simaan General Contracting Co v Pilkington Glass Ltd (1987) 84(11) L.S.G. 819 ("without prejudice" negotiations inadmissible on an application for security for costs) should not be referred to by either party at any stage in the proceedings without the consent of the other and should not influence the award of costs: (Stotesbury v. Turner [1943] K.B. 370) an offer, a Calderbank offer as it is called, which is made without prejudice save as to costs can be referred to. Such an offer reserves the right to bring the offer to the attention of the arbitrator or judge for the purpose of dealing with costs once all other matters have been dealt with: (Calderbank v. Calderbank [1975] 3 All E.R. 333; see also Computer Machinery v. Drescher [1983] 3 All E.R. 153; McDonnell v. McDonnell [1977] 1 All E.R. 766; Cutts v. Head [1984] 2 W.L.R. 349 at 363, 365).

Such an offer is appropriate in court proceedings when it is an offer to settle a claim which is more than a simple money claim so that a payment into court is not an appropriate means of settlement. Then it can be expected to result in an award to the party making the offer of all his costs from the date when the offer ought reasonably to have been accepted by the other party. A payment into court can safeguard a defendant or a plaintiff faced with a counterclaim. A Calderbank offer can similarly be made by either party. Thus in Cutts v. Head [1984] 2 W.L.R. 349 it was the plaintiff who had made the Calderbank offer to the defendant. A Calderbank offer is not a substitute for a payment into court where such a payment can be made: (Cutts v. Head [1984] 2 W.L.R. 349 at 365, 369; and Corby D.C. v. Holst & Co Ltd [1985] 1 All E.R. 321. Contra Messiter v. Hutchinson (1987) 4 A.N.Z. Insurance Cases para 60-802).

An offer must be structured to allow the arbitrator to compare his award with the offer. In any money award there will be three elements, the debt or principal, interest and costs. A comparison between the two can only be valid if it takes account of how both have dealt with these three elements.

### 3.4.5 The need for reasons in departing from the general rule

The fact that an award deprives a successful party of his costs will not of itself be enough to establish a failure to act judicially but where the award departs from the English Rule without recognising it and without giving a sufficient reason for the departure there will, prima facie, be evidence of a failure to act judicially. Even more so if the reasons given are unintelligible: (Heaven & Kesterton Ltd. v. Sven Widaeus A/B, [1958] 1 W.L.R. 248 at 257).

## 4. The future

The future of the English Rule in court proceedings is in question. As arbitration has become judicialized and arbitrators have been discouraged from seeking to ensure that each party bears his or her own costs or from making awards other than in accordance with the English Rule arbitration has had to share these criticisms.

There is probably a greater use of arbitration clauses and commercial arbitration in the construction industry than in any other industry but there has been clamorous criticism of recent times that arbitration is failing to provide a cheap and efficient resolution of construction industry disputes. A recent world-wide survey by the Australian Federation of Construction Contractors (A.F.C.C.) of the phenomenon of claims and disputes in the construction industry concluded:

> "It is a reasonable perception that arbitration has broken down as a cheap and efficient means of resolving construction disputes, albeit that the cause may be the strenuously adversarial manner in which the disputants themselves pursue the arbitral process. The increasing extent of arbitrations which involve a determination of the parties' legal rights and obligations, rather than a determination of matters such as quality is an important factor in the

> deterioration of the efficiency of the arbitral process in the industry.
>
> Whatever the cause, there have been situations in recent times where parties have felt obliged to settle, as they could not afford to continue with the arbitral process, given the costs of doing so compared with the amount in dispute. There have also been situations where a claimant has spent more on legal costs than the amount claimed, which to the casual observer appears as the ultimate absurdity": (<u>Strategies for the Reduction of Claims Disputes in the Construction Industry - A Research Report</u> (1989) Australian Federation of Construction Contractors, Sydney p.70).

The reference here to parties being forced to settle should be noted. The right to costs incident to the English Rule can carry with it the right to have those costs secured and a security for costs application in arbitration proceedings as in litigation, has, routinely proved to be decisive of them.

The costs of litigation and arbitration has encouraged world wide the development of Alternative Dispute Resolution (A.D.R) procedures. In such procedures the usual principle regulating the incidence of costs is the American Rule.

Commonly claims are made that these procedures are extremely successful and that their cost (borne equally by the parties) is a mere fraction of the cost of litigation or arbitration. The A.F.C.C. report, for example, said that the United States experience and the experience of the Australian Commercial Disputes Centre (A.C.D.C.) suggested that the costs of mediation could be as low as approximately 3% of the likely costs of litigation or arbitration: (Id., p.71) and that the experience of the Australian Commercial Disputes Centre: (A.C.D.C.) was that mediation had close to a 100% success rate: (Id., p. 70)

There is currently on foot in Australia a wide ranging Parliamentary inquiry into the cost of justice, its accessibility and any practical alternatives. The Secretary-General of the A.C.D.C. told the Committee conducting this inquiry, the Senate Standing Committee on Legal and Constitutional Affairs, that the average cost of mediating "a million dollar type of claim" for a building dispute would be $7 - 10,000, (about 3,500-5,000 pounds Sterling) plus preparation: (See (<u>Costs of Legal Services and Litigation Discussion Paper No. 4, Methods of Dispute Resolution</u> (September 1991) para. 6.14 p.60). The A.C.D.C estimated that the cost to each party in using its mediation service (each party again bearing its own costs) would normally be 5% of the cost of litigation, although this figure would increase to 25% when the parties were assisted by their lawyers: (Ibid.)

It is remarkably hard to prove empirically these claims for ADR procedures. Although there are a number of studies being conducted which may help in this it may be that conclusive evidence on the relative costs of the different methods is simply not obtainable if only because the confidentiality of these procedures prevents it. The Senate Standing Committee on Legal and Constitutional Affairs was inclined to the preliminary view that by encouraging the use of a range of A.D.R. procedures there would be at least some scope for reducing the financial cost to the parties to a dispute: (Costs of Legal Services and Litigation Discussion Paper, No 4. Methods of Dispute Resolution (September 1991) para. 6.11, p.59).

It is also remarkably difficult to justify the English Rule except by reference to principles such as fairness or fault or its long history and general acceptance: (See Pfennigstorf "The European Experience with Attorney Fee Shifting" (1984) 47 Law & Contemporary Problems 37). It seems, however, that the English Rule will survive if arbitration itself survives.

Paralleling the attempts to utilise A.D.R. procedures there have been repeated experiments with the American Rule by English and Australian courts particularly in non-adversarial litigation such as matrimonial disputes. As we have seen in Australia, s.117 of the Family Law Act 1975, No. 53 (Cth.) states expressly that each party is to bear his or her own costs and in England O.62 r.3(5) RSC states that the English Rule does not apply to proceedings in the Family Division of the High Court. The Australian experience with s.117 has been that such qualifications have had to be made to the express statement that it is closer to the truth to say that with qualifications the English Rule applies than it is to say that the American Rule applies. It is possible, too, to see in the recent development of the American Rule in the United States the first faltering steps towards the adoption of the English Rule. That development has been characterised by increasing numbers of exceptions to the American Rule, increasing volumes of litigation about when an award of costs should be made and how it should be calculated and considerable academic comment critical of the American Rule: (For a review of this development see Dobbs, "Awarding Attorney Fees against Adversaries: Introducing the Problem" (1986) Duke Law Journal 436)

# CONSTRUCTION CONTRACTS: TOWARDS A NEW RELATIONSHIP

BARBARA COLLEDGE
Leeds School of the Environment, Leeds Polytechnic, Leeds, England

Abstract
This working paper comprises preliminary research into the use of law in the construction industry. It seeks to explore the nature of commercial relationships which exist and develop a framework for the drafting and selection of construction contracts. The primary objective is to investigate the extent to which greater recognition may be given to the commercial relationship between parties to the transaction as a means of reducing or managing conflict. Specifically, an analysis will be made of the nature of construction contract transactions and this matched with classifications developed by those such as Beale and Dugdale (1975), Macaulay (1963), Macneil (1978) and Williamson (1981) as to the way in which law is used in commercial relationships.
<u>Keywords:</u> Construction Contracts, Legal Frameworks, Governance Structures, Conflict Management.

## 1 Introduction

The potential for conflict is inherent in any relationship and often it is a breakdown in communication between the parties that results in its failure. Parallels may be drawn in the UK construction industry with the frequency and nature of conflict between the parties resulting in proposals for new approaches to the management of the construction process (NEDO 1988, NCG 1990). The resolution of conflict through the adoption of alternative dispute resolution techniques provides a mechanism for the management of conflict which recognises the commercial relationship between the parties and provides an opportunity for maintaining communication. If, by sustaining a "good" relationship, failure in the achievement of the parties objectives is less likely to occur, then avoidance of conflict, through the planned application of contractual mechanisms which give greater emphasis to the relational aspects of the agreement, might provide an alternative to the adversarial traditional contractual relationships.

This paper explores the necessity for changes in contractual frameworks in the construction industry and argues for greater recognition of the specific nature of the commercial relationship. An analysis is undertaken of the way in which the law is used to regulate commercial relationships and a conceptual classification of contract law usage is applied to construction contracts. From this, it is argued that a matrix of key characteristics of the transaction might be developed for use in the selection of an appropriate legal framework. This matrix will permit closer attention to be given to the nature of the commercial relationship and provide a framework for the development of appropriate contract conditions. By using such an analysis it is argued that informed decisions may be taken regarding the drafting and selection of contractual arrangements which recognise the importance of the commercial relationship.

## 2 The use of law to regulate commercial relationships in the construction industry

### 2.1 The legal framework and the role of the contract

The legal framework provides for the management and regulation of the relationship between the parties to a contract. In construction, this is most frequently evidenced by reference to the selection by the client of a procurement method or path and the adoption of a contractual arrangement, (such as lump sum, remeasurement, cost reimbursement), very often facilitated by the use of a standard form of construction contract, which brings with it a predetermined allocation of risk. However, the legal framework is not restricted solely to the contract itself but may be more broadly defined to include associated bodies of general law which impinge on and influence the performance of the contractual agreement eg contract law, the law of tort, statute. 'Indeed, there are few contracts today which are not governed by specific rules which in some measure derogate from the general law' (Atiyah 1986).

Fuller (1981) identified two categories of principles of social ordering, or rather a matrix of principles which guide and influence the nature of interactions in society, (including business or commercial relationships): (a) those which operate vertically such as legislation or managerial direction (b) those which operate horizontally such as custom or contract (Atiyah 1986). Contract as a means of regulating future human interaction was seen as a dominant principle, although custom should not be ignored. From this, an explanation of the role of the contract can be developed.

In simplest terms, contracts could be defined as 'devices for conducting exchanges' (Macaulay 1963). However, the distinction between a contract and an exchange is made by Atiyah (1986) in his criticism of Fuller:

> '...contract differs from mere exchange because it contains an element of futurity - what Ian Macneil calls "presentiation". Contracts (or anyhow many contracts) bind people to future performances. Exchange can be a purely present transaction...'

Equally, Macaulay makes a similar distinction, seeing contract as not 'synonymous with an exchange itself, which may or may not be characterized as contractual' (Macaulay 1963). It is perhaps an element of benefit or detrimental reliance apart from mere voluntary conduct which characterises a contract from an exchange. However, given the long term nature of construction projects, it is rare for the transaction to conform to instantaneous exchange, as depicted in the classical model. Consequently, futurity or presentiation is inherent in the relationship and contract law 'affords a legal framework within which parties can plan' (Feinman 1983). The contract thus performs two primary functions (Macaulay 1963):

**(a) Creation of an exchange relationship**

Planning for and regulation of the transaction for example with regard to contingencies or unforeseen circumstances, remedies for defective performance.

**(b) Solution of problems arising during the course of the contract**

Provides sanctions to induce performance or to compensate for non performance.

The broader legal framework (that of Fuller's vertical principal of social ordering) supports 'the network of private, unregulated transactions' (Friedman 1983) by

(a) providing a legal framework within which parties can plan;
(b) assuring enforcement of provisions;
(c) performing a remedial function in filling gaps in planning or resolving disputes as to the parties intentions.

(See Beale and Dugdale 1975, Feinman 1983, Macaulay 1963).

The contract and the legal framework both serve to assist the governance of the commercial relationship between the parties for a particular transaction. The nature of the agreement clearly will be influenced by the relationship between the parties at the time of formation. Existing commercial interests or long standing business relations with a company may affect the approach adopted in the formation of further agreements. What is needed for effective governance is an appropriate legal structure which recognises the context of the commercial relationship and the specific characteristics of the transaction.

The contract itself will form a dominant part of this structure, providing planned contractual mechanisms for regulation and performance. The nature of contractual relationships for construction project transactions will be considered next to assess the extent to which such planning is undertaken.

## 2.2 Contracts in the construction industry

Contracts used for construction project transactions invariably reflect the procurement path whether an Employer - Contractor or Contractor - Sub-contractor relationship is considered. A distinction should be made between the procurement path, (determined by the nature of the process used eg management contracting, design and build, traditional) and the contractual arrangement, (such as lump sum, remeasurement and cost reimbursement), which will be influenced by the former but could be the same for differing procurement paths. The agreement itself may take a variety of forms and clearly will reflect the procurement path and contractual arrangement. However, the precise terms are not determined by such approaches in themselves. Rather, they are the result more or less of conscious planning and agreement of the terms (to a greater or lesser extent) by the parties to the contract.

In construction, this is often achieved by the use of standard forms of contract which cater for a variety of contractual arrangements and procurement paths. Certainly this is borne out by the results of the 1989 RICS survey of contracts in use (RICS 1989) which indicated that in the region of 90 percent of all construction contracts were so let. Even after allowance is made for inaccuracies and limitations of the survey the results are still highly significant. Similar conclusions have been drawn by Hibberd, Merrifield and Taylor (1990). This is not, however, intended to be a treatise on the merits or otherwise of standard forms of construction contract. The fact that they are so widely used in itself suggests that they do have merit even though questions remain as to their interpretation[1]. What is important is the degree to which the parties to a construction contract use the law to regulate their relationship.

In 1974, Lord Diplock provided an insight into the purpose of the standard form of contract and the use of law in A Schroeder Music Publishing Co Ltd v Macaulay (1974) 3 All ER 616, House of Lords. The following two kinds of standard form were identified:

(a) those where 'the standard clauses...have been settled over the years by negotiation by representatives of the commercial interests involved and have been widely adopted because experience has shown that they facilitate the conduct of trade. Contracts of these kinds affect not only the actual parties to them but also others who may have a commercial interest in the transactions to which they relate, as buyers or sellers...If fairness or reasonableness were relevant to their enforceability the fact that they are widely used by parties whose bargaining power is fairly matched would raise a strong presumption that their terms are fair and reasonable';

(b) those where 'the terms...have not been the subject of negotiation between the parties to it, or approved by any organisation representing the interests of the weaker party...To be in a position to adopt this attitude towards a party desirous of entering into a contract to obtain goods or services provides a classic instance of superior bargaining power'.

Whilst the majority of construction contract standard forms are of the first variety and negotiated on behalf of those who use them, this serves to illustrate the purpose of the form itself in the conduct of the commercial relationship. They 'have been widely adopted because experience has shown that they facilitate the conduct of trade' (Lord Diplock op cit.). The use of a standard form is traditional in facilitating a discrete

1 For example see Lord Reid in North-West Metropolitan Regional Hospital Board V T A Bickerton & Son Ltd (1970) 1 AllER 1039 commenting on the RIBA standard form and Wallace I N D (1978) "Appendix C" in the ICE Conditions of Contract Fifth Edition, A Commentary, Sweet & Maxwell, London.

transaction,[2] where '...no duties exist between the parties prior to the contract formation and in which the duties of the parties are determined at the formation stage...'(Goldberg 1976). In reality, the existence of these transactions is rare, with exchanges between parties displaying relational characteristics[3], for example, being of significant duration, enabling future co-operative behaviour, sharing of benefits or burdens, friendship - see Macneil (1978) (although in construction, the purchase of commonplace materials on a supplier's standard terms might be considered relatively discrete where no prior commercial relationship exists or in a market of many suppliers).

'Most actual exchanges are at least partially relational...it is...more useful to think of transactional and relational characteristics as creating a spectrum...As one moves towards the relational end of this spectrum presentiation plays a relatively smaller role, since increasing aspects of the relation must be left to future determination...' (see Macneil 1978). In other words, it becomes impossible to finalise all aspects of the contract before commencement. Such 'presentiated' contract provisions are displaced by those which seek to provide mechanisms or procedures to be followed at some future date, for example clause 13 of the JCT Standard Form of Building Contract 1980. In general, despite the fact that many construction projects are of a 'one-off' or idiosyncratic nature, the standard forms facilitate longer term exchanges, providing rules to govern the interaction between the parties for the duration of the contract (and sometimes beyond). These rules have developed over time related to practices in

---

2 Macneil I R (1978) at page 856 states that "We do find in real life many quite discrete transactions: little personal involvement of the parties, communications largely or entirely linguistic and limited to the subject matter of the transaction, the subjects of exchange consisting of an easily monetized commodity and money, little or no social or secondary exchange, and no significant past relations nor likely future relations. For example, a cash purchase of gasoline at a station on the New Jersey Turnpike by someone rarely traveling the road is such a quite discrete transaction."

3 Macneil I R (1983) at page 345 considers that "it is readily apparent that even a transaction deliberately chosen for its discreteness is deeply embedded in a wide range of interconnected relations". In footnote 12, Macneil summarizes these relational elements to include, for example: primary personal relations; multiple participants; some utility that is difficult or impossible to measure or specify; extended periods of commencement, duration, and termination; planning for change; circumstances where future co-operation will be essential to the relation. See also Macneil I R (1980) the new Social Contract: An inquiry into Modern Contractual Relations, pp 23-35.

the industry and driven by economic forces (Uff 1989). However, not all trade customs and 'unwritten laws' are incorporated into the express provisions of an agreement. Beale and Dugdale (1975) discovered a 'positive resistance to the idea' and that 'these customs were felt to operate more satisfactorily and flexibly in a purely commercial context than they would in the formal context of legal rights and duties'.

Consequently, the extent to which the users of standard forms of construction contract consciously plan the remedies and provisions within them is unclear. Admittedly the forms on the whole result from compromises made by representatives of various factions within the industry but very rarely would this specifically encompass the parties to a contract themselves [(4)]. Whether in fact these standard forms are then meeting the objectives set, for example to provide an effective legal framework for the regulation of the agreement, is an important question. The degree to which in such circumstances planning of the contract has been undertaken must surely be the primary responsibility of the negotiating bodies or drafting principles [(5)], not the individual parties concerned. However contrary this might seem in relation to meeting the needs of a specific project and the parties to that exchange it is generally accepted that the use of a standard form presupposes an acceptance of the contract planning and remedies embodied therein. It would be naive to suggest that parties to a contract do not consider the nature and content of their agreement: to contract on such a basis would be wholly inadvisable. Rather the selection and agreement to the use of a particular standard form indicates that the parties have inevitably gone through a process of planning. Furthermore, it is not uncommon for standard form provision to be the subject of amendment, (see Hibberd, Merrifield and Taylor 1990). Therefore it may be concluded that some degree of planning of construction contract provisions is undertaken by the parties but the full extent and nature of such planning, particularly that of a transaction-specific nature, is largely unknown. The same could equally be said in relation to non-standard forms of construction contract.

Macaulay's research into contractual relations of manufacturing organisations (Macaulay 1963) indicated

---

4 This does not include Government contracts such as GC/Works/1, which would be an exception.

5 Such as the Joint Contracts Tribunal (JCT); The Institution of Civil Engineers (ICE); the Federation of Civil Engineering Contractors (FCEC); British Property Federation (BPF); Association of Consultant Architects (ACA).

that 'businessmen often fail to plan exchange relationships completely and seldom use legal sanctions to adjust these relationships or to settle disputes. Planning and legal sanctions are often unnecessary and may have undesirable consequences. Transactions are planned and legal sanctions are used when the gains are thought to outweigh the costs'. Similar conclusions were drawn by Beale and Dugdale (1975) in their study of engineering manufacturers. When the risk was considered to justify it, careful planning on a particular issue was carried out.

Further empirical research in the construction industry is needed to validate the relevance of these findings for construction project transactions. However, Hibberd, Merrifield and Taylor (1990) identified a set of key factors which were addressed in construction project transactions: time, payment, certification, quality. Furthermore, for issues such as payment, design liability, insurance, sub-contracts and time, standard form conditions were frequently amended to respond to legal decisions, change the apportionment of risk or for clarification (op cit p22; see also School of Business and Industrial Management 1991). This does provide evidence that some planning of the contractual relationship is undertaken. However, given a predominant reliance on JCT standard conventional forms, it is difficult to establish the true extent of planning of the contractual relationship.

Given therefore the commonplace use of standard forms of construction contract, the extent to which it might be said that the parties plan the contents of their agreement for a specific project appears (superficially at least) to be largely limited to the selection of an appropriate procurement path and standard form of contract. Consequently, the commercial relationship is assisted by a standard set of familiar conditions which provide for a pre-agreed allocation of risk. These have developed to cater for a wide variety of contractual arrangements and procurement paths. Given, in some cases described as adhoc, developments in standard forms McDonagh (1990) has proposed that "the fundamental factors...of a contractual relationship for construction procurement" be reconsidered in the context of a specific procurement method. Hibberd, Merrifield and Taylor (1990) have contributed to this review and highlighted a number of problems with existing contracting methods. Whilst their research has identified key factors which parties seek to address in contractual arrangements, they 'ignored the fundamental question: were these key contractual factors appropriate for the particular circumstances of the particular job ?' (Hibberd, Merrifield and Taylor op cit.). Certainly the extent of real evaluation of a procurement path before decisions are made has been questioned by Hibberd and Taylor (1991) who considered that 'the procurement advice proffered by consultants is more often given on the basis of "gut feeling" rather than real knowledge based upon a scientific evaluation of antecedents'.

A number of studies have been undertaken in order to establish criteria for the selection of an appropriate contractual framework or procurement path (see Hibberd, Merrifield and Taylor 1990, pages 6-12 for a review). However, these do not address the selection of the terms of contract themselves, (although they will clearly have an influence) and are in general, limited to criteria which consider the objectives of the client for a specific project. Hibberd, Merrifield and Taylor (1990) concluded that further research was needed in providing definitions for procurement paths and options, and in the development of frameworks for flexible contractual relationships and the evaluation of procurement paths. It is submitted that closer attention needs to be given to the fundamental principles of contractual frameworks in order to develop policy for the drafting and selection of construction contracts. Attention must be given to the nature of the commercial relationship (evidenced by research undertaken by Macaulay 1963 and Beale and Dugdale 1975) and the role which the law and the contract are to perform. This inevitably involves closer attention being given to economic models of contracting behaviour which seek to analyse more sympathetically than a pure neoclassical market orientated approach (with its emphasis on discrete exchanges) the way in which inter-company economic activity is organised: via markets, hierarchies and networks.

It is suggested by Uff (1989) that 'the aspect of standard forms which is least developed yet most capable of creating beneficial impact on a project... is their ability to regulate and modify the position of the parties, primarily the contractor, and thereby to influence the performance of the work itself'. It is submitted that change in isolation will make a limited contribution to Uff's proposition unless appropriate legal frameworks are devised and/or selected which relate more closely to the nature of inter-company economic activity. Draft proposals for the development of a model for matching economic transactions with approaches to contract drafting are discussed in Section 3.

## 3 Developing a new relationship

### 3.1 The use of law: theoretical concepts

Macneil (1978) proposes a three-way classification of the

way in which law is used in contracts by analysis of twelve key concepts (see Fig.1). Williamson (1981) considers that 'serious problems of recognition and application are posed by such a rich classificatory apparatus' and argues for a simpler three-way classification initiated by Macneil: Classical, Neoclassical, Relational. In developing this three-way classification, Williamson focuses on the transaction itself and matches governance structures (what Williamson describes as the institutional matrix within which transactions are negotiated and executed) to different categories of transaction. The objective is to devise an appropriate framework for the conduct and regulation of transactions which gives due recognition to the nature of the transaction and the relationship between the parties. Thus an attempt is made to 'harmonise interests or at least achieve order where otherwise there would be conflict' (Williamson 1981). The legal framework is an important aspect of governance but the commercial relationship itself will determine the significance given to the legal framework and the remedies which it provides.

Williamson (1981) recommends that for a predictive theory of contract, criteria for the description of a transaction need to be identified and matched with governance structures in a discriminating way. Clearly, in order to develop a matrix for use as a predictive model in the selection and drafting of construction contracts, a detailed analysis of the nature of the transaction and contract provisions would need to be undertaken. This would accord with Macneil's "twelve concepts" approach. However, before this is undertaken, a broader generic preliminary framework will be developed so that governance structures, such as the three-way contract classification model, can be tested and matched to the general nature of the transaction. Consequently, Williamson's three-way classification has been adopted and will be tested against construction project transactions.

### 3.2 Constructing a Model

The basis of Williamson's theory and the contribution made by others in this area (Goldberg 1976, Macneil 1978) is an economic analysis of contractual relations. More recently, similar approaches have been adopted in other construction related fields (Lewis 1982, Flanagan & Norman 1989, Winch 1989). This permits a specific construction project transaction to be analysed in economic terms and appropriate mechanisms for the regulation and conduct of the activity to be devised. Three broad classifications of economic model are apparent in these theories and may be referred to by

| NUMBER | CONCEPT |
|---|---|
| 1 | Overall relation type |
| 2 | Measurability and actual measurement of exchange and other factors |
| 3 | Basic sources of socio-economic support |
| 4 | Duration |
| 5 | Commencement and termination |
| 6 | Planning |
| 7 | Future co-operation required in post commencement planning and actual performance |
| 8 | Incidence of benefits and burdens |
| 9 | Obligations undertaken |
| 10 | Transferability |
| 11 | Number of participants |
| 12 | Participants' views of transaction or relation |

Source: Derived from Macneil I R (1978) Contracts: adjustment of long-term economic relations under classical, neoclassical and relational contract law, Northwestern University Law Review, 72, no 6, 854-905

Fig.1: Macneil's twelve key concepts

adapting the concept of markets, networks and hierarchies:

Markets: ie the discrete transaction - linked historically to neo-classical economics;

Networks: ie the development of longer term commercial relations between autonomous parties;

Hierarchies: ie the internal organisation of the transaction -vertical or horizontal integration;

which are matched with models of governance structures: Classical - Neo-classical - Relational (see Fig. 2).

Based on this assumption of efficiency, the nature of the legal framework appropriate in these circumstances to govern the transaction is assessed resulting in a general theory of contracting, see Fig.2. In this way, transactions can be analysed and matched with an appropriate economic model and governance structure.

The above model provides a general theory for classification of transactions. However, the nature of the transaction needs to be analysed more closely to develop criteria to assist a more accurate classification. Williamson (1981) identifies three critical "dimensions" or criteria for describing contractual relations: uncertainty, frequency and investment idiosyncracy. Uncertainty is considered separately, as a feature of all contracts and a matrix based on the other two criteria is developed. The following features of these criteria have been identified by Williamson:

(a) **Frequency (of buyer): Occasional or Recurrent**

This determines the characteristic of the commercial relationship. One-time and occasional transactions are not distinguished.

(b) **Investment characteristics (of supplier's services)**

| **Nonspecific** | **Mixed** | **Idiosyncratic** |
|---|---|---|
| Standardised service | Non standard service but not highly specialised | Highly specialised service |

This determines the individuality of the transaction and will affect the extent to which standardised governance structures are adopted.

From this a matrix of possible transactions is developed and matched with the economic model/governance structure. Specific features of the transactions may be identified when the above criteria are combined in the matrix (see Fig. 3). Increasing the level of uncertainty has little effect on standard transactions as parties can easily be replaced. However, where the transaction is less standardised, greater flexibility to cater for change must be built into the governance structure. This provides a starting point for the development of a similar model for construction project transactions. In Section 3.3, criteria for selection of standard forms of

| ECONOMIC MODEL | GOVERNANCE STRUCTURE | FEATURES |
|---|---|---|
| Markets | Classical Contracting | Reliance on the market; discrete transactions; adherance to legal framework; use of legal remedies; standardised contract planning |
| Networks | Neo-classical Contracting | Longer term relationship begins to assume more importance; development of relational tendencies; contract provisions cater for flexibility |
| Hierarchies | Relational Contracting | The commercial relationship assumes equal or greater importance compared to the legal agreement; significant sharing of benefits and burdens; bilateral or unified governance(6) |

Fig.2: Economic models and governance structures

construction contract for use between the client and contractor will be reviewed and matched with Williamson's model to evaluate (albeit in a cursory way at this stage) whether this could be applied to a construction context.

6 See Williamson O E (1981) at page 52: "Two types of transaction-specific governance structures for intermediate-production market transactions can be distinguished: bilateral structures, where the autonomy of the parties is maintained; and unified structures, where the transaction is removed from the market and organized within the firm, subject to an authority relation (vertical integration)."

| INVESTMENT TYPE | FREQUENCY | |
|---|---|---|
| | OCCASIONAL | RECURRENT |
| Nonspecific | market governance (classical contracting)<br><br>less reliance on previous experience<br><br>reputation governed by market | market governance (classical contracting)<br><br>greater reliance on past experience and relationship that has developed<br><br>reputation through direct experience |
| | market alternatives protect each party against opportunism by his opposite<br><br>concentrated efforts to sustain the relation are not made because the relation is not independently valued | |
| Mixed | stronger incentive to see contract through to completion: not so easy to obtain a replacement<br><br>cost of transaction specific governance structure is prohibitive but market governance provides no incentive to sustain relationship<br><br>mechanisms to resolve future disputes are introduced eg third party assistance<br><br>trilateral governance (neoclassical contracting) | greater incentive to sustain relationship<br><br>primary reliance on market is unreliable; cost of specialised structure can be recovered<br><br>bilateral structure where autonomy of parties is maintained (relational contracting) |
| Idiosyncratic | as mixed occasional<br><br>---------------------<br>--transition to unified structure as transaction becomes more idiosyncratic | as mixed recurrent but bilateral structure is replaced with unified structure ie where project is undertaken jointly (relational contracting) |

Source: Derived from Williamson O E (1981)

Fig.3: Features of commercial and governance structures

### 3.3 Theory into practice

It has been argued that, before a detailed analysis of construction contracts is made on the basis of Macneil's twelve concept approach, a broader generic framework needs to be developed to match contract classifications with the transaction. In this section, Williamson's criteria are used to allocate standard forms of construction contract to the matrix.

Standard forms are classified as to their frequency and nature of investment using as evidence, a variety of texts/sources which provide guidance on the selection of an appropriate form. This results in a preliminary classification of standard form construction contracts.

### 3.4 Criteria for selection of construction contracts

Current sources of data on the selection of an appropriate form of construction contract were used to classify standard forms according to the Williamson model. (See for example JCT 1988; BEC 1987; Powell-Smith and Chappell 1990).

Definitions of Williamson's "Frequency" and "Investment" criteria had to be developed for application in a construction context in order to classify contracts in accordance with this model. Frequency criteria were considered to be acceptable but for a more detailed model, one-off transactions would need to be considered. The Investment criteria presented greater difficulties given that construction work is generally of a transaction specific nature. A definition therefore needed to be arrived at for "nonspecific" and "mixed" investment. In a strict sense, nonspecific construction investment ie standardised transactions for which continuity has little value, does not easily translate to a construction context apart from where it can be applied to the purchase of traditional building materials or the hiring of non specialist plant. However, it could be argued that a standardised construction project, such as a "package deal" or repetitive housing construction would qualify for this classification. Furthermore, it is also possible to argue that any construction work of a minor, non complex nature or standard repetitive maintenance work might equally be considered non-specific. It is this broader definition which has been adopted.

### 3.5 Classification of standard forms

A pattern of investment classification - nonspecific, mixed, idiosyncratic - for projects where standard forms might be used was established on the basis of the following synthesised set of characteristics developed from an analysis of the sources referred to above:

Complexity of work
Duration
Estimated contract value
Certainty

These characteristics of the project (or transaction) were, in general, factors which governed the selection and use of a particular standard form. The following assumptions were made:

(a) Low complexity work is indicative of non-specific investment and vice versa.
(b) Projects of long duration are more likely to indicate that the works are of a large or complex nature resulting in increased uncertainty and a mixed or idiosyncratic investment.
(c) Projects of a high estimated contract value will tend to be classified as (b) but exceptions to this may occur, for example, with a highly standardised package deal.
(d) The extent to which design is certain will determine the extent of uncertainty and hence flexibility required.

Details of contract specific criteria were used to determine an appropriate classification. The contracts were then classified and inserted into Williamson's matrix structure (see Fig.4).

It may be concluded that on the basis of Williamson's frequency and investment criteria, the spectrum of classifications of construction contracts bears similarity to the Williamson model. A more detailed theoretical and empirical study of construction contracts is needed to refine the model and validate its suitability for development into a framework for drafting and selection of construction contract provisions. The extent to which these and other construction contracts adopt Williamson's governance structures - Classical, Neoclassical and Relational Law approaches - in these circumstances needs to be evaluated and will form the basis of further research. This will enable clearer guidelines to be developed to permit the circumstances and context of a transaction to be addressed. A closer analysis of key criteria of the transaction will necessarily be involved. Whilst Williamson's frequency and investment model provides a tentative framework, it is insufficiently detailed to make distinctions between groupings of forms in each category. For this, criteria such as those identified by Macneil (1978) - see Fig.1 - will need to be developed and used to analyse contract provisions along the discrete-relational spectrum.

| FREQUENCY | INVESTMENT | | |
|---|---|---|---|
| | NON SPECIFIC | MIXED | IDIOSYNCRATIC |
| OCCASIONAL | 1. JCT MINOR WORKS<br>2. JCT IFC 84 drwgs + spec<br>3. JCT 80 drwgs + spec | 1. JCT IFC 84 drwgs + spec/ bq<br>2. JCT 80 drwgs + spec<br>3. JCT 80 drwgs + approx quants<br>4. JCT 80 drwgs + bq<br>5. JCT IFC 84 drwgs + bq<br>6. JCT C DES SUPP<br>7. ICE 5th & 6th<br>8. JCT MAN CONTRACT | 1. JCT 80 drwgs + approx quants<br>2. JCT 80 drwgs + bq<br>3. JCT C DES<br>4. ICE 5th & 6th<br>5. JCT MAN CONTRACT<br>6. JCT FIXED FEE |
| RECURRENT | 1. JCT MINOR WORKS<br>2. JCT IFC 84 drwgs + spec<br>3. JCT 80 drwgs + spec<br>4. JCT MEASURED TERM | 1. JCT IFC 84 drwgs + spec/ bq<br>2. JCT 80 drwgs + spec<br>3. JCT 80 drwgs + approx quants<br>4. JCT 80 drwgs + bq<br>5. JCT IFC 84 drwgs + bq<br>6. JCT C DES SUPP<br>7. ICE 5th & 6th<br>8. JCT FIXED FEE PC<br>9. JCT MEAS TERM<br>10. JCT MAN CONTRACT | 1. ADHOC or OWN STANDARD FORM<br>2. Use of 1-6 above<br>3. Joint ventures / consortium |

Fig.4: Classification of standard forms of construction contract according to Williamson's frequency and investment criteria

## 4 Conclusions

The UK construction industry has suffered much criticism in recent years with respect to its performance and achievement of time and cost objectives (NEDO 1988, NCG

1990). Furthermore, as the existance of a wider European industry is recognised, traditional UK practices (not so extensively favoured in the rest of Europe) are being critically reviewed and influenced. This is assisted by the growth of "non-traditional" contracting methods such as management contracting, construction management and design and build which attempt to provide greater flexibility and consistency in achieving a client's objectives. A recent report on the UK industry, "Building Towards 2001" (NCG 1990) called for 'a new form of contract...that arranges the parties to a construction contract in such a way that they all have identical goals of a timely, economical, profitable and high quality product'. A reason given for this is that 'adversarial roles are part and parcel of the traditional contract structure, with confrontation often leading to greater reward than might arise from pursuing the objective of economical construction of the required quality'.

Since this report, the industry has seen the introduction of "The New Engineering Contract (NEC)" (ICE 1991) which, although derived from an engineering background, is intended for use on a broad range of engineering and building projects. The philosophy behind its introduction and that of recent empirical research undertaken into contractual relationships (Hibberd, Merrifield and Taylor 1990) is that there is a need to establish clear, flexible and effective means of contracting which result in meeting a client's objectives and a reduction in the incidence of disputes. Whether this will be achieved by the use of the NEC is outside the scope of this paper. However, the commercial relationship envisaged by this contract, irrespective of the procurement path selected, is one in which risk is clearly allocated between the parties and each is motivated to contribute efficiently to serve the interests of the employer. It is argued that this emphasis on a "working relationship", with the introduction of provisions which seek to enhance and maintain the relationship, is fundamental to the philosophy of the new form and to improvements in existing contractual frameworks.

This paper has explored the application of an economic model of contractual relations which might be developed for use in the selection of an appropriate legal framework. The model gives recognition to the nature of the transaction between the parties and enables a transaction-specific approach to be adopted. Preliminary research indicates that analysis and classification of Employer-Contractor construction project transactions might be assisted by the application of such a matrix.

The next stage of the research will seek to develop a more detailed model using as references Macneil's "twelve concept" approach, recent empirical findings into the nature of contracting and criteria developed for selection of an appropriate procurement path. The refined model will be tested both theoretically and empirically against current industry practices. The research will lead to a greater understanding of the nature of contractual relations in the construction industry and provide guidance for the drafting and selection of construction contracts.

## 5 References


Atiyah, P. S. (1986) **Essays on Contract**. Clarendon Press, Oxford.

Aubert, V. (ed) (1969) **Sociology of Law**. Penguin, Harmondsworth.

Beale, H. and Dugdale, T. (1975) Contracts Between Businessmen: Planning and the Use of Contractual Remedies. **British Journal of Law and Society**, 2.

Beale, H. G. Bishop, W. D. & Furmston, M. P. (1990) **Contract Cases and Materials**. Butterworth, London.

BEC (1987) **Simple Guide and Flow Chart for Selecting the Appropriate JCT Form of Contract**. Building Employers Confederation, London.

Bingham, T. (1989) New service charges may prove hard to follow. **Building**, 6 October, 43.

Burrows, P. and Veljanovski, C. G. (1981) **The Economic Approach to Law**. Butterworths, London.

Cheetham, D. W. Carter, D. J. Lewis, J. and Jaggar, D. (eds) (1989) **Contractual Procedures for Building**. Proceedings of the International Workshop held on 6th and 7th April, 1989, at the University of Liverpool, CIB.

CIB W92 Proceedings (1990) **International Symposium on Procurement Systems**. Zagreb, Yugoslavia, 10-13 September.

Feinman, J. M. (1983) Critical Approaches to Contract Law. **UCLA Law Review**, 30, 848.

Flanagan, R. and Norman, G. (1989) Pricing Policy in Hillebrandt, Patricia M. and Cannon, Jacqueline, **The Management of Construction Firms: Aspects of Theory**. The Macmillan Press Ltd, London.

Friedman, L. (1983) in Feinman, J. M. Critical Approaches to Contract Law. **UCLA Law Review**, 30.

Friedman, Lawrence M. and Macaulay, Stewart (1977) **Law and the Behavioural Sciences**, Second Edition. The Bobbs-Merrill Company Inc, New York.

Fuller in Kenneth, I. Winston (ed) (1981) **The Principles of Social Order**. Duke University Press, Durham N C.

Goldberg, V. P. (1976) Towards an Expanded Economic Theory of Contract. **Journal of Economic Issues**, Vol X

No 1, March.
Harris, D. and Tallon, D. (1989) **Contract Law Today, Anglo-French Comparisons**. Clarendon Press, Oxford.
Hibberd, P. Merrifield, D. and Taylor, A. (1990) **Key Factors in Contractual Relationships**. Royal Institution of Chartered Surveyors, London.
Hibberd, P. and Taylor, A. (1991) Key Factors in Contractual Relationships. **The Quantity Surveyor Bulletin**, October, Royal Institution of Chartered Surveyors, London.
Hillebrandt, Patricia M. and Cannon, Jacqueline (1989) **The Management of Construction Firms: Aspects of Theory**. The Macmillan Press Ltd, London.
Institution of Civil Engineers (1991) **The New Engineering Contract - Consultative Document**. Thomas Telford, London.
JCT (1988) **Practice Note 20 Deciding on the appropriate form of JCT Main Contract**, Revised July 1988. RIBA Publications Ltd, London.
Lewis, R. (1982) Contracts Between Businessmen: Reform of the Law of Firm Offers and an Empirical Study of Tendering Practices in the Building Industry. **Journal of Law & Society**, Vol 9 Nr 2, Winter.
Macaulay, S. (1963) Non-Contractual Relations in Business: A Preliminary Study. **American Sociological Review**, Vol 28.
Macaulay, S. (1969) Non-Contractual Relations in Business: Planning and the Use of Contractual Remedies. **British Journal of Law and Society**, 2.
Macneil, I. R. (1983) Values in Contract: Internal and External. **Northwestern University Law Review**, Volume 78.
Macneil, I. R. (1981) Economic analysis of contractual relations in Burrows, P. and Veljanovski, C. G. **The Economic Approach to Law**. Butterworths, London.
Macneil, I. R. (1978) Contracts: adjustment of long-term economic relations under classical, neoclassical and relational contract law. **Northwestern University Law Review**, 72 Nr 6, 854-905.
McDonagh (1990) Foreward in Hibberd, P. Merrifield, D. and Taylor, A. **Key Factors in Contractual Relationships**. Royal Institution of Chartered Surveyors, London.
National Contractors Group and Centre for Strategic Studies in Construction (1990) **Building Towards 2001**. Building Magazine, London.
NEDO (1988) **Faster Building for Commerce**. NEDO, London.
Powell-Smith, Vincent and Chappell, David (1990) **Building Contracts Compared and Tabulated, second edition.** Legal Studies and Services (Publishing) Ltd, London.
School of Business and Industrial Management (1991) **Amending the Standard Forms**. Conference Proceedings, 7

June, London.
The Royal Institution of Chartered Surveyors, Junior Organisation (1989) Contracts in Use 1987. **Chartered Quantity Surveyor**, January.
Uff, J. Capper, P. (eds) (1989) **Construction Contract Policy, Improved Procedures and Practice.** Centre of Construction Law and Management, King's College, London.
Uff, J. (1989) Origin and Development of Construction Contracts in Uff J, Capper P (eds) **Construction Contract Policy, Improved Procedures and Practice.** Centre of Construction Law and Management, King's College, London.
Wallace, I. N. (1978) **The ICE Conditions of Contract Fifth Edition A Commentary.** Sweet & Maxwell Ltd, London.
Williamson, O. E. (1981) Contract Analysis: the transaction cost approach in Burrows, P. and Veljanovski, C. G. **The Economic Approach to Law.** Butterworths, London.

# CONSTRUCTION CONTRACTORS LIABILITY IN SAUDI ARABIA

SADI A. ASSAF AND ABDULMOHSEN AL-HAMMAD
King Fahd University of Petroleum & Minerals
Dhahran, Saudi Arabia.

**Abstract**
This paper highlights construction liability for public projects in Saudi Arabia. It focuses on the Saudi Standard Public Works Contract which governs all public projects in Saudi Arabia. The main issues discussed are liability for contracts, withdrawal of contract, change orders, delays, damages, subcontractors liability, contractor insurance liability, and settlement of disputes.
*Keywords*: Contractors Liability, Saudi Arabia, Public Projects, Conflicts, Construction Projects, Delays, Change Orders, Insurance.

## 1 Introduction

Over the past two decades Saudi Arabia has experienced a major construction boom, whch included the construction of major infrastructure of roads, airports, seaports, hospitals, schools, universitities, and a large number of residential houses and buildings. Many foreign and local designers and contractors were involved in this massive construction, and many problems were encountered due to different contracts used and different construction methods and specifications. Many claims were raised by contractors against government agencies. In order to minimize conflicts and to establish standards which govern all public projects, the Saudi government enacted, the Standard Public Works Contract on February 1, 1988. This law specifies the liability of different parties in the construction process especially the construction contractor. This paper discusses the main issues raised by this law and its relationship to liability of contractors. The issues discussed include the definition of contract, withdrawal of contract, change orders, delays, damages subcontractors liability, contractor insurance liabilitity, and settlement of disputes.

## 2 Definition of contracts

Contracts as specified by the Saudi Standard Contract includes the following documents:

(a) Principal Contract documents
(b) Special conditions
(c) General conditions
(d) Special specificaitions
(e) Plans and drawings
(f) General specifications
(g) Bill of Quantities and Price schedules
(h) Letter of intent or acceptance of proposal.

The Saudi law states that whenever a contradiction occurs between the rules of the contract documents, the rules of the documents preceeding in the order shown above shall supercede.

## 3 Withdrawal of contract

The owner has the right to withdraw the work from the contractor under the following circumstances:

(a) If the contractor delays the work beyond reasonable period set by the owner.
(b) If the contractor quits the work or engages subcontractors without the knowledge or permission of the owner.
(c) If the contractor violates any of the terms of the contract and refuse to remedy the situation as specified in the contract.
(d) If the contractor gives the owners representative a present, gift, loan or reward.
(e) If the contractor becomes bankrupt.

The consequences of the withdrawal of work are the following:

(a) The owner has the right to give the work to the next bidder at the same prices tendered by him. If the next bidder refuses to take the work, the owner will negotiate with other bidders for undertaking the work.
(b) Invite other contractors for part or all of the unfinished work.

All extra expenses incurred are at the expense of the withdrawn contractor. The owner also has the right to sieze and sell materials, machines, and equipments belonging to the contractor, to pay for losses incurred to withdrawn work. The contractor has the right after settlement with the owner to remove the siezed property owned by him from the site of work.

**4 Change orders**

The owner has the right to initiate a change order at his request. This change order's value should not exceed 10% of the original contract value or decrease the contract value by a maximum of 20%. In such cases the contract should be modified to reflect such changes.

**5 Delays**

The contract duration start from the day the site is officially handed over to the contractor and a period of days is set for the duration of contract. The delay fine for the contractor will be calculated as follows:

(a) For the first fifteen days or 5% of the duration whichever is greater, the contractor pays a fine of one fourth the average daily cost times the number of days delayed. The average daily cost is the contract value divided by the contract duration.
(b) For the next fifteen days or 10% of the duration of the contract whichever is greater, the contractor pays a fine of one half the average daily cost times the number of days delayed.
(c) After a delay of 30 days or more than 10% of the contract times the number of days delayed.
(d) The total fines for delay should not exceed 10% of the contract value.
(e) The contractor pays for supervision fees incurred by the owner due to delay of work completion.

**6 Damages**

The Saudi Public Works Contract states that the contractor shall be responsible for all losses and damages incurred by persons and property as a result of the execution of the work. The contractor shall not be responsible for:

(a) Losses and damags that are inevitable result of work performed.
(b) Losses and damages due to errors and negligence of the owner.

**7 Subcontractors liability**

The contractor liability regarding the utilization of subcontractors on their jobs include the following:

(a) The contractor cannot subcontract all of the work.
(b) The contractor can subcontract part of the work with the consent and approval of the owner, and it should be mentioned in the contract documents which specific parts will be subcontracted.
(c) The contractor is legally liable for all work performed by the subcontractors and their employees.

**8 Contractors liability insurance**

There are many types of insurance that can be used by the contractor to cover liability in Saudi Arabia and policies available are of the same international standards as available in the United States and United Kingdom and other developed countries. Liability occurs due to the following reasons:

(a) Injury to persons and damages to property not part of the construction project.
(b) Indirect liability due to contractor association with others such as subcontractors, material supplier etc.
(c) Contractor liability due to entrance in the contract.
(d) Liability after completion of work.
(e) Liability due to operation of contractor or subcontractors equipments.
(f) Liability from the use of professional services needed for completion and excution of the work.

The Saudi public contract specifies that the contractor must purchase insurance to cover all risks on construction project where the contract value exceeds 10 million Saudi Riyal (3.75 Riyals = 1 US$).

## 9 Settlements of disputes

All disputes that cannot be resolved mutually due to interpretation of contrat will be referred to the board of grievences (Diwan Al-Mathalem) for its final judgement.

## 10 Conclusion

It can be concluded that liability is an important issue for construction contractors in Saudi Arabia and foreign contractors should familiarize themselves with the contract laws and how it relates to their execution of work. This paper highlights some of the aspects of contractor liability in Saudi Arabia and discussed the risks involved in liabilities for contracts, withdrawal of contracts, change orders, delays, damages, subcontractors liabilities, contractor insurance liability, and settlement of disputes.

An awareness of construction contracts and the risks involved in performing construction work in Saudi Arabia will lessen many of the problems the contractor could face.

## 11 Acknowledgement

The authors wish to express their appreciation to the King Fahd University of Petroleum & Minerals, Dhahran, Saudi Arabia for providing the time and facilities to carry out this research.

## 12 References

Assaf, Sadi A. (1982) **A Systematic Approach for Selection of Risk Retention Plan,** Ph.D. Thesis, University of Illinois, Urbana, USA.

Council of Minister Resolution No. 136, dated February 1, 1988, **Saudi Economic Survey,** April & May 1988.

Dong Sae, Cho (1982) **Risk Management Decision Making: An integrated Approach,** Ph.D Thesis, University of Illinois, Urbana, U.S.A.

Erickson, Carl, A. (1971) **Risk Sharing in Construction Contracts,** Ph.D Thesis, University of Illinois, Urbana, USA.

Mehr R., and Hedges, B. (1974), **Risk Management Concepts and Applications,** Richard D. Irwin, Honewood, Illinois, U.S.A.

# THE ROLE OF INTEGRATED COST AND TIME MODELS IN CONFLICT RESOLUTION

P H McGOWAN, R M W HORNER and R ZAKIEH
Dept of Civil Engineering, University of Dundee, UK

D JONES and P A THOMPSON
Dept of Civil and Structural Engineering, UMIST, UK

**ABSTRACT**
Fair settlement of contractual claims requires cooperation and communication between the parties. It also requires the contract to provide a clear statement of risks and a mechanism for dealing fairly with associated cost and time effects. A significant proportion of claims in the construction industry relates to the incidence of delays, disruptions and other variations which have cost and time implications. Existing project models fail to reflect the interactive link between cost and time in construction contracts. This failure leads to an unsystematic evaluation of the effects of change, and frequently, to bitter disputes. There is an urgent need for a simple model integrating time and cost within the contract so that the cost and time effects of variations, delays and disruptions can be systematically and objectively analysed.

New models have been developed which contain fewer than 20% of the items in traditional models, yet which can accurately predict construction costs and durations. The new models are based on operations which realistically reflect a contractor's site operations. They provide him with the flexibility to offer the optimum construction solution within a standardised work breakdown structure. It is hoped that further development of these "resource-significant" models will yield a new, systematic framework for evaluating the effects of change.
Keywords: Resource Significance, Cost and Time Modelling, Delay and Disruption, Claims Evaluation.

## 1 INTRODUCTION

### 1.1 Variations and claims
Contractual claims are a common feature of the UK construction industry. Under most conditions of contract in general use, they arise when the work the contractor

is required to execute changes in a way which he could not reasonably have been expected to foresee at time of tender. The changes may concern the nature or extent of the work, or the conditions under which the work has to be carried out. The extent to which the liability for variations and claims rests with the client, and the level of time and cost compensation to which the contractor may be entitled are governed by the terms of the contract.

Variations and claims are not synonymous. Hughes (1985) defines a claim as "a demand or request or application for something to which a contractor considers, believes or contends he is entitled but in respect of which agreement has not yet been reached". By this definition, a claim remains at large as long as agreement has yet to be reached. Disagreements may relate either to the grounds of a claim or its evaluation. The incidence of claims is minimised when contracts provide a clear allocation of risk, and a swift, equitable and rational means of settlement once liability is admitted. This paper reports progress on a research project whose objective is to provide a more systematic basis for the valuation of claims.

### 1.2 Interdependence of cost and time

Construction costs are a function both of the material element of the project, and of the labour and plant resources required to operate on those materials. Thus, when an estimator prices a contract, he is concerned not only with the quantity of materials, but with all those factors which may affect the cost of labour and plant. In addition to the unit cost of each resource, these include project duration, the method of construction, productivity, the scope of the work, continuity, the timing of activities, resource and sequence constraints. The same parameters dictate to the planner the method of execution, timing and duration of operations, expressed in the programme and method statement for the works.

The programme, method statement and estimate form the basis of the contractor's tender. Any subsequent changes to the works must be evaluated in the light not only of any changes in quantities, but in any of the factors which affect the cost of labour and plant. The interdependence of cost and time is irrefutable.

### 1.3 Shortcomings of traditional contractual frameworks

There exist more than 30 so called "standard" forms of construction contract in the UK alone (McGowan et al 1991). The proliferation simply reflects how different clients seek to allocate risk in different ways. It also reflects the many ways available to evaluate risk. In this context, risk evaluation is concerned with the contract procedures for cost and time compensation, and any contractual aids to that process.

Despite the interdependence of cost and time, traditional forms of contract continue to isolate them. Cost is modelled separately from time. Under most contracts the powers of the Engineer or Architect to instruct commonly extend to programme, sequence and method of execution of the works, execution of work by other contractors, postponement, acceleration, access to and possession of the site, and other variations. In addition, the client usually carries the risk of adverse ground and physical conditions and late issue of information. All of these events can give rise to a change in the use of resources. Good contracts and contractual procedures will deal swiftly and fairly with claims. Unfortunately, traditional forms of contract, far from encouraging agreement, are adversarial in their very nature (McGowan et al, 1991; European Construction Institute, 1991). Their failure to provide systematic procedures for evaluating the effect of changes on cost and time promote protracted disputes. Time limits for the notification, submission or settlement of claims are usually ambiguous or inadequate. Rules and procedures for cost and time compensation are totally separate. The contractor is forced to produce unsubstantiated details of tender build-ups and programme assumptions in order to demonstrate entitlement.

Forms of contract such as JCT80, ICE6 and GC/Works/1 continue to separate time and cost in both risk allocation and risk treatment. It is then little wonder that claims for delay and disruption are so frequently and bitterly disputed.

## 1.4 New forms of contract

More recent and proactive forms of contract have broken with tradition. The most notable are the New Engineering Contract (NEC), GC/Works/1 Edition 3 and the ACA/BPF forms of contract. Common to all three is a recognition of the importance of quantifying the effect of delay and disruption at the time when a variation occurs. This should reduce the potential for protracted disputes. The failure of bills of quantities to satisfactorily model construction projects has lead to their replacement in the ACA/BPF and NEC by an activity schedule. Both NEC and GC/Works/1 have promoted the programme to a central supporting role.

NEC has deliberately set out to minimise the effects of risk and to promote cooperation between the parties through good management practice. It aims to offer a sound basis for the swift and equitable settlement of claims. The contract allocates risk clearly by reducing subjective judgement and recognising the interdependence of cost and time. Each compensation event is subject to early warning and quotation where possible. Delay and disruption are automatically considered with each event, and the cost and time models are revised in the wake of each event. NEC reimburses delays, disruptions and other variations on a cost plus basis - a marked departure from traditional procedures.

### 1.5 Objective

This paper outlines recent research aimed at developing simple, integrated time and cost models which can be incorporated in the contract. It also seeks to provide a contractual framework specifically designed to recognise the interdependence of cost and time, thereby facilitating the swift and equitable settlement of claims.

## 2 PRINCIPLES OF PROJECT MODELLING

### 2.1 Purposes

Project models may serve a variety of purposes. They provide a vehicle for predicting costs and durations, a means of control, both for the client and contractor, a system for valuing the work done, and for evaluating the effects of any changes; they are indispensable for planning and for predicting cash flow (Skinner, 1981; Pasquire and Tyler, 1987).

#### 2.1.1 The model's role in dispute resolution

A model is a representation of the construction process and of the factors which affect it. An effective model allows us to understand the complex interrelationship between cost, time and the environment within which the works are executed. Only if these interrelationships are clearly exposed is it possible to systematically evaluate the effects of change on cost and time. It follows that cost and time must be linked dynamically within the contract if equity is to be achieved and the potential for dispute reduced.

#### 2.1.2 The model's role in control

Figure 1 illustrates the classical control loop. Throughout the life of a project, cost and time must be predicted, monitored, analysed and controlled. To close the loop, historical information is fed back into the database. At each point in the loop, time and money are closely interdependent, and must be controlled together. This is particularly pertinent when claims proliferate and cashflow is a major determinant of the project's profitability. The whole system is defined absolutely by the characteristics of the model. It is only possible

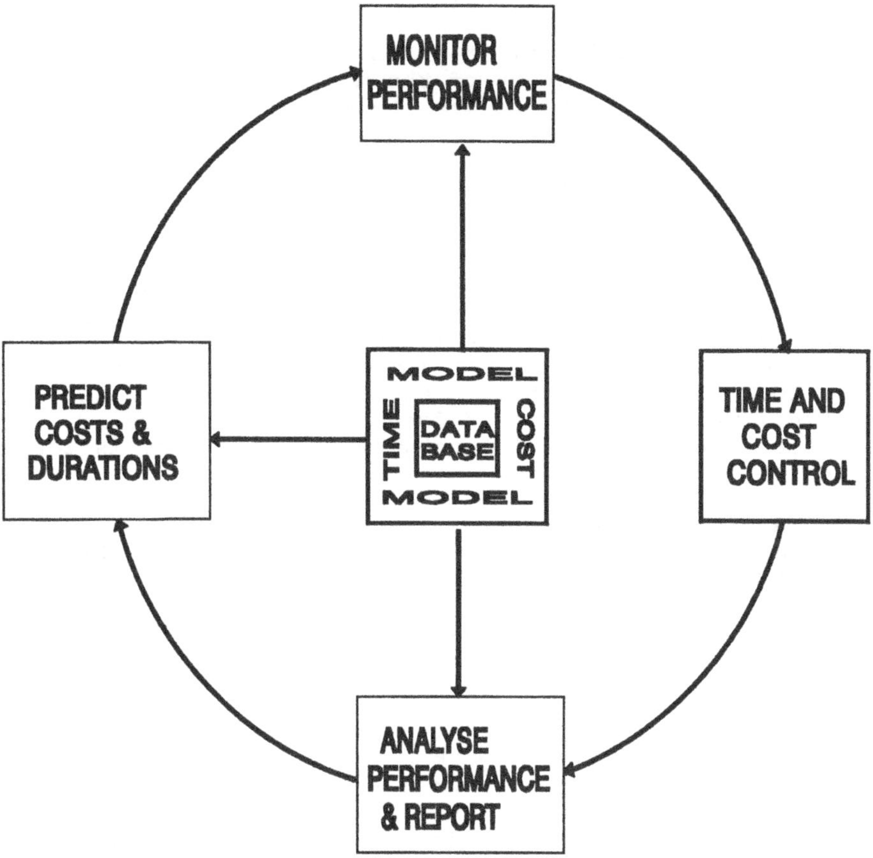


**Figure 1**
**Classical control loop**

to control those elements which are expressed within the model; those same elements define the nature and scope of the data which has to be collected. The quality of the model and the accuracy of the feedback affect the usefulness of the database. If the model realistically represents the construction process and the factors which affect it, and if the quality of information in the database allows the relationship between cause and effect to be quantified with confidence, the scope for dispute is significantly reduced.

### 2.2 Characteristics of the ideal model

There are four major considerations:

- the level of detail at which the process is expressed;
- the form and content of the model elements;
- the way in which the link between time and cost is provided; and
- the potential conflict between flexibility and the need for standardisation.

#### 2.2.1 Level of detail

The best solutions are often the simplest. In project modelling, this is no less true. Over-itemisation and unnecessary detail inhibit efficiency not only in the settlement of claims, but in the wider field of valuation and control. They make it impossible to monitor progress or to collect meaningful productivity data (Paulson, 1976); the allocation of feedback data becomes inaccurate, and the link between cost and time is lost (Mair, 1991).

#### 2.2.2 Form and content

We can only control that which we can measure. If measurement and data feedback are to be effective, the model must be simple. It must also reflect realistic site operations. It is pointless therefore to incorporate elements like "formation of cavities" or "formwork less than 300mm wide to edges of slabs". Since we cannot measure the time it takes to carry out such work, it follows that we cannot isolate the costs associated with it. Why then should we model it and ask a contractor to price it when its duration and cost is impossible to predict? The cavity is part of an operation "construct wall"; the slab edge formwork is part of an operation "formwork to slab".

Construction is principally concerned with the incorporation of materials into the works through the execution of a series of operations. It is desirable that operations are defined in a way which allows a single unit rate to be applied to each one. Thus an operation must involve the work carried out by a single operative or gang of operatives so that it can be characterised by a single productivity. It must be continuous and take place in a single location.

#### 2.2.3 Link between cost and time

Operations translate into programme activities. An activity either "installs, removes, modifies or tests a particular component" of a facility (Echeverry et al, 1991). Activities are characterised by work content and location. Thus an operation may represent a single activity, part of an activity or a number of activities.

The link between time and cost depends on the use of the same operations in the cost model as in the time model.

#### 2.2.4 Conflict between flexibility and standardisation

For programming purposes, the fundamental requirement when defining the model of site operations is flexibility to allow the contractor to choose the most effective construction solution within the boundary conditions imposed by his own resource constraints. The imposition of a rigid activity structure may impair the ability of the contractor to optimise his solution. However, effective feedback depends on a comparison of like with like both within a project and from project to project. The ideal model must therefore strike a balance between flexibility and the capability of producing standards which can be applied to future projects and which can help to quantify the relationship between cause and effect.

### 2.3 Model specification

From the foregoing, the ideal model must be sufficiently simple for effective control, yet sufficiently accurate to generate confidence in the predictions of cost and duration. At its heart must be a set of work packages which relate to realistic site operations and which are capable of site measurement for monitoring, control and feedback. They must lie within a standard framework, yet provide sufficient flexibility for the contractor to identify and implement the optimum construction solution. There must be a close correspondence between the elements of the cost model and the activities in the programme to provide that dynamic link between cost and time which is the key to the resolution of claims for delay and disruption.

## 3 APPRAISAL OF CURRENT MODELS AND THEIR PREDECESSORS

### 3.1 Bills of Quantities

The principal attraction of the bill of quantities is the facility it provides for contractors to tender on the same basis. It is essentially a shopping list of materials to be incorporated in the works (Horner , 1991), against which the contractor enters the labour, plant and materials costs. It is however ill equipped to satisfy the needs of a control model, for it totally fails to reflect the contractor's site operations. As a result, feedback to the database is practically impossible. Worse than that, it fails to expose the resources on which the contractor's bid is based and is quite independent of any programme, whether that programme forms part of the contract or not. It is therefore ill-suited to the evaluation of change. Indeed, part of the reason that it continues to find favour with contractors is the opportunity it provides for creating a smoke screen around the contractor's original intentions. Thus, front end loading may go undetected, and new rates may be negotiated almost from scratch. It is this lack of transparency that is a major contributor to the problems associated with valuing claims for delay and disruption. The premise that cost is proportional to quantity alone fails to recognise that many variations result in a delay to the programme, or a drop in productivity, or both. Neither of these effects can be valued by the use of the tender unit rate.

### 3.2 Operational Bills

Although labour and plant may account for a smaller proportion of costs than materials, their costs are subject to much greater variability. This is a direct result of their sensitivity to factors other than quantity. Such variability requires control. In a direct attempt to slant the bill of quantities towards planning and control, Skoyles (1965) introduced the Operational Bill. Its structure was based on an activity network prepared by the client's consultant. The contractor prepared his estimate by inserting the appropriate labour and plant cost as a sum against each activity. Despite offering advantages to planning, it was perceived as lacking in flexibility, even though the use of the network was not mandatory, and failed to gain general acceptance.

### 3.3 Method Related Charges

The introduction by Barnes and Thompson (1971) of the method related bill was another attempt to model the interdependency of cost and time. It is a simple idea which seeks to create a model which is sensitive to cost factors other than the quantity of the work. The method related charge (MRC) is defined as a charge entered by the contractor "to cover items of work relating to his intended method of executing the work, the costs of which are not to be considered as proportional to the quantities of other items". (Barnes, 1986). Evidence suggests that the MRC operates well where it is adopted positively or where its use is mandatory (Fraser, 1991). The philosophy is embedded in the CESMM, and to a lesser extent in SMM7. Because their use is discretionary in the UK, MRC's rarely feature in modern bills of quantities.

## 4 SIGNIFICANCE MODELLING

### 4.1 The 80/20 Rule

The difficulty in creating suitable models of construction projects is to strike the right balance between the level of detail and the required accuracy. If the model is too detailed, it becomes unmanageable; if it is too coarse, it lacks sensitivity to change. It has been known for a long time that 80% of the value of a bill of quantities is contained in only 20% of the items (Figure 2). There seems obvious merit in directing management's attention and seeking to control the whole project through the significant items (Crow, 1992; Horner 1991). This implies that the insignificant items, representing some 20% of the total project must be either ignored, accounted for within the significant items, or wrapped up into a single "non-significant" item. Which of these alternatives offers the best solution is yet to be determined.

Early work by Saket (1986) showed how significant items were simply those whose value was greater than the mean.

### 4.2 Cost modelling

More recent work by Asif (1988) and Zakieh (1991) has shown how this finding can be used to create simple models of the construction process. It has proved possible to categorise construction projects in such a way that within any category, the cost-significant items are roughly the same. Because of the way traditional bills of quantities have developed, it is necessary to create a new item whenever there is a change in circumstances which might affect the cost of the existing items, no matter how small that change might be. As a result, many items carry either the same unit rate, or unit rates which are insignificantly different one from the other. Because contractors cannot consistently price individual items with an accuracy of better than ±100% (PSA, 1980), it is perfectly sensible to combine such items into a single cost package.

Furthermore, Zakieh (1991) has shown that whilst operations such as steelfixing are spread over many bill items, there is a strong linear relationship between cost and quantity of those items involved in the same operation (Figure 3). Thus the quantity of one item, representing a single material component or class of material frequently predominates over the quantities of all similar items. This class of material is said to characterise the operation. For example, reinforcement may be characterised by 24mm $\phi$ high yield steel bars. If the unit rate associated with the predominant item is used to price all other items within the same operation,

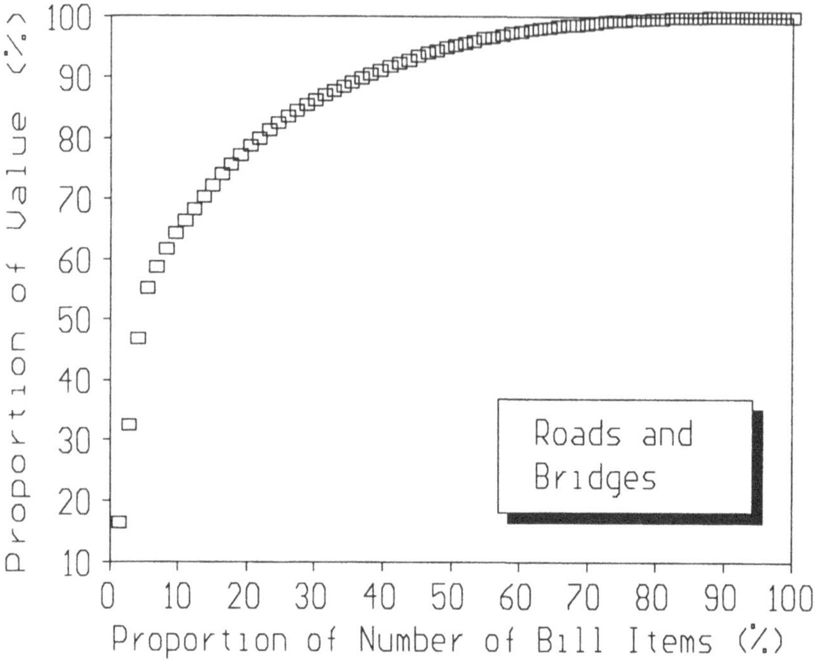


**Figure 2**
**80/20 Relationship**

a negligible error results.

By applying these and similar techniques, it is possible to derive a series of cost packages which reflect site operations, which relate to a single material, which involve a single trade and a single productivity, and whose value represents a consistent proportion (called the cost model factor, usually close to 0.8) of the total value. The operations can be priced using a single unit rate, and typically number some 10% of the items in a conventional bill of quantities. The value of a project can be estimated simply by pricing these cost-significant work packages and dividing by the cost model factor.

## 5 LINKING TIME AND COST

### 5.1 Resource significance

Integration of cost and time requires a model that is sensitive to both. It must therefore represent the consumption of plant and labour resource as well as

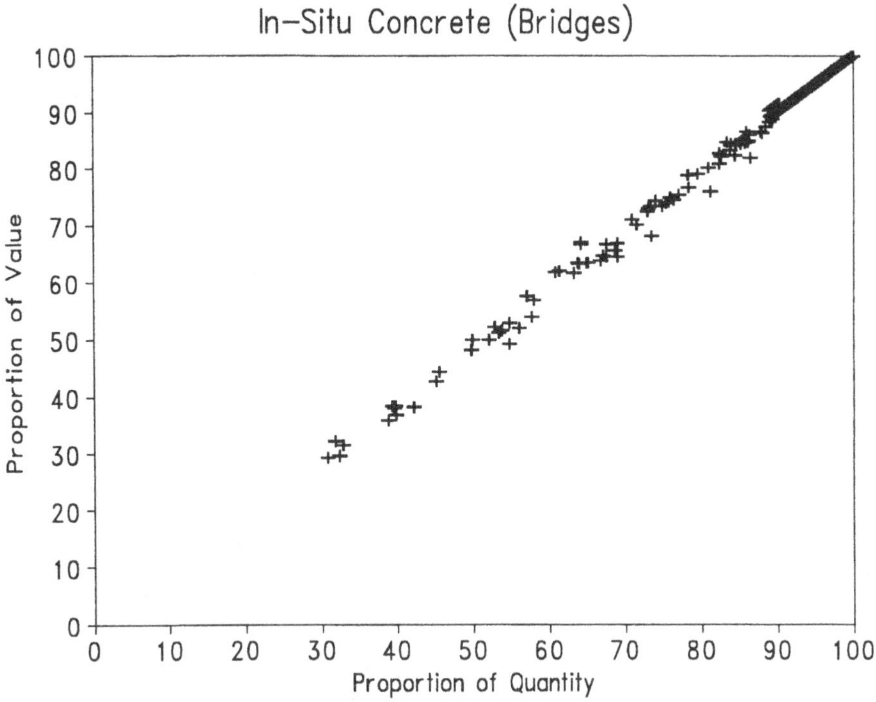


**Figure 3**
**Relationship between cost and quantity**

materials. Analysis similar to that described in section 4.2 has demonstrated that work packages can be used not only to predict total costs but to predict the cost of each resource separately, and the plant and labour hours as well. Models have been developed for roads and bridges. They predict costs to an accuracy of 4% (coefficient of variation) and resource hours to 6%. The operations can be translated into programme activities which take account of those variations in location, sequence and timing which must be incorporated in a realistic project network. Thus, we are now in a position to link time and cost models through a single set of work packages which properly reflect site operations.

### 5.2 Application in practice

Although the models specify the work breakdown structure for the project, they do not prescribe the interrelationships between the various operations, or indeed, the way in which the operations translate into activities. Indeed, it is important that the contractor is provided with sufficient flexibility to allow him to construct the project in the way which best suits his resources. It is also important to allow him sufficient discretion to adjust his tender in the light of the risks he is required to accept and the prevailing commercial conditions.

The European Construction Institute (1991) has suggested that the ideal approach is one in which the client recognises "that the contractor is endeavouring to achieve an equitable level of profit commensurate with the risk involved, and should be prepared for him to receive such profit subject to satisfactory performance."

In order to ensure that the effect of changes can be properly evaluated, it is necessary in the model to separate those costs which are time sensitive from those which are not. In order to do this, material costs must be separated from resource costs. This may be accomplished by requiring the contractor to break down the cost of each operation into three components:

quantity proportional (primarily materials);
time proportional (primarily labour and plant); and
fixed (typically setting up and subsequent removal costs).

Obliging the contractor to provide this breakdown ensures that there is a suitable basis from which to measure the effect of change. It neither requires him to divulge his method of working nor ties him to decisions made at time of tender which might subsequently seem inappropriate. It simply sets the boundary conditions within which he is expected to work. It is not a perfect solution, but it seems to us to be a much more systematic one than any others which currently exist.

## 5.3 Future work

A number of problems remain to be solved. The current research project is confined to a study of roads, bridges and water treatment works. It will be necessary to develop models for many other types of projects. It will also be necessary to define with precision the conditions which define the boundaries between different categories of project. The bill compiler must know with certainty to which category a new project belongs.

In addition, decisions must be made about how non-significant items are to be dealt with. This will have greatest effect on the final account. Should the contractor be allowed to choose the value of the cost model factor, or should it be specified in the instructions to tenderers? Should the cost and durations of the operations within the model be increased to allow for the non-significant items, or should non-significant items be valued separately? If they are valued separately, how are changes to be dealt with? In the remainder of this project, computer simulations of completed contracts will be used to determine which of the various alternatives leads to the most equitable outcome.

## 6 SUMMARY

If contractual conflict is to be reduced, it is necessary to model construction projects in a way which maximizes the scope for systematically evaluating the effects of change. Application of the philosophy of resource significance to defined categories of projects produces simple yet sufficiently accurate models of the construction process. The models incorporate work packages which accurately reflect site operations, and which allow sufficient flexibility for a contractor to define those activities and their interrelationships which best represent his preferred method of working. The requirement to separate material costs from resource and fixed costs establishes through the programme that link between cost and time which it is hoped will provide an equitable method for evaluating the effects of variations, delays and disruptions. The NEC provides the sort of contractual environment which would allow such a system to operate successfully.

## 7 REFERENCES


Asif, M. (1988) **Simple Generic Models for Cost-Significant Estimating of Construction Project Costs.** Ph.D. Thesis, University of Dundee, Scotland.

Barnes, M. (1986) **CESMM2 Handbook.** Thomas Telford, London.

Barnes, N.M.L. and Thompson, P.A. (1971) **Civil Engineering Bills of Quantities.** Report No. 34, CIRIA, England.

Crow, T.W. (1992) Information Management Comes of Age. **Proc. 3rd Intl. Conf. Modern Techniques in Constr., Engrg. and Proj. Mgmt.**, Singapore, pp. 25-31.

Echeverry, D. Ibbs, C.W. and Kim, S. (1991) Sequencing Knowledge for Construction Scheduling. **J. Constr. Engrg. & Mgmt.**, ASCE, 117, 1, pp. 118-130.

European Construction Institute (1991) **Client Management and its Role in the Limitation of Contentious Claims.** Client Management Task Force, Loughborough, England.

Fraser, R.M. (1989) The Palmiet Pumped Storage Scheme: The construction phase. **The Civil Engineer in South Africa**, 31, pp 417-425.

Horner, R.M.W. (1991) Fundamentals of Construction Project Cost Control Systems in **Proc. 4th Yugoslav Symp. on Org. and Mgmt. in Constr.** (ed. Prof. dr. Marko Zaja), Gradevinski Institut, Zagreb, pp. 437-445.
Hughes, G.A. (1985) **Building and Civil Engineering Claims in Perspective.** 2nd Ed., Longman, London and New York.
Mair, A.J. (1990) **Estimating and Control of Construction Projects using Cost Significant Work Packages.** Ph.D. Thesis, University of Dundee, Scotland.
McGowan, P.H. Horner R.M.W. Jones D. and Thompson P.A. (1991) **Allocation and Evaluation of Risk in construction Contracts.** Occas. Paper No. 52, C.I.O.B., England. In press.
Pasquire, C. and Tyler, A. (1987) Bills of Quantities - Are they needed? **Building Technology and Management,** June/July, pp. 16-23.
Paulson, B.C. (1976) Concepts of Project Planning and Control in **J. Constr. Div.,** ASCE, 102, CO1, pp 67-80.
Property Services Agency (1980) **Construction Cost Database.** Second annual report on behalf of the P.S.A. by the Dept of Const Mngmnt, University of Reading.
Saket, M.M. (1986) **Cost Significance Applied to Estimating and Control of Construction Projects.** Ph.D. Thesis, University of Dundee, Scotland.
Skinner, D.W.H. (1981) **The Contractor's use of Bills of Quantities.** Occas. Paper No. 24, C.I.O.B., England.
Skoyles, E.R. (1965) **Introduction to Operational Bills.** Design Series 32, Building Research Station, England.
Zakieh, R. (1991) **Quantity Significance and its application to Construction Project Modelling.** Ph.D. Thesis, University of Dundee, Scotland.

## 8 ACKNOWLEDGEMENT

The work described in this paper has been financed in part by the Science and Engineering Research Council, whose support is gratefully acknowledged.

# THE POSITION OF MATERIALS RE PAYMENT AND OWNERSHIP IN CONSTRUCTION PROJECTS IN THE UK

G. BOWLES AND H.A. GOW
Dundee Institute of Technology, Scotland

Abstract
The paper investigates the position of materials on and off-site with reference to advance payments, ownership and transfer of title to the employer. Respective duties, attendant risks and alternative arrangements are reviewed along with the differing legal structures in Scotland and England.
Keywords: Payment of materials, Ownership, Title, Building Contracts, Insolvency.

## 1 Introduction

Construction is a complex production system (1) involving at least three and as many as seven main consultants and a main contractor with perhaps seven or eight domestic sub-contractors and often as many nominated (2) sub-contractors; with both of the latter categories also employing sub-contractors and certainly involved with numerous suppliers of goods.

The traditional structure involves an operation as described above while current approaches have as alternative methods of operation the consultant (5) engaging the contractors or a management contractor operating with a consultants design or an in-house design (design and build) and employing (and managing) the various separate contractors either traditionally or in accordance with SMM7 in work sections (parcels of work designated in accordance with the publications of the Committee on Co-ordinated Project Information(3)).

The chain of operations between client/consultants and contractors can be long and often is not direct. This itself raises legal problems as to who has contracted with whom and what may seem to be simple solutions to these problems are often not an acceptable way to proceed given the timescales being operated by the parties to the project.

There are three main forms of contract used by the major clients in the UK today. The government uses GC/Works/1 (4) and the civil engineering industry uses the ICE General Conditions of Contract (5) and the building industry largely operates under the JCT Standard Form of Contract 1980 (6). There are other lesser forms but the paper concentrates on the JCT Standard Form as the legal position from an operational point of view is similar to all.

## 2 Legal overview

The contentious issue regarding payment for materials stored off-site is in who actually owns the title for the goods after payment has been certified. The position would appear to be clear cut in the light of JCT 80 clauses 16, Materials and Goods Unfixed or Off Site and clause 30.3, Certificates and Payments - Off Site Materials or Goods.

Clause 30.3 gives the Architect discretionary power to include in interim certificates the value of goods or materials intended for the works before their actual delivery to the site. The reason for this clause is to help maintain a healthy cash-flow position of the main contractor and his sub-contractors (7) in the modern and specialist environment of the construction industry, where much fabrication is done off site. There are certain restrictions prescribed in the clauses specifically aimed at preventing the architect from including in his certificate materials off-site for which the original supplier has retained title until payment. Cl 30.3.4, 5 and 6 expressly provide that a contract for the supply of goods exists in writing and that property therein shall pass to the contractor before their value is included in a certificate.

Construction contracts and sub-contracts are contracts of work and materials and not contracts of supply. They are not subject to the provisions of the Sale of Goods Act 1979, where property in the goods pass at the time when parties intend it to pass - irrespective of whether payment has been made. In domestic sub-contracts there is no contractual relationship between the sub-contractor and the employer and no passing of ownership of the goods to the contractor. The contractor does not at any point buy, or agree to buy, the materials which a sub-contractor supplies. Hence, the main contractor does not at any stage acquire property in the goods and so has nothing to pass on to the client. The general rule is that an unpaid sub-contractor can reclaim the goods until materials are incorporated in the building; at which point they become property of the Employer. It is essential, therefore, that the Employer enters into a separate contract of supply with the Contractor and the sub-contractor or supplier as the case may be in order to effectively transfer the right of property in the materials concerned. This action would be required under JCT 80 to enable the architect to use his discretionary powers when considering payment for materials stored off site. Conflict over ownership is not uncommon in cases where materials have been paid for without a separate contract of sale, whether stored on or off site.

According to Cl. 16 of JCT 80 ownership of materials off site passes to the client when the main contractor is paid the value of the goods in an interim certificate. These conditions cannot, however, bind suppliers who are not parties to the main contract. This notion of ownership has been seen to have no substance in common law in cases of insolvency of the main contractor.

Under the JCT form (6) Parris reported the case of Dawber Williamson Roofing v Humberside County Council (8) in which the main contractor subcontracted the roofing work to DW on a JCT standard form of sub-contract which stated that property of materials which had been delivered to site passed to the client when it paid the value of goods as certified by the architect. DW delivered slates to site and the MC was paid for them in the next valuation. The MC subsequently went into liquidation before DW had been paid for the slates. It was held that the clause regarding property of the goods passing to the client when they were paid for did not operate against DW, who retained ownership of the slates.

This case is important under both Scots and English Law where the Employer does not have full rights in the materials unless the MC has proper title to the materials which he can transfer to the Employer. Interestingly, the Employer owns the materials once incorporated into the works whether paid for or not.

The case of Stirling County Council v Official Liquidator of John Frame Ltd (9) based on the Scottish National Building Code Regulations relates to the ownership of materials stored off site. In 1952 The company contracted to carry out work for SCC on a housing scheme. As space was limited on site certain materials were stored off site with SCC's consent. The company went into liquidation and the county council claimed the property in the materials on the grounds that they were constructively on site and Condition 5 of the contract provided all materials delivered by the contractor for the execution of the works "shall become and be the absolute property of the Employer." It was held that because there was no contract of sale and title of the goods could not pass Condition 5 of the main contract could not operate. In the opinion of the case Sheriff-Substitute the condition did "seem to be of some foreign system of law."

The implications of the decision of this case have been far reaching. Twenty years later the Scottish branch of the RICS advised all its members of the QS section to highlight the case to their clients when passing the first certificate for payment. At that time many local authorities were still making payments of up to 90% of the value of materials placed on sites, presumably without investigating who has title on any of the goods.

### 3 Position of the employer

Under JCT 80 it would appear that the Employer bears less risk for materials paid for and stored off site than for materials stored on site, if the conditions are rigidly applied. Cl.16.2 provides that the value of materials stored off site included in an interim certificate are paid in accordance with Cl 30.3 i.e a contract of sale which transfers property to the MC has been entered into (in Scotland this will be the SBCC Contract of Purchase (10)). In turn the MC automatically passes ownership to the Employer when paid for in a certificate. In contrast for unfixed materials on site, Cl.16.1 states that where the value of materials have been included in an interim certificate, such materials and goods shall become the property of the Employer. Reading this clause the Employer would quite reasonably, but erroneously, believe ownership of such materials would be his. This assumption that materials on site unreservedly become the property of the Employer is repeated in many standard forms of contract but it does not conform with common law as illustrated by case law (see Dawber Williamson v Humberside County Council).

The point at which goods/materials become the property of the Employer, unless specific provision is made otherwise, is when they have been incorporated into the works regardless of whether the Employer has paid for them or not. It is clear then that the Employer bears the risk for materials in the period between paying for them and having them included in the construction.

### 4 Position of main contractor and sub-contractor/supplier

The position of the main contractor is defined by the Conditions of Contract . He has an obligation to submit to the architect/quantity surveyor claims at valuation time (usually monthly) for work done and materials available (on and off site) to carry out the works. If paid under Clause 30 for work and materials made available by subcontractors then the main contractor is paid within 14 days and in turn has to pay the sub-contractor in accordance with the subcontract conditions. Only the nominated sub-contractors and suppliers have the right to request direct payments for their claims, other domestic subcontractors and suppliers have no such arrangements.

Under common law there is a distinction between the position of the sub-contractor, who supplies goods and labour, and the supplier, who supplies only goods. It is often a matter of chance, though, whether a particular input supplier is classified as a sub-contractor or supplier. In the case of a sub-contractor who hasn't been paid by an insolvent MC, the sub-contractor can bring a direct claim against the client without using a reservation of title clause. However, a supplier can only claim if it can show its contract includes such a clause.

## 5 Position of consultants

The position of the consultants appears ambiguous. The quantity surveyor in computing the valuation will draw the architects attention to those materials which are on-site and those included which are off-site. The architect is allowed discretion in the certification of the latter (JCT 80 Cl.30.3) but no longer has the protection at law as a quasi-arbitrator inferred in 1901 by the Court of Appeal, confirmed in 1957 by the House of Lords but overturned in the judgement in Sutcliffe v Thackrah 1974 (10). The poor architect is faced with upsetting necessary cash-flows, hindering certification by requesting proof of ownership or leaving the possibility of being sued for negligence by an irate and out-of-pocket client. At least in court the JCT could be called an expert witness to verify their 1978 notice in which they discounted a universal administrative procedure to check actual ownership to counter-act the verdict in Aluminium Industrie Vaassen B.V v Romalpa Aluminium Ltd. in which an agreed clause allowed transfer of title "on receiving payment in full" (11). This leaves the problem of title to goods on-site and to goods off-site and the risk factor in agreeing to a payment "on-account." The reporting of Dawber Williamson Roofing Ltd. v Humberside County Council 1979 raises a fine point in contract administration as three months passed between the depositing of the materials and the contractors insolvency. A tight reporting of cash distribution to all sub-contractors would have produced the information that the roofer had not yet been paid. It could be inferred another interim payment was probably made before the insolvency. This mechanism would have reduced the risk factor to activities within a monthly period for the supply and payment of materials.

## 6 Risk management

The construction industry as an industry where performance of contractual obligations is projected well into the future and is thus largely concerned with risk management. This task is made difficult due to the large number of independent, but interdependent, contributors to the construction process as outlined in the introduction. The problem of entering into separate bills of sale for materials, whether on or off site, to provide proof of ownership is that it can be a complex, time consuming exercise for the main contractor. This can ultimately reflect in a higher price to clients in tender levels. In avoiding litigation in the case of a main contractor insolvency between receiving payment for goods from the Employer and passing money on to the supplier there remains the problem of which of the inconsistent contracts prevail regarding ownership of the goods. Rational apportionment of risk is made problematic since no direct contractual link exists between client and supplier/domestic sub-contractor. G.Antoinette Williams (1) writing in the Construction Law Journal suggests economic theory may inform a decision as to who ought to bear the risk. The question of how this could be achieved though remains.

## 7 Conclusions

The industry has traditionally operated without insurance bonding as a fall-back protection and has therefore saved thousands if not millions of pounds in oncosts for such insurance. The careful selection of contractors cannot guarantee that insolvency will not happen and problems only occur in insolvencies.

The mechanisms necessary to ensure a minimal problem for the few would create major administrative problems for the many - to no great purpose in most cases.

Careful monitoring of all operations and a common sense approach to particular problems should minimise risk and neutralise a negligence claim. Where large sums are involved and it becomes imperative to pay monies for off-site materials then the client would be more fully covered by entering into a contract to purchase the goods. V Powell-Smith and J.Sims (12) suggest a pro-active contractor should furnish a written statement and enclose proof of ownership! For the moment the introduction of a bond has not been suggested except, perhaps, on very large projects. For the consultants the timespans of often less than a month between valuations will minimise the risk management of the problem and for clients in general a careful monitoring of materials and components and their incorporation in the building by the consultants is their best safeguard. For the particular client, his advisers and their contractors there must be risk, there must be loss and sometimes there must be scapegoats. Usually we hope its someone else -insurance costs money!

## 8 References


1 G.A Williams, Reservation of Title in the Construction Industry, **Construction Law Journal,** Volume 3, 252 - 257, Sweet and Maxwell 1987.

2 Nominated Sub-contractors Clause 35, Joint Contracts Tribunal, **Standard Form of Building Contract,** RIBA Publications 1980.

3 Committee for Co-ordinated Project Information, various publications

4 **General Conditions of Government Contracts for Building and Civil Engineering** Works 2nd Ed., HMSO 1977

5 Joint Contracts Tribunal, **Standard Form of Building Contract - Private Edition with Quantities**, RIBA Publications 1980

6 Joint Contracts Tribunal, **Practice Note 5**, RIBA Publications 1981

7 Parris. J, (1982) **Retention of Title on the Sale of Goods,** Granada

8 Stirling County Council v Official Liquidator of John Frame Ltd, (1951), **The Scots Law Times,** April 14, 37 - 39.

9 Sutcliffe v Thackrah [1974]1 **All LR** 860
10 Aluminium Industrie Vaassen B.V v Romalpa Aluminium Ltd. [1976]2 **All LR** 552
11 Powell-Smith, V. & Sims, J., **Contract Documentation for Contractors** (2nd. Ed. by Powell-Smith), BSP Professional, 1990

# STATISTICAL MODELLING OF CLAIMS PROCEDURES AND CONSTRUCTION CONFLICTS

D.DALTON AND N.SHEHADEH
Department of Civil Engineering, University of Leeds, England

Abstract
Conflict can occur at any stage before the contract is agreed. This paper considers the conflict that occurs after the contract has been formed and why this conflict should be so evident in constriction contracts. Indication of conflict can be cost and time over-run. Claims are clearly correlated with cost and time over-run.

Certainly claims are frequent features of conflict as seen in conciliation, arbitration and litigation. In this paper some of the results of a statistical investigation into the causes of claims are given. Factors involved in claims procedure are shown to be correlated with the number of claims, value of claims as well as cost and time over-run. These factors are dependent on the Client, Designer, Supervisor, and could also include type of contract, authority given to the Engineer, Client's involvement etc. Knowing the various factors for a particular project, and by using various statistically derived models, it is possible to predict the number of claims and the value of claims. It is possible to consider changing some of the project factors and being able to predict the effect this will have on claims and hence conflicts.
Keywords: Claims Procedures, Claims Management.

## 1 Conflicts in construction contracts

In any potential purchase and where there is no agreement, i.e a contract, there can be a conflict of interests between the purchaser and the supplier. Once the contract has been agreed, conflict occurs when the contract fails to live up to the expectations of one or both of the parties to the contract.

The likelihood of conflict is greater when there is uncertainty as to the details of the subject of agreement i.e where there are different expectations. This is common in construction where projects are large, complex, and many parties are involved. This is added to by the large number of uncertainties, such as ground conditions, which are part of these projects.

### 1.1 Cost overrun, time overrun and claims vs. conflict

In construction conflict regularly occurs either when the client feels that he is having to pay more than he expected, or having to wait longer than he expected; or when the contractor feels he is having to carry out extra work that he did not envisage having to do, and for which he may not be paid. In such a case he may raise a claim to recover money or time. These are reflected in cost overrun, time overrun and claims. These three elements will be taken in this study as dependent variables that work as indicators for the degree of conflict.

## 2 Definitions

In the context of this paper four terms need to be clearly defined: conflict, claim, cost overrun and time overrun:

(a) **Conflict** is not getting what you expect, either having to pay more or failure to obtain what you feel you bargained for. Conflict is where the project fails in some way to live up to expectations.

(b) **Claim** has a rather more specific meaning in a construction contract when compared to that in general English. In this paper a claim will be defined as, "a situation where the Contractor, rightly or wrongly, requests an adjustment in the original contract time/cost".

(c) **Cost overrun** is defined as:

$$\frac{\text{Contract final sum}}{\text{Original tender sum}}$$

(d) **Time overrun** is defined as:

$$\frac{\text{Contract actual duration}}{\text{Original contract period}}$$

## 3 Statistical analysis model

Statistics can play a major role in the functions of construction management. By adopting statistical techniques, one can derive forecasting models from historical data that can assist in the decision making process.

In this paper a model that describes the relationships between conflict and a number of independent variables is investigated. The prime objective of the models is to analyze the trend pattern of conflict. Investigation into conflict using historical data and statistical techniques can lead to advantageous results in the form of a practical prediction trends models.

## 4 Method of development

Two main questionnaires were developed to collect the data required for the statistical analysis, one for the contractors and the other for the consultants or clients who may have an in-house design and/or supervision capability. The questionnaires were carefully designed to ensure maximum response.
175 questionnaires were despatched to contractors, with 101, (58%), of those returned found usable. 127 of the consultant's questionnaires were dispatched, with 89, (70%), of those returned being usable for analysis. Considering the well-known low rate of response to questionnaires in the construction industry, the length of the questionnaires, as well as the complexity and confidentiality of the subject addressed by the study, this response indicates the significance of the subject to the construction industry and the effort dedicated to the following up and collection of the data.

## 5 Claims procedure variables

Conflict was considered as being measured by the variables of cost over-run, time over-run, number of claims, value of claims, extension of time etc. Mathematical models were developed that linked the conflict variables with a number of independent variables. The independent variables were related to the Client, Contractor and Consultant.

The mathematical models were developed by statistical techniques using data obtained from questionnaires. Separate questionnaires were developed for collection of data from contractors and clients with 'in house' design capability.

A large number of independent variables were considered for possible correlation with the conflict variables.

Certain independent variables were found to be significantly correlated with conflicts. The degree of significance is shown in Table 1. The relationship between the variables found to be strongly correlated and the conflict variables is shown in Table 2.

Independent variables considered for possible correlation with claims are described in the following sections. The reasons for possible correlation are explained.

### 5.1 Previous experience of contractor

One of pertinent factors that can serve as an indicator of the firm's capabilities is its previous experience. The Contractor's capabilities can have an influence on his performance in a project, particularly his management of claims. Henceforth, the Contractor's track record for similar projects was regarded as one of the variables that could have an effect on the claims in a project.

Table 1. Degree of significance of independent variables

| Independent Variables | Highly Significant | Significant | Insignificant |
|---|---|---|---|
| Previous experience of contractor | | | * |
| Type of contract | | * | |
| Communication (By Engineer) | | * | |
| Coordination (By Contractor) | * | | |
| Decision making on site (Client's involvement) | * | | |
| Type of supervision | * | | |
| Previous experience of supervision | | | * |
| Type of Client | * | | |
| Client's cooperation and attitude in supervision | * | | |
| Standard conditions of contract | * | | |
| Extent of Engineer's authority | * | | |
| Engineer's adherence to the conditions of contract | | * | |
| Claims management: | | | |
| (a) whether claims were an issue of discussion | | * | |
| (b) frequency of claims' status report | | * | |
| (c) steps followed when claim was received | | * | |
| Method of settling claims | | | * |

Table 2. Relationship between strongly correlated and conflict variables

| Factor | Parameters | Cost Overrun 0.5 1.0 1.5 2.0 | Time Overrun 0.5 1.0 1.5 2.0 | Number of Claims | Value of Claims Raised | Time Claimed |
|---|---|---|---|---|---|---|
| Coordination (by contract) | 1 | | | | | |
| | 2 | | | | | |
| | 3 | | | | | |
| | 4 | | | | | |
| Clients involvement in decision making | Yes | | | | | |
| | No | | | | | |
| Type of Supervision | Part of Clients organisation | | | | | |
| | Independent | | | | | |
| | Government Body | | | | | |
| Type of Client | Ministry | | | | | |
| | Financial Organisation | | | | | |
| | Private | | | | | |
| Clients cooperation & attitude in supervision | Yes | | | | | |
| | No | | | | | |
| Standard Conditions of Contract | Conditions of Contract 1 | | | | | |
| | Conditions of Contract 2 | | | | | |
| | Conditions of Contract 3 | | | | | |
| | Conditions of Contract 4 | | | | | |
| | Conditions of Contract 5 | | | | | |
| | Conditions of Contract 6 | | | | | |
| Extent of Engineers Authority | Little | | | | | |
| | Too much | | | | | |
| | None | | | | | |
| | Too Little | | | | | |

## 5.2 Type of contract

The type of contract for a project carries a probable effect on the construction conflicts. This can be concluded from the direct and firm relationship between the type of contract and the following parameters:

(a) risks
(b) methods of dealing with cost overrun or time overrun
(c) incentives

These parameters are now explained in further detail:

(a) Risks: each type of contract is associated with a certain degree of risk. Each party to the contract is allocated a proportion of the risks. The more risks forced upon the Contractor coupled with less chance for him to consider them in his tender, the more likely we should expect an influence on claims.

(b) Methods of dealing with cost overrun or time overrun: the different types of contract handle cost overrun and time overrun in different ways. The more restrictions on changes in the original cost and time of the project coinciding with one or all of the following:

- Exceptional number of requests for variations
- Disruption to progress caused by Client or his agents
- Conflicts and ambiguities in contract documents

the more likely we expect an effect on claims.

(c) Incentives: Different types of contract offer different degrees of incentive to the Contractor. This may affect his attitude towards claims.

## 5.3 Communication (by the Engineer)

Some disputes on projects occur because communication was not maintained. The Engineer and Client should encourage the Contractor to raise any problems he may face. It is through this that the Engineer will be able to identify early potential delays or extra expenses and seek to avoid them. Conflicts can be minimized through a cooperative atmosphere in dealing with Contractors. The variables that are considered to measure the effect of this factor are the extent of ease for the Contractor to communicate his ideas to the Client, the Engineer, and the Engineer's Representative and the mean of these extents, as assessed by the Contractor.

## 5.4 Coordination (by the Contractor)

The Contractor's contractual obligation of coordination may not be limited to coordination with the Engineer. By virtue of his contract with the Client, the Contractor may have to coordinate with other parties such as; the Client's representative, independent supervision office, designer, other Contractors, specialized Consultants and quantity surveyors. This increases the burden of responsibility for the Contractor and this may affect his performance. If the Contractor fails to exercise a skillful management practice in coordination, he may suffer delays, disruption and incur extra expenses. The number and nature of the parties with whom the Contractor has to coordinate as part of his contractual obligations were considered as variables that may generate conflicts.

**5.5 Decision making on site (Client's involvement)**
The decision making process on site can have a significant influence on conflicts' occurrence. One of the critical issues is the existence of more than one decision maker that can issue instructions to the Contractor. In most cases it is the Engineer's Representative who is the sole decision maker on site. Serious difficulties occur when the Contractor is confronted with multi-decision makers. This has happened for example, where the Client decided to appoint a representative on site and failed to define his authority adequately. This lead to contradictions in decisions received by the Contractor which caused disruption. The Contractor has claims as his last resort. The number and nature of the decision makers on site are two further independent variables.

**5.6 Type of supervision**
Communications and understanding of the Client's requirements tend to be more effective where the supervisor is part of the Client's organisation. The independent supervisor is expected to be more productive and efficient in terms of quality of work, time, and cost. The probable effect these issues have on conflicts is examined. In addition, the view of the Contractor towards the role of the Engineer as an impartial party in both types is different.

**5.7 Previous experience of supervision**
The number of previous projects supervised by the office/department can serve as a useful variable for measuring the experience of the office that might be related to the performance on projects of a similar nature to that under investigation. It is also thought that the more projects supervised for the same Client would lead to more successful performance of future projects, particularly in terms of time and cost. Using the same logic, more projects supervised with the same Contractor could mean mutual understanding and well-set communication channels. As a result, this variable may influence conflicts.

**5.8 Type of client**
The data collected for this research was obtained from Kuwait. Construction projects are promoted in Kuwait by Ministry, government bodies, financial organisation or individual investors. Each Client is characterized by a financial capacity, and technical and managerial capabilities. These will be reflected through a number of aspects, inter alia, method of design (in-house or private),

effect of cost and time over-run, quality of contract documents, cooperation and understanding. All these aspects can have an implication on the conflicts.

**5.9 Client's cooperation and attitude in supervision**

The Client can affect claims during construction. His attitude and cooperation towards the supervision team is vital.

**5.10 Standard conditions of contract**

The particular type of conditions of contract may influence the whole contract administration procedure, including conflicts.

The standard conditions of contract in this study are identified by codes: COC1, COC2 etc. The cost over-run varies with the type of conditions of contract. This applies also to the time over-run with COC1 and COC5 having the highest cost overrun and time over-run. The high cost over-run and time over-run for COC1 is associated by the highest number of claims and the value claimed and the second highest time claimed. Therefore, one can conclude the adopting COC1 in a project may lead to more conflicts than any other conditions of contract. In contrast COC4 has the lowest number of claims, value claimed and time claimed and a relatively low cost over-run and time over-run compared to other conditions of contract.

Turning to COC5 and COC6, both seem to follow different trends. COC6 has a lower cost over-run and time over-run than COC5, but has a higher number of claims and value claimed and time claimed.

The value claimed and the time claimed varies considerably between the type of conditions of contract. In contrast the number of claims does not vary considerably.

**5.11 Engineer's authority**

During the construction stage the Engineer has to perform a multitude of duties towards the realization of the project. These duties flow from both the agreement he has with the Client and from the construction contract between the Client and the Contractor. Although the Engineer acts as the agent of the Client, or may be an employee in the Client's organisation, he should not accept instructions by the Client on the attitude he is to adopt in the discharge of his duties under the contract. The Engineer must act independently and impartially and must show no bias towards either party. If the Engineer is not fully authorized to perform any of his duties, his role as an independent party may not be accepted by the Contractor. This may influence the Contractor's attitudes, particularly towards claims and hence conflicts.

**5.12 Engineer's adherence to the conditions of contract**

The conditions of contract say how the Engineer should carry out his responsibilities. Ideally, the Engineer should adhere to these conditions. Deviation from these conditions causes conflict.

To study the effect of this issue on the claims occurrence in a project, five key procedures were identified. These are:

(a) Issue of variation orders.
(b) Approval of work or materials.
(c) Giving decisions regarding claims.
(d) Approval of shop drawings.
(e) Preparation and issue of payment certificates.

The extent to which the Engineer adhered to the conditions of contract when performing them was tested for possible correlation with conflicts.

**5.13 Claims management**

The management of claims may have a strong relationship with the frequency and severity of claims. This includes the notification, submission procedure and processing by the Engineer. The following variables take account of the possible effects of this factor:

(a) Whether claims were an issue of discussion,
(b) Frequency of claims' status report,
(c) Procedure (steps) for processing claims.

The steps considered and their sequence differ from one Engineer to another, and from one project to another. The way the Engineer manages the claims will not just affect the occurrence and value of claims, but it may also affect the method of settlement.

**5.14 Method of settling claims**

Different means exist for resolving construction disputes such as; negotiation, mediation, arbitration and litigation. Arbitration and mediation are increasingly used in the construction industry.
The major elements that differentiate between these four means are:

(a) Legal implications.
(b) Expense.
(c) Time.
(d) Level of technical knowledge.

With arbitration and mediation, disputes are more likely to be resolved within a shorter time than litigation. In addition, disputes are resolved at reasonable cost, (Yanoviak 1987), (Hoellering 1984), and through experts who have more technical knowledge of construction than lawyers. Arbitration, compared to mediation, provides a means of adjudication. A mediator only participates impartially in the negotiations, advising the various parties involved.
It is obvious that negotiation offers the most economical, quickest and highest level of technical knowledge, provided highly qualified staff are employed.

From the above, it was concluded that the choice of the method will affect the following:

(a) The decision of the Contractor to raise a claim.
(b) The value of such claim, since this should include part of his expenses on any method adopted.
(c) The extent to which the Contractor will decide to proceed with his claim.
(d) The value of claim awarded.

Some of the above four variables are part of the dependent variables considered.

## 6 Conclusions

Claims procedures are a major factor in the success of a project. Conflicts in a construction project, as measured by the dependent variables, are shown to be a predictable characteristic. These variables have particular trends affected by the independent variables related to the claims procedure.

The prediction obtained by these relationships are associated with a certain degree of probability to account for the uncertainty in the construction industry.

In addition to the claims procedures, other factors can also have an effect on construction conflicts (Shehadeh 1990).

Some of the variables were found to be only significant or insignificant in terms of their effect on construction conflicts.

## 7 References

Hoellering, M.F. (1984) Construction arbitration and mediation. **Journal of the American Waterworks Association,** 76/2, 34-38.

Shehadeh, N. (1990) **Factors Influencing Construction Claims in Developing Countries, The Case of Kuwait**, Ph.D. Thesis, Leeds University, Department of Civil Engineering, Leeds.

Yanoviak, J.J. (1987) Improving project management for claims reduction. **Proceedings of the 31th Annual Meeting of the American Association of Cost Engineers, Atlanta (U.S.A)**, H.2.1-H.2.5.

## Part Four

Alternative Dispute Resolution

New methods of resolving disputes have appeared in many countries as alternatives to litigation and arbitration. The papers in this section investigates the experiences and the techniques.

'Mediation, the experience in the United States' (Cooper) discusses the process of mediation and how it differs from arbitration, the unique application of mediation to construction disputes is considered.

'Alternative dispute resolution - a far east perspective' (Houghton) considers the ADR techniques employed by the construction industry, the experience in Hong Kong and the pacific rim is outlined.

'Alternative dispute resolution and construction disputes' (Mackie) considers the essential approach of ADR as an attempt to return to more appropriate methods

'The problem of using ADR in the construction industry (Miles) considers the problems of implementation of ADR in the UK, the experience in the USA is considered.'

'Mediation and mini-trial of construction disputes' (Stipanowich and Henderson) provides an empirical study of the role of mini-trial (executive tribunal) and mediation in north american construction disputes.

'The dispute resolution adviser in the construction industry' (Wall) describes the role of the dispute resolution advisor and provides a detailed exposition of the operation of DRA in Hong Kong.

'Whither small value residential dispute settlement in Australia?' (Eilenberg) considers a system which provides for quick but effective dispute resolution of small value claims.

'Peace, love and harmony' (Nicholson) attempts to identify areas of conflict and to offer alternatives of contract methods and procedures.

'The use of mini-trials to resolve construction disputes' (Siedel) considers the place of mini-trials in the taxonomy of alternatives to litigation via: dispute prevention; dispute management and dispute resolution.

'FIDIC study on amicable settlement of construction disputes' (Hollands) provides a progress report on the work of FIDIC's ADR committee, the study considered amicable settlement processes.

# MEDIATION, THE EXPERIENCE IN THE UNITED STATES

CHARLES A. COOPER
Regional Vice President, American Arbitration Association, San Francisco, California.

Abstract
This paper discusses the process of mediation and how it differs from arbitration. It is specifically oriented to the use and development of mediation in the United States. Where appropriate, the paper discusses the unique application of mediation to construction disputes.
Keywords: Mediation, Negotiation, Caucus, Mediator Functions, Personal Qualities, Benefits.

## 1 Introduction

Mediation is a process in which disputants voluntarily choose to select a person or persons to work with them in an effort to resolve the dispute in a mutually acceptable manner by fashioning a mutually acceptable remedy. The mediator has no authority to impose a settlement. Thus mediation does not replace negotiations. It is an extension and enhancement of the negotiations process. A party using a mediator must also actively engage in the normal process of bargaining. The mediator, through techniques discussed below will facilitate the crafting of an agreement unique to the dynamics of the particular dispute, having in mind the bargaining strengths of the parties involved. This is not to say that the mediator will not use issues such as law external to the contractual relationship, or the consideration of the precedents involving similar disputes.

There is a classical model of the process used by mediators working under the rules of the American Arbitration Association, AAA. This process consists of the filing and exchange of pre-mediation papers by the parties. These papers discuss the facts and the law of the dispute as viewed by each party. The mediation itself begins with the expression in a joint session, of the position and interests of each party. Generally the other parties and the mediator are allowed to ask clarifying questions. Often a discussion arises between the parties which helps to educate both parties in the facts of the dispute. The joint session continues until the mediator judges that progress is diminishing. He or she will then move to a period of separate caucuses with each party. In these private meetings, the mediator will ask more direct questions in an effort to understand more fully the nature of the dispute and the underlying interests of each of the participants. It is generally toward the end of the period of caucusing that the mediator is given authorization by the parties to begin carrying offers of settlement between the parties. There may be many rounds of caucusing both before and after this point in the mediation is reached. A successful mediation ends with a written and signed document setting forth all of the necessary points of agreement.

**2 Limitations**

The limitations on mediation stem from the voluntary nature of the process. The mediator by definition has no authority to compel the parties in any way. Unlike an arbitrator the mediator has no authority to render a binding decision. In fact he or she has no power to make any decision affecting the merits of the dispute. The mediator can only listen, advise, suggest, persuade, reason with, question, and in some cases recommend. Perhaps the most dynamic power a mediator possesses is the ability to question a party about the facts of their position or the interpretation of laws affecting the dispute. The mediator has no formal tenure of position. In the most classic sense, a mediator can be replaced by the whim of any of the parties to the dispute. Because mediation is a completely voluntary process, the mediator must be acceptable to all parties to a dispute. He or she must also appear to be impartial and objective. In construction disputes this can be a major factor in selecting the mediator because many problems involve a large number of parties, each one of which must be satisfied of the mediator's honesty and integrity.

**3 Benefits of mediation**

Under most, if not all statutory schemes in America, settlement discussions are privileged, and thus are not subject to disclosure in subsequent legal actions. Parties may thus participate in a mediation, which is a form of settlement discussion without fear that an admission during the course of the mediation will be used against their interest in subsequent litigation.

Mediation also offers a number of incentives to the parties who approach settlement discussions in good faith. A trained mediator will strive to translate the concerns of one party into language which the other party can more readily understand. This greater understanding by both sides promotes the maintenance of relationships. In the construction industry, parties will often have the opportunity to work together in the future. The ability to negotiate voluntary acceptable settlements enhances the ability of parties to contemplate future cooperative ventures.

It is often possible to obtain the services of a mediator with knowledge of the subject matter of the dispute. For example the AAA Center for Mediation has trained architects, engineers, general contractors, sub-contractors and construction attorneys in the skills of mediation. These experts can often be more helpful in particular disputes then someone with only a knowledge of the process of mediation.

Mediation helps parties avoid the negative consequences of disputes. Confrontation, hostile negotiations and lawsuits do not benefit business relationships. By preventing the breakdown of communications, or by repairing communications after a breakdown has occurred, a skilled mediator can create positive communication patterns.

Occasionally it is necessary during the course of negotiations for one party to be able to establish that its position is more appropriate empirically. A skilled mediator with subject matter expertise can be very helpful in convincing one's counterpart that their position is more appropriate. It can be especially helpful if the mediator's professional training is in the same discipline as that of the other party. For instance, the American Arbitration Association, the largest supplier of mediation services in the United States, has many examples in its files demonstrating that, for example, in a dispute between architect and owner, an architect-mediator can more easily facilitate the architect-party's appropriate acceptance of responsibility. Of course this example also applies to contractors, owners, and others in the construction relationship.

There are times when one party has taken a public position which it understands is not correct. Yet that party cannot easily accept a change in position, perhaps because of the very public nature of its original statement. Mediation can be very useful in this situation, providing cover for someone's change of position.

## 4 Optimal prerequisites for mediation

Staff at the American Arbitration Association are often asked if there are any cases which are not subject to resolution through mediation. At one time the author maintained a list of such cases which now, for the most part, I no longer believe is accurate. What is required is a good faith effort to settle. After more than ten years of overseeing construction mediations, we know that a skilled mediator can even help create, in an otherwise recalcitrant party, enough good faith to settle the dispute.

If a party is truly unwilling to settle the dispute on any terms other than its own, than this is probably the only substantive reason to decline to participate in a mediation. An example might be a defendant who is facing a number of potential plaintiffs, and that defendant is more interested in establishing a precedent in order to create a guide for future settlement discussions. Actual examples of this situation in the construction industry are very rare because of the unique relationship which exists in each separate contractual relationship.

### 4.1 Procedural prerequisites

There are several procedural prerequisites which, while not reasons to avoid mediation, might be grounds for delay in initiating the process.

For a mediation to move smoothly and quickly, it is necessary that the issues be appropriately formulated and reasonably defined. If a party truly does not understand what the dispute involves, that party is not ready to bargain, and thus mediation is not yet appropriate. One method which the AAA has developed to test the preparedness of the parties is to require the filing of a short pre-mediation paper covering the facts of the dispute and any legal issues involved. These papers are exchanged between the parties and supplied to the mediator in advance of the mediation. Parties can quickly determine whether the other party has, at a minimum, organized its positions regarding all items of concern to any participating party. A pre-hearing conference or conference call can generally establish whether a party is prepared to begin bargaining. These preliminary steps are particularly important in complex construction disputes which often involve a multitude of issues and claims.

A party's organization is of interest, not only to itself, but to the other parties as well. It is best to engage in mediation only when there is sufficient settlement authority present in the other teams. When negotiating against insurrors, it is necessary to outline in broad terms the level of authority of the insurance company's representative. When negotiating with a public corporation it is best that the representative of the corporation be someone whose recommendation will likely be accepted by the board of directors. This recommendation is equally important when negotiating with governmental bodies whose staff report to a public body.

### 4.2 Inappropriate attitudes

Very rarely a party will engage in mediation for purposes other than to settle the dispute. Occasionally, in the American context, a party will agree to mediate in order to obtain free discovery. Most parties and most mediators believe that this is an inappropriate use of the mediation process. If it is suspected by the other party, they will generally question the good faith of those engaging in such tactics. It will almost certainly delay the process and may lead to a withdrawal from mediation of the party who suspects that their opponent is engaging in such manipulation. Most mediators will terminate a mediation which they believe is being manipulated by a party for reasons other than a good faith effort to settle the dispute.

### 4.3 Definition of good faith

As used in this paper, the term good faith bargaining or good faith negotiations implies a willingness to listen to and reason with the other party or parties in an effort to educate or be educated as to the interests involved, in an effort to resolve the dispute. It does not imply a requirement that a party surrender an appropriately held interest.

## 5 Roles of the mediator

As has been said, the mediator has no authority to impose a solution on the disputants. The role flows from a model of negotiations between the parties and enhances those negotiations. The main task of the mediator is to see to it that the parties achieve a settlement of the dispute. The mediator will therefore play many different roles to ensure the success of the parties.

### 5.1 Interpreter and translator

It is often the case in a construction dispute for the parties to each analyze the problem from their own perspective. Each party has an obvious interest in maximizing their own profits and in escaping as much responsibility for flaws as is possible. There are other barriers to effective communication. Construction projects are cooperative ventures by people and organizations with diverse orientations. The outlook of any participant is often markedly different than that of the other parties to the project. Owners, designers, and construction experts each have different orientations, and most probably, have different motivations regarding their involvement in construction. One role of the mediator is to translate the different motivations and outlooks of each disputant to the other parties. After lengthy discussions in a private caucus, the mediator is able to understand the motivation of a party. The mediator may be in a better position to communicate that motivation or interest than the party himself.

### 5.2 Facilitator

By his very presence the mediator transforms the negotiation process. Once a problem has reached the stage where parties realize that they have a dispute, they often stop talking to each other. The AAA has found that the mediation is often the first time that the parties have talked directly with each other since the dispute arose. By his very presence the mediator thus often facilitates the reestablishment of effective communication between the parties.

### 5.3 Agent of reality

Using the model of mediation promoted by the AAA, the mediator will often, during the caucus, question a party about the viability of a position espoused by that party. In other words, one role of the mediator is to suggest that a party question its position in light of external forces. It would be difficult to overemphasize this role. No dispute can settle until one or both parties begin to question their belief in their own position.

### 5.4 Resource expander

If the parties are availing themselves of the services of a mediator with subject matter expertise, they will be employing someone who can help them expand the number of available options. The mediator can draw upon his or her own construction experience to suggest possible remedies which the parties have not discovered. By using his own expertise, the mediator has expanded the number of options which the parties can use to create their own remedy.

## 6 The functions of the mediator

One other way to view the office of the mediator is to examine it in light of the functions performed by the mediator.

### 6.1 Procedural functions

The mediator performs a number of procedural functions. She calls the parties together by scheduling the date and time of the mediation conference. Note that this role is often performed by an administrative agency. Before the parties reach this point the mediator, or the agency, may have to educate the parties of the value of mediation. Once the mediation has commenced, the mediator is the one who decides that the joint session has reached a point of diminishing returns, and that the period of private caucuses should begin. It is the mediator who determines when, or whether, the parties should come back together for more joint meetings. It is often the mediator who is best able to determine that the parties will benefit form a period of study, away from the negotiation, and it is the mediator who is best able to call for a return to the mediation-negotiation process.

### 6.2 Communication functions

Many people when engaged in a dispute, have a tendency to discuss the problem as they affect themselves, rather than from the viewpoint of the listener. During the joint sessions, the mediator enhances communication because each party is talking to the mediator, in an effort to educate the mediator of their interests and real or potential harms. As this communication proceeds in the presence of the other parties, they are required to listen, perhaps for the first time to a cogent rendition of the dispute as described by their opponent. One other role of the mediator is to facilitate effective communication by serving as the filter through which communication passes during the caucus period. The mediator is able to translate the interests of one party into terms which the other parties can more readily understand.

### 6.3 Substantive functions

While a mediator who is unfamiliar with the subject matter of the dispute may be able to facilitate a settlement, it is the AAA's experience that, particularly respecting construction matters, someone trained in mediation skills who also has subject matter expertise can more easily facilitate a settlement. An expert in construction is able to bring a greater range of possible solutions to the negotiation table than someone who is not familiar with the industry.

## 7 The personal qualities of the mediator

As the director of the AAA's largest mediation program, I am often asked to define the qualities which make a good mediator. Several of these qualities can be drawn from what I have said about the factors which make mediation successful. A candidate must be trained in the mediation process and must believe that facilitated negotiations will almost certainly result in a settlement of the dispute. Many people have inherent mediation skills which they are unable to put to effective use because they have not organized those skills in such a way as to produce results. An important component of the mediation process is the ability of the mediator to listen to what someone is saying in such a way as to make the speaker believe that they are being heard in a sympathetic manner. As stated above it is our unscientific conclusion that a knowledge of the field of the dispute can be very helpful in crafting creative remedies. The mediator must be intelligent, clear thinking and spontaneous in his or her communication. Flexibility is very important. The process of mediation belongs to the parties and the most successful mediators are people who approach the problem with an open mind, without having determined how the successful remedy will be fashioned. Good mediators are patient. The process of negotiation takes time. It often takes time for one party, or for both parties, to realize that a change in position is warranted. Once the realization is made, still more time is often required before the change can be manifested. The mediator must have an understanding of the importance of timing. Timing is obviously important when determining when to schedule the mediation conference and when to break into caucus groups. Equally important is knowing when to make a proposal, when to listen, when to challenge a statement.

While the positive interpersonal communication skills just discussed are important, the ability to confront people is also sometimes necessary. Good commercial mediators must be able to both present the objections of the other party **presented in an acceptable way** and to be able to directly confront a cherished belief or desire held by a party. If a mediator is silent regarding critical feedback the party is denied information which may be necessary to their understanding of the total dispute.

In conclusion, we are looking for individuals who are intelligent, quick thinking, sensitive listeners who are able to challenge others in ways which still allow the mediator to maintain the relationship.

## 8 Overcoming objections to mediation

Most civil disputes in the United States are settled between the parties without benefit of lawyers or the litigation process. The overwhelming majority of civil cases filed in our courts are settled before trial. This fact is the most important argument in overcoming objections to mediation. The question: "why don't you want to settle this case?" is almost certain to produce the answer: "of course we desire to settle the matter, but...." The parties are now in the position where the recalcitrant one can be educated to the benefits of mediation. Set out below are a few of the most common objections to mediation, and examples of successful information which educates the objector and can bring them to the table.

**8.1 The other side is too upset or hostile**

The use of a neutral mediator tends to reduce the hostility of a party because they are able to present their case to someone in apparent authority. In addition, one of the roles of the mediator is to translate the interests of the parties to each other. Therefore, the mediator will help the upset party to understand the legitimate interests of the other side.

**8.2 We do not trust the other side**

The mediation process is both confidential and voluntary. You have control of the outcome. The mediator will hold your position in strict confidence and will only disclose what you specifically authorize him to disclose. If you do not like the way the mediation is progressing, you have the right to withdraw at any time. Most disputes settle. Mediation gives you an extra tool because the mediator, who will be an expert in construction matters, will be challenging both of you regarding your stated interests. If you are honest the mediator will soon know that. If you are not trustworthy the mediator will probably discover that too.

**8.3 The other side is uncooperative**

Mediation is designed to solve just this sort of problem. The mediation process requires them to disclose to the mediator, in confidence, aspects of their case. The mediator will do two thing with that confidential information. She will educate them as to why she should be allowed to share it with you, in order to educate you as to their legitimate interests. She will also confront them, challenge them with both reality as she understands it and she will restate your objections in a way which may be more acceptable to them, if only because some neutral person is presenting the message.

## 9 Conclusion

The use of mediation in the construction industry in the United States is growing because of its proven track record. Statistics at the AAA indicate that eighty-five per cent of construction mediations lead to a mutual agreement. Because mediation can occur as soon as construction executives know enough about the dispute to make rational settlement plans, the savings in transaction costs are very great as compared to litigation. Savings occur in discovery costs, expert witness fees, savings in executive time, and often in attorney fees.

Any negative consequences of mediation, such as dilatory tactics practiced by the other party, will soon be discovered. The voluntary nature of the process allows a party to act as its own policeman, and the mediator will not be interested in participating in a frivolous endeavor.

The AAA and elements of the construction industry are so impressed with the benefits of mediation that they created the Dispute Avoidance and Resolution Task Force, DART, to educate the construction industry about mediation. Participants include owners, insurance companies, general contractors, sub-contractors, architects, and engineers.

# ALTERNATIVE DISPUTE RESOLUTION - A FAR EAST PERSPECTIVE

ANTHONY HOUGHTON
Barrister-at-Law, Hong Kong

Abstract
This paper considers the ADR techniques employed by the construction industry. The experience in Hong Kong is described and some outline of the wider Asiatic is given.

## 1 Introduction

"Alternative Dispute Resolution" has become a fashionable cure for a common disease in the construction industry during the last few years, namely disputes. To be pedantic, the term "alternative dispute resolution", however, is something of a misnomer, since what is in fact being put forward as the alternative is not the type of dispute but rather the method of resolution of the dispute.

Unfortunately for the building industry and fortunately for claims consultants and lawyers, the disputes which usually arise are in fact the same as they have been for many years. Ground conditions consistently fail to live up to expectations, Architects will inevitably be delayed in issuing drawings for various reasons, contractors will run into difficulties and delays caused by their sub-contractors.

In Hong Kong at least, (and I suspect elsewhere), the repetitiveness of these disputes may have much to do with the repeated use of standard forms of contract which make detailed provision for the recovery of loss by the employer in the event of default by the contractor (by way of liquidated damages) but which at the same time leaves claims by contractors for the almost inevitable defaults on behalf of the employer to be dealt with on general principles, as and when they arise. While possible alternatives to this approach are known, they are rarely explored.

I am in no better position than anyone else to resolve the problems which give rise to these regular defaults by both the employer and by the contractor. What I would like to offer however are a few observations as to the ways in which disputes between contractors, employers and sub-contractors have been dealt with previously in my part of the world, Hong Kong and South East Asia, and briefly to provide an overview of the current position.

## 2 Hong Kong

In Hong Kong, construction is a major part of the business environment with an annual gross value of construction work performed in the region of £4.7 billion. Property related stocks account for a significant proportion of the value of quoted shares on the Hong Kong stock market, and Hong Kong developers are active in many other parts of South East Asia. It is impossible to say that Hong Kong businessmen are anything other than hard-nosed in their approach to business, but nevertheless, major conflicts in the construction industry are a relatively recent phenomenon, and have evolved from a restricted base.

One reason for this is that the territory is small. Correspondingly the number of professionals engaged in the construction industry is also small, and news, both good, but especially bad spreads quickly. If a contractor, sub-contractor, or employer is minded to pursue claims to the bitter end then this will quickly become known to the entire industry. The effect of being branded, fairly or unfairly as either a "claims conscious" contractor or a "difficult" employer are obvious.

The benefit that has traditionally accrued therefore from the comparatively small environment in which Hong Kong contractors and developers operate was that there was a positive incentive to reach a compromise.

## 3 The region

This incentive to compromise is in line with old established traditional values in Asian cultures which see compromise as the natural first solution to most contractual disputes. In other parts of Asia compromise has also traditionally been seen as the best solution to a problem. Japanese businesses are well known for the long term view of their business relationships which they take. When a dispute does arise, the claim resulting from it will often be treated to a similar long term approach, Japanese contractors often being willing to sustain a loss on one contract with a view to recouping it on the next. This is of value of course only where the chance of subsequent (profitable) work does exist. Dealt with in this way it will often be unnecessary to fully determine the rights and wrongs of a dispute, and such compromise solutions if they are to be reached at all can usually be reached very quickly. This approach has obvious economic benefits for both parties, once again supplying the incentive element, and by not forcing the claim to a conclusion leaves "face" intact on both sides.

In the wider Asian context it is surprising that disputes on major construction projects have not been more prevalent. The region as a whole is in the course of rapid economic growth and developers have been happy to invest in countries other than their own. Contractors have followed on, and Japanese, Korean, European and American names all appear on the tender lists of major projects, competing with local contractors.

The successful contractor will then face not only the usual complexities of completing a large and difficult building project on time and within budget, but also the possible additional hazards of a developer and sub-contractors schooled in entirely different approaches to construction. Despite the potential pitfalls however the fact remains that such projects comparatively rarely result in either litigation or arbitration.

Building contracts in China, which almost invariably involve at least one Government or quasi-Government party, will usually contain a provision, which seems strange Western lawyers, to the effect that the parties must attempt a settlement between themselves before a matter can be referred to arbitration. As I say, the concept does appear to be strange to the Western eye, where it may be expected that such an attempt will have been made in any event, however, when viewed from the Chinese perspective of compromise as the solution of first choice, then it may be seen that a serious attempt at an agreement should be a precursor to any third party settlement of the dispute.

In many countries throughout the region there are also significant practical reasons behind an approach which relies on the parties themselves to settle their dispute. Quite simply, unlike Hong Kong, and certainly unlike the United Kingdom, the laws, and the mechanisms of a legal system are under-developed. If a compromise cannot be reached, then resolution by other means will be, at best, a very lengthy process.

There is no doubt however that throughout South East Asia, with states such as Hong Kong and Singapore in the forefront, there is a tendency to follow the current western thinking, and to have disputes resolved by third party intervention. Also in line with the western approach, the standard methods, litigation and arbitration, are giving way to some extent to the so called alternatives, namely mediation, conciliation, and mini-trials.

For example mediation clauses are now standard in all Government Building Contracts in Hong Kong, including those relating to the new airport project. In the standard Government contract in Hong Kong, mediation of a dispute is to be adopted only in the event of both parties agreeing. If mediation is considered to be a form of structured comprise of a dispute with the assistance and guidance of a third party then this consensual approach seems necessary. However, under the Airport contracts if one party alone requests mediation, then the other party is obliged to take part in the process. If the mediator is unable to obtain an agreement between the parties, then he is to continue to act as an adjudicator, and to make a decision which will be binding on the parties for the duration of the remainder of the contract. Such a decision of an adjudicator can only be overturned by arbitration.

One approach which has been adopted on one or two large contracts with which I have been involved, has been the adoption of a named adjudicator to deal with disputes between the contractor and the architect arising during the currency of the contract. I have to say that on long projects whatever faith the contractor may have had in the adjudication process at the commencement of the contract has usually evaporated by its conclusion.

As formal processes these are without doubt relatively new and unusual concepts in the region however.

Hong Kong, Singapore and Malaysia have full staffed Arbitration Centres able to advise parties on how to instigate or conduct an arbitration or any of the other forms of ADR. Indonesia and other countries have their Chambers of Commerce which provide similar services, publishing their own rules for the conduct of arbitration. These centres are able usually either to provide names of qualified and suitable arbitrators, or alternatively to actually retain individuals to conduct the arbitration. It is of course the intention, and sometimes even the result, that these individuals will be able to deal with a dispute under the relevant system quickly and fairly. Unfortunately, speed is not helped by the fact that once it has become necessary for third party intervention to help resolve a dispute, it is usual that third party will be that helps resolve the dispute!

## 4 Summary

In summary there are I think lessons which may be learned from the Asian experience, and which may from time to time be forgotten in the usual run of disputes.

In terms of a long term business relationship a settlement between the parties themselves is probably more satisfactory than success in arbitration litigation, or even via mediation. It will almost certainly be cheaper in those disputes whose there is some element of right and wrong on each side. If the intervention of a third party is necessary however, it will be clear that the current "alternative" forms of dispute resolution are perhaps a closer approach to this eastern ideal than are the "traditional" forms of dispute resolution. Only the parties to a dispute are able to evaluate the success of a particular compromise in terms of dollars and cents however, and where it is felt that the rights are all on one side, then it may be that the traditional methods are the ones to be preferred.

# ALTERNATIVE DISPUTE RESOLUTION AND CONSTRUCTION DISPUTES

KARL J MACKIE
Centre for Dispute Resolution (CEDR), London, England

Abstract
This paper considers the essential approach of ADR as an attempt to return to more appropriate methods of resolving disputes with a number of benefits compared to existing methods. It explains that ADR has had to challenge existing culture and practices, but that change has been made possible with the development of CEDR and others campaigning for ADR use. A growing track record of ADR successes and contractual approaches confirm the success of the new methods. Finally the author suggests that ADR can also be perceived as part of the search for management of quality by way of reducing the dispute cycle and by preventing disputes from emerging or from escalating once they do emerge.
Keywords: ADR, Dispute Resolution, CEDR, Culture, Quality, Dispute Prevention.

## 1 Development of ADR

Why was ADR needed? A simple story demostrates the point. Once upon a time a mother came home to find her two young daughters arguing over who should have an orange. "I want it" one daughter was yelling. "No,it's mine" cried the other one, "you had an orange yesterday and you're always so greedy anyway." And so the bickering escalates and the relationship deteriorates.

The mother could act in the role of professional adviser or consultant to one of the parties and explain how best to persuade the other daughter to hand over the orange or how to use power to get it. Or she could act as a judge or arbitrator and decide who deserves the orange most. She could split the orange in half in an attempt to be fair, with costs on both sides. Or she could go for the perfect litigation Lose-Lose outcome and refuse them both a piece for bad behaviour and share the orange with professional advisers instead.

In this case, the mother did none of these things. Rather she turned to one daughter and asked why she wanted the orange. "To eat it of course" came the reply. She asked the other daughter the same question. "I need to bake a cake for school tomorrow and the recipe says I have to use orange peel".

You can guess the terms of the settlement of this dispute. The story illustrates on the one hand how adversarial negotiations can develop into impasse through assumptions and misunderstanding with consequent poor communication and deteriorating relationships - all adding to an increased sense of grievance and self-justification that further impede problem-solving. On the other hand many adjudications may miss the point, may feel to meet the real concerns of the parties and may ultimately divide the parties even further after one or both loses. Resorting to professional advice in order to substantiate one's best case and undermine the other's may assist but it also will often only succeed in helping one or more parties to construct an expensive and time-consuming edifice that is built on sand and is most successful only in obscuring the light for all the parties involved.

Of course many disputes are built on genuine grievances or are cases where the parties really do lay claim to the same orange. However even in those there is no effective mechanism that bridges the gap between delicate negotiating (whether directly between the parties or between their professional advisers) and the stage of arbitration or litigation.

## 2 Benefits of ADR

It was to alter this situation, to fill this gap that Alternative Dispute Resolution - ADR - came on the scene. ADR stresses that parties can add value to difficult negotiations by bringing in a skilled third party - not to make a binding legal award, but to help the parties make their negotiations more productive. By doing so, ADR also achieves a number of benefits that tend to be lost in the context of adjudication proceedings and which are frequently referred to in the ADR literature (Mackie, 1991a; 1991b):-

* better communications
* continued business relationships
* active management of the dispute
* more options for settlement
* speed
* reduced costs in achieving settlement
* confidentiality
* control of the outcome and the process.

ADR achieves these benefits not because of any magic but because (a) third party intervention alters the dynamics of any dispute or negotiation, and (b) skilled third party intervention ensures that the parties begin to communicate effectively, begin to focus on problem-solving and settlement, not on acrimonious debate and point-scoring.

## 3 ADR techniques

In the main ADR approach, mediation, the mediator takes a high-profile, active part in negotiations - meeting parties jointly and separately until a binding agreement between the parties is reached or until the mediator or the parties decide that they are not going to reach agreement. If agreement is reached, signature to a document setting out the terms ensures a binding written contract which the parties can enforce. If no agreement is reached, the parties will at least go away with a clearer idea of what the dispute is about and what will be needed to settle it in the future. The evidence is that most mediations conducted by skilled neutrals are successful in achieving agreement.

It is of course vital that mediators are skilled, with sufficient training and experience to understand the process and to be able to act as a neutral additional negotiator working in the interests of all parties in the dispute. Mediation is not a soft option - the parties are usually in an impasse before they feel ready to call in a third party. However the fact of previous negotiations often helps ensure that mediation can be a quick process - most mediations are completed in a day or a few days.

Mediation (sometimes called conciliation - see Mackie 1991a for a discussion of the difference) is the technique closest to negotiation in filling the gap between negotiation and adjudication. Other ADR techniques can be developed as appropriate to the needs of the dispute and the parties - indeed the 'A' in ADR is often best thought of as standing for 'Appropriate' more than 'Alternative'. Particularly since ADR can be conducted alongside preparation for trial or arbitration. The parties are free to return to those options if ADR fails. ADR offers a 'window of opportunity' for parties to seek a more amicable and effective resolution at a stage earlier than judgment or award. Other techniques which are used as appropriate move closer than mediation to the adjudication end of the spectrum of dispute resolution methods. For example the executive tribunal or mini-trial, where senior executives hear a presentation of each of their company's cases before sitting down to negotiate (with the assistance of a third party neutral); expert opinion or appraisal where parties agree to be advised or to be bound on a restricted technical aspect of the dispute; or fast-track adjudication , where the parties agree to have an advisory award in their dispute or an award on which they agree to act until the completion of a contract or until one party overturns the award in subsequent formal arbitration or litigation.

## 4 Achieving change in the management of disputes

ADR can be seen as a return to commonsense approaches to dispute resolution rather than the highly developed formalistic options offered by the advance of professional arbitration and litigation proceedings. Many informal examples of ADR can be found in past practice and indeed have sometimes been associated with the engineer, surveyor or architect role. (Changing employment relations have been sometimes seen to undermine the 'neutrality' of those roles and therefore the capacity of holders of the role to act as independent third parties where they are employed in the project involved.) However many of those who have participated in or observed the management of conflict and dispute in the construction industry (and not only that industry), appreciated that more was needed to 'shake up' existing atitudes, assumptions and to alter current practice. It is never easy to call in an outside source of assistance and there are always questions of who makes the suggestion first, how to agree who should be called in, and their role in relation to existing advisers and proceedings.

What was needed was a powerful campaign to achieve a change in the 'mind-set' or dispute culture of the industry, alongside an effective and straightforward mechanism by which skilled neutral ADR services could be offered. It was for those reasons that the Centre for Dispute Resolution was launched in November 1990 at the Confederation of British Industry. CEDR's objectives are to research and promote the use of ADR ,and to ease that use by providing model ADR contract clauses and dispute services via references to trained neutrals. Our activities have also supported or been assisted by a range of other organisations which have recognised the value of ADR.

I believe the activities of CEDR and other supporters of ADR has already begun to create some Important achievements. Apart from a growing use of ADR contract clauses, and a growing track record of successful ADR references in disputes of all shapes and sizes, the promotion of ADR also helps the construction industry question the wisdom of adversarial approaches to negotiation and dispute. The use of ADR clauses and processes of aids dispute prevention since disputes are resolved at an earlier stage before they escalate out of the parties's control. Just as a commitment to quality in industry can reduce the construction or 'manufacturing cycle', so can ADR commitment reduce the dispute cycle in business relationships to feed back into overall quality of performance, not least more constructive relationships in an industry which has always been renowned for its propensity to disputes and claims. That is an achievement worth striving for.

## 5 References


Mackie, K.J. (1991a) **A Handbook of Dispute Resolution**, Routledge/Sweet and Maxwell, London, England.

Mackie, K.J. (1991b) **The ADR Route Map**, Centre for Dispute Resolution, London, England.

Wilkinson, J.H.(Ed)(1990) Donovan Leisure Newton & Irvine **ADR Practice Book**, John Wiley:New York.

# THE PROBLEMS OF USING ADR IN THE CONSTRUCTION INDUSTRY

DAVID R MILES
Glovers, London, England

Abstract
This paper considers the problems of implementation of ADR in the UK construction industry. The experience in the USA is considered. Attitudes towards ADR, notably by the UK legal profession are discussed.
Keywords: ADR, ADR in US, Implementation, Construction, Sub-Contractor, Main Contractor, Employer, Insurers, Institutions.

## 1 Introduction

Despite a concerted effort by the organisations involved, it is still fair to say that throughout commerce generally and the construction industry in particular, there is still a general lack of awareness of what ADR is and what it seeks to achieve. Even in the legal profession, a number of eminent lawyers consider ADR to be some form of arbitration, involving a traditional adversarial approach and an adjudicatory award of some nature.

Perhaps the lack of awareness is not surprising. Even in the United States, where ADR methods have now been practised for nearly 15 years, there is still a lack of awareness. Bearing in mind that general lobbying for the ADR movement only began in this country some three or so years ago and that, for example, CEDR was only launched under 2 years ago, perhaps this is not surprising.

It is, of course, up to the ADR providers to spread the word. This has been done by CEDR through a number of regional launches and seminars, through the industry sector working groups of CEDR, through arrangements for affiliation with professional bodies and through such events as the Chartered Institute of Arbitrators ADR Road Show.

## 2 Implementation

Once there is an awareness and an understanding of the process, it is easier to encourage parties to use the procedures. Probably the first hurdle that has to be overcome in the construction industry is the traditional "macho" approach to resolving disputes. Whilst the majority of disputes still are resolved through commercial negotiations, there is an incorrect perception that the suggestion of ADR could be taken as a sign of weakness. That perception ought to disappear with a better understanding of the process and with the establishment of a track record of successful settlement in the construction field. Further, of course, the need for ADR processes to be suggested at all can be overcome by the inclusion of ADR clauses in standard forms of construction contracts (see below)

ADR has considerable advantages for the construction industry, especially if a dispute occurs early on as it is vitally important for harmonious relationships to be maintained between the parties if at all possible. In most construction contracts the contract period is of a significant length and in the major construction projects can go on for many years. Major dam projects, the Channel Tunnel and bridge contracts are supreme examples. If the parties fall out with each other early on, it really is not satisfactory from either parties' point of view for a contractor either to walk off site or to be kicked off. The parties have to learn to live with each other and co-operate with each other to see the project through to a successful completion. The existence of a dispute which may not be resolved until many years after completion of the project will not help relationships on site. If the dispute can be nipped in the bud early on there is less scope for parties to take up entrenched positions, the retreat from which may be perceived as being a sign of weakness or loss of face. If the parties can establish a precedent of settling disputes, hopefully this can only enhance the relationship on site so that a constructive atmosphere of the parties working together to achieve completion can be established rather than what is often the traditional acrimonious and adversarial atmosphere on site. Admittedly the British form of construction contract and approach does not help matters - but that is a topic for discussion on another day.

Whilst there are always points of law to be argued in construction cases, more often than not the issues between the parties narrow down to those of a technical nature, be they architectural, engineering or quantity surveying. The reinvolvement of the businessman in the ADR process, therefore, brings back into the arena those most naturally suited and trained to argue matters of a technical and commercial nature.

The pressures in today's construction industry, the drive fast track building, coupled with the endless arguments over collateral and sub-contract warranties, mean that inevitably some contracts are signed long after the workmen enter the site. Sometimes no contract is signed at all. A contractor, and sometimes an employer, can find himself in a situation where the forum for his dispute is a mixture of litigation and arbitration. ADR enables the parties to come together in a common arena regardless of the requirements of the contract.

Whether the economy is booming or in recession, continuing relations between the parties are still important. Whilst it may be thought that employers only rarely find themselves signing up building contracts, there are many government bodies, institutions, investment funds, developers and large retail groups that regularly enter into construction contracts. Such organisations prefer to deal with contractors whose method of operation they know. Similarly, contractors prefer to work for employers they know and trust. ADR works hard at preserving the business relations and often can produce a settlement involving a future contractual relationship from which both parties will benefit.

Anyone who has had experience of a construction dispute that has involved the formalities of instructing solicitors and barristers, be it in litigation or arbitration, will be aware that the costs of even a simple dispute can escalate rapidly. A glance round any Court or arbitration room easily illustrates the costs clocking up. Whilst in Court both the Judge and the room are paid for by the Crown, in an arbitration the Arbitrator is on an hourly or daily rate. There will be one barrister (at least) and one solicitor (at least) on both sides. Inevitably there is also at least one expert. Add to that the cost of management time and lost earning opportunities by the attendant witnesses together with the cost of the room and very quickly it can be seen that the overall cost of running even the simplest of construction cases can easily run to between £1-2,000 per hour. Expenses on such a scale soon call into question the cost effectiveness of disputes of a value below £50-100,000.

The engineering industry in particular is a component manufacturing industry. Many components go into the end product. The components may of themselves be of low value, but of great importance in the functioning of the end product. Such disputes therefore often involve significant factual disputes which need the assistance of experts. The cost of sorting out technical disputes is therefore sometimes far greater than the value of the end product. Further, because the industry is component based, ongoing business relations are very important.

Finally, apart from the multi party aspects, there is often a multi-national dimension. This can lead to further complicated and technical arguments between the parties as to the proper forum and jurisdiction. The application of negotiation and mediation techniques can sometimes overcome these impasses.

## 3 The parties

### 3.1 The sub-contractor

The common perception is that it is the sub-contractor who is likely to be an enthusiast of ADR procedures. This is often because he perceives that it is a quicker way to his money, as indeed it may well be. The construction industry has for many months been full of reports about the trend of increasingly lengthy payments, such that, not only in the construction industry, the Government has been lobbied to introduce legislation encouraging prompt payment. Further, the early ADR trade association schemes have all largely been concerned with specialist sub-contractor associations.

The advantage to a sub-contractor in embarking upon ADR is that this may give him a route to the ultimate payer, the employer, in circumstances where no direct route exists, as his claim can only be brought in contract against the main contractor. Often an employer will make deductions against a main contractor for reasons which the main contractor then seeks to pass on to the sub-contractor. The unfortunate sub-contractor, being kept out of his money, has no means of persuading the employer to pay. It may even be that the main contractor has the benefit of a "pay when paid" clause in his contract with the sub-contractor.

The sub-contractor may not necessarily always be in favour of ADR. He may feel he has an open and shut case. A straight forward entitlement to a set sum of money. He will perceive ADR as a pressure upon him to make concessions and to compromise. However, it should be remembered that even where there is an indisputable sum due, there may still be advantages in negotiation. There is no point in obtaining judgment against a party for a sum if that party effectively is likely to go into liquidation if judgment is executed. In such circumstances there is still scope for negotiating payments by instalments thereby easing the pressure upon the payee's cash flow.

**3.2 The main contractor**

The main contractor's principal resistance to ADR is that he perceives that the procedures will merely result in his having to make an earlier payment to his sub-contractor than he would otherwise have to by forcing the sub-contractor to go through the costs, delays and expenses of arbitration or litigation. Rather like the insurance company, if the main contractor holds out for long enough there is always the hope that either the sub-contractor will go into liquidation or accept a vastly reduced sum later down the litigation or arbitration track merely to overcome cash flow problems. If the litigation or arbitration process goes on for some time, even the prospect of having to bear the costs may not be so dreadful as a "fighting fund" may have been built up upon the interest earned on the sums in dispute.

Of course, the main contractor will see the benefits of ADR so far as any dispute he has against an employer. The benefits of an early private resolution of a dispute will also have an attraction to a contractor who does not wish to gain the reputation of being claims orientated.

**3.3 The employer**

Once again, although initially the prospect of an early payment may not seem attractive to an employer, no defendant or respondent can afford merely to forget about a claim secure in the knowledge that the day of judgment is some day off. Provision and sometimes qualification may still need to be made in the accounts. Whilst every effort is made to be as accurate as possible, with the uncertainties of litigation or arbitration, no professional advisor can ever be 100% accurate. Such qualifications or provisions can have a material bearing on how outsiders view the position of the company. A number of bodies regularly involved in the construction process now feel that the litigation and arbitration costs have rendered such an approach uneconomical.

Employers, more than most bodies in the construction industry, have the opportunity to introduce ADR. This is because may standard forms now contain such clauses. The 6th Edition of the ICE Conditions contain a conciliation clause, the FIDIC Form requires the parties to attempt amicable settlement. The South African Institution of Civil Engineers has introduced a mediation procedure and in Hong Kong the Government has introduced a set of mediation rules for use in its Government construction contracts. In the US, at the end of 1991, the standard dispute clause in federal procurement contracts allowed contractors and the Government, by mutual consent to agree to use ADR. Even if the standard form does not contain an ADR clause, there is nothing to stop employers introducing one and incorporating arbitration rules which allow for conciliation procedures such as those of the ICC and UNCITRAL.

### 3.4 Insurers

Insurers have traditionally been resistant to early settlement. However, it is significant that in the US, together with in-house Counsel, insurers have been amongst the greatest backers of the ADR movement. Recently, they have produced statistics showing that of sums paid out in settlement, 60p in the £ goes towards the costs of fighting the case. Insurers will, of course, inevitably be involved behind the professionals indemnity policy. Therefore, there is a further attraction to ADR because of the confidential aspects of the procedure. Few professional advisors want their mistakes aired in Court.

### 3.5 End users

A significant construction development can often be faced with protest groups. As those involved in the business began to appreciate the wide range of application of ADR processes, it was soon realised that there was ample scope for involving neighbourhood and environmental interests in the consultation process using ADR methods. ADR has thus overcome the need to lie down in front of the bulldozer.

## 4 The institutions

How have the institutions reacted to the ADR movement? On the whole they have acted favourably and, in some instances, with considerable speed. The Chartered Institute of Arbitrators has produced its own guidelines and rules for mediation and conciliation. It proposes the introduction of the teaching of ADR procedures onto the arbitration courses. Whilst initially not training mediators itself, it is proposed to appoint mediators who have been trained by recognised bodies.

### 4.1 The Law Society

The Law Society produced a report upon ADR in July of 1991. Latterly it has proposed the introduction of ADR teaching as one of its priority areas for a period of twelve months beginning on 1st November 1992.

### 4.2 The British Academy of Experts

The BAE has for some time seen the use of ADR methods as a logical addition to the function of an expert and the efficient resolution of disputes. In addition to training mediators, it will, if approached, make appointments.

## 5 The courts

It is known that some senior members of the judiciary, amongst them the Lord Chancellor, feel that there is a role and place for ADR. Not all members of the judicial establishment are convinced. The Master of the Rolls, in a speech to the London Common Law and Commercial Bar Association delivered in June 1991 stated:

> "ADR is a PR man's dream. It conjures up visions of a factor 'X' which will do for dispute resolution what it is said to have done for washing powders and petrol. The truth is that there is no factor 'X'. Indeed, I rather doubt whether there is any such thing as ADR. It is simply an umbrella term or 'buzz word' covering any new procedure or modification of old procedures which anyone is able to think up."

Notwithstanding his scepticism, the Master of the Rolls ended by suggesting an equivalent of civil JPs to compliment the traditional approach to criminal disputes. Whether individuals will feel it their public duty to assist in disputes between two commercial bodies to the extent that they do in respect of the Magistrates Court, remains to be seen.

Notwithstanding these remarks, a number of proposals have been put for Court annexed schemes. In September 1990 the London Common Law and Commercial Bar Association put forward a proposal for a pilot scheme in the County Court. Although the scheme was not adopted by the Lord Chancellor's Department, various discussions took place with County Court Judges in London in the hope that it would be possible to implement a scheme on a trial basis at little or no cost. The scheme envisaged mediation by lawyers of some experience.

In June 1991 Philip Naughton QC (who had been involved in the preparation of the LCLCBA report) delivered a paper to the Official Referees Users Committee suggesting a pilot scheme similar to that proposed for the County Court.

An interim report by the Lord Chancellor's working group on ADR in July 1991 also considered Court annexed ADR. It recommended further study into the perceived benefits of ADR and whether the use of voluntary and commercial ADR services could be encouraged by introducing a recognition of them in Court proceedings.

In October 1991 the Beldham report was produced under the chairmanship of the Right Honourable Lord Justice Beldham commissioned by the General Council of the Bar. Amongst the distinguished members of the committee were Anthony Scrivener QC, Philip Naughton QC and Chris Chandler of the Law Society. The report suggested the introduction of a scheme along the lines of the London Common Law and Commercial Bar Association proposal.

## 6 The lawyers

In any process which seeks to be an alternative to the traditional methods of dispute resolution, be they litigation or arbitration, it is inevitable that consultation with lawyers will be necessary. In a high value claim it is unlikely that any company executive will take the decision to use ADR of his own initiative without prior consultation with his advisors.

It follows that the role of lawyers in the ADR process is therefore crucial. It is probably even more important to educate lawyers and ensure that they understand the process if there is to be any prospect of them either approving or indeed recommending ADR to their clients. Some ADR organisations are purely lawyer based, such as the commercial organisation IDR which recruits as its members and trains as mediators only solicitors. CEDR takes a different view and seeks to be an organisation for commerce led by industry. However, over 50% of CEDR's members are lawyers. Further, it is probably fair to say that a similar proportion have trained as mediators. Barristers chambers have joined the organisation and the London Common Law and Commercial Bar Association has drawn up a special training programme on ADR. In addition, the Official Referees Solicitors' Association is producing a list of approved mediators. The MSc course on construction law and management run by Kings College includes three hours tuition upon ADR.

Finally, it is not only those lawyers involved directly in litigation and arbitration that should be concerned with ADR. The starting place for consideration of ADR is when the contract is negotiated. It is therefore of importance that the non-contentious lawyers appreciate ADR methods and encourage the parties to insert a suitable ADR clause. It makes sense that when the parties are negotiating and are essentially in a co-operative frame of mind, a clause suggesting that any dispute be resolved through settlement negotiations in good faith will meet little or no resistance.

Are lawyers prepared to back ADR as a concept; after all a substantial piece of construction litigation can involve steady work over a period of two to three years followed by intense activity (and therefore the opportunity for substantial fees) during the course of the case. Unfortunately, there will still be lawyers who perceive privately, if not publicly to their partners, that there is more money in it for them if the case is to fight rather than settle. However, as is evidenced by the 50% membership of CEDR, there are some lawyers who are questioning the effectiveness of our traditional means of dispute resolution.

Whilst undoubtedly there will always be a need for the Courts and arbitration and the procedures offered by them, such as the need for injunctive relief, declarations, precedents and publicity, it must also be remembered that something in the region of 80-90% of cases settle - a large proportion of which do so at the steps of the Court. Does it not make sense if settlement is highly likely, to do so in such circumstances, sooner rather than later.

Even in instances where a case runs the full breadth of the legal process from the Official Referees Courts to the House of Lords, by the time it reaches that distinguished Court the costs involved often have far outstripped the sums at stake. D&F Estates concerned a piece of defective plastering, Donaghue -v- Stephenson a decomposed snail, yet they both ended up in the House of Lords.

Ultimately lawyers need satisfied clients if they are to obtain repeat business. A client is not readily going to subject himself to the litigation process if he has had bitter experience of going through the mill and losing or possibly even winning and having to pay substantially for the privilege. In my view, if there is room for improvement in the system, then it is incumbent upon the lawyers to put their creative talents towards exploring the options and possibilities. ADR is one such option.

It has been suggested that lawyers support of ADR is merely a matter of "defensive marketing". There are two rules to defensive marketing:

**Rule 1**:The Strategy - keep the old products (and the fees that go with them) while manipulating the image to suggest the new service is on offer.

**Rule 2**: The Tactic - suggest to the client that nearly every case is suitable for ADR except the one he has in front of him where unique circumstances apply requiring "that old adversarial magic".

Recent research conducted for the Legal Skills Research Group has found that failure to advise effectively on ADR emerged as the least satisfactory area of performance amongst outside lawyers. Outside lawyers were also rated poorly in terms of their understanding of the real costs to a company of litigation in terms of management time lost and other factors.

## 7 Conclusion

Slowly the perception of the advantages of ADR is changing. With the introduction of ADR clauses into standard forms of construction contract, the need for the suggestion of "ad hoc" ADR will diminish - thus overcoming the fear of the "who blinks first" attitude to the suggestion of settlement discussions. In the early days there is a need for the maximum publicity to ADR to the successful use of the process in construction cases of all types in order to convince the sceptics. "Actions speak louder than words".

# MEDIATION AND MINI-TRIAL OF CONSTRUCTION DISPUTES

THOMAS J. STIPANOWICH and DOUGLAS A. HENDERSON
University of Kentucky College of Law, Lexington, Kentucky, U.S.A.

**Abstract**
Drawing on more than 500 responses to the 1990 American Bar Association Forum on the Construction Industry Survey, this paper investigates the role of mediation and mini-trial in construction disputes. Addressed in the paper are specific attitudes and experiences of construction lawyers using these techniques. The paper represents one of the first empirical investigations of non-binding dispute resolution in the construction field.
Keywords: Mediation, Mini-trial, Summary Jury Trial, Survey Analysis, Construction Disputes.

## 1 Introduction

The modern construction attorney is confronted with a bewildering array of tools and techniques for resolving disputes outside the courtroom. The traditional industry stalwart, binding arbitration, is gradually giving way to a whole host of alternatives for addressing claims and controversies--including partnering, dispute review boards, mediation, and mini-trial.

For too long, the unacknowledged main source of information regarding dispute resolution alternatives has been anecdote and hearsay. Attorneys seeking to select or implement such techniques have found it necessary to rely on the often limited experience of others. The result has been that an alternative may be misapplied, or ignored in favor of a less appropriate method. For example, one popular notion is that proposing to mediate a dispute reflects lack of confidence in the strength of one's case. Another is that mediation and other dispute resolution alternatives do detriment by revealing trial strategies. Neither of these propositions, however, has ever been tested empirically. Few researchers confront these issues, despite their practical relevance, and few writers attempt to ground their conclusions in empirical observation.

This paper addresses these issues, by examining the 1991 American Bar Association (ABA) Forum on the Construction Industry survey on non-binding dispute resolution. Increasingly, construction lawyers seem to be exploring a range of ADR processes aimed at settling controversies without the courts and without binding arbitration. Results from the Forum survey, summarized in detail here, provide a current picture of these key professional perspectives on ADR and furnish a broad base of data from and for the field. The 1990-91 survey, extending an earlier ABA survey focusing exclusively on construction arbitration, canvassed the construction bar's attitudes and experiences with mediation, mini-trial, and, to a lesser extent, summary jury trial, non-binding arbitration, and other ADR processes. From this research, practitioners and academics alike should be able to identify, based on something more than intuition, significant guideposts for designing and implementing settlement-oriented procedures.

## 2 About the Survey

In 1985-86 the Forum on the Construction Industry and the ABA Litigation Section co-sponsored a survey on arbitration under the American Arbitration Association (AAA) Construction Industry Arbitration Rules. The survey produced a wealth of valuable information regarding construction arbitration, received international attention, and played a significant role in revision of AAA procedures.[1]

The success of the arbitration survey inspired the 1990-91 survey of Forum members regarding mediation, mini-trial, and other settlement-oriented procedures used in construction disputes. Forum sponsorship was premised on the notion that many practitioners currently lack sufficient information or experience to make knowledgeable decisions regarding the use of such procedures. The survey, developed with input from representatives of the Forum, the AAA, and the Center for Public Resources, was intended to supplant anecdote and hearsay with data reflecting members' collective attitudes and experiences in the field.[2]

---

1. For a detailed discussion of the construction arbitration survey, see Thomas J. Stipanowich, Rethinking American Arbitration, 63 Ind. L. J. 425 (1988).
2. The survey instrument was developed at the University of Kentucky with assistance from the Survey Research Center and Computing Center.

The questionnaire contained two sections. The first sought information regarding the respondent's perceptions of a range of dispute resolution processes. The second section collected information concerning the respondent's actual experiences with a range of dispute resolution techniques. Special emphasis was placed on mediation,[3] defined in the survey as "a private, informal process in which disputants are assisted by one or more neutral third parties in their efforts toward settlement"; and mini-trial,[4] defined as "a private process in which counsel for the opposing parties present their cases in condensed form in the presence of designated representatives for each side who have authority to settle the dispute . . . [and, usually,] an impartial third party advisor."[5] Selected questions addressed the use of summary jury trial[6] and non-binding arbitration.[7] The questionnaire was distributed to approximately 5,400

---

3. For analysis of mediation research generally, see Kenneth Kressel, Dean G. Pruitt and Associates, Mediation Research: The Process and Effectiveness of Third-Party Interventions (1989) and James A. Wall, Jr., Mediation: An Analysis, Review, and Proposed Research, 25 J. Conflict Resol. 157 (1981).

4. The classic description is given by Eric Green, Jonathan Marks and Ronald Olson, Settling Large Case Litigation: An Alternative Approach, 11 Loy. L.A. L. Rev. 493 (1978). Another comprehensive source is provided by James Davis and Lynne Omlie, Mini-Trials: The Courtroom in the Boardroom, 21 Willamette L. Rev. 531 (1985). In addition, Ronald Olson, Dispute Resolution: An Alternative For Large Case Litigation, 6 Litigation 22, 59 (1980), and Lester Edelman and Frank Carr, The Mini-Trial: An Alternative Dispute Resolution Procedure, 42 Arb. J. 7 (1987) provide useful summaries.

The most detailed evaluation of the mini-trial is presented in ABA Sub-Committee on Alternative Means of Dispute Resolution, Committee on Corporate Counsel, The Effectiveness of the Mini-Trial in Resolving Complex Commercial Disputes: A Survey (1986), which qualitatively reviews 19 mini-trial experiences.

5. "After the presentation, the parties' representatives meet to discuss settlement prospects. At some point, the third-party advisor may offer certain non-binding conclusions regarding the probable adjudicated outcome of the case and may assist in negotiations."

6. As explained in the survey, "Summary jury trial is similar in concept to mini-trial, but involves condensed presentations before a jury which draws nonbinding conclusions regarding issues in dispute. It is utilized by some courts as a means of facilitating pretrial settlement of legal actions."

7. The survey states: "Non-binding arbitration, like mini-trial and summary jury trial, usually involves condensed case presentations before one or more third persons who draw non-binding conclusions regarding issues in dispute.

Forum members in late 1990; 552 completed surveys were ultimately returned and coded for analysis.[8]

## 3 Perceptions and Processes

### 3.1 General Experience with Various Processes

#### 3.1.1 Binding arbitration

According to Forum participants, binding arbitration remains by far the most widely used form of ADR in construction cases. The great majority of the lawyers completing the survey (81.5%) had participated in at least one binding arbitration; more than half (55.0%) had arbitrated five or more times, and one quarter (25.2%) had 10 or more arbitration experiences.[9] Approximately three quarters (72.4%) of the respondents had arbitrated during the previous two years.

#### 3.1.2 Mediation and mini-trial

Although the survey revealed less breadth and depth of experience with mediation, responses indicated such procedures are being extensively employed. Nearly two-thirds (64.2%) of the respondents had participated in at least one mediation, and most of those (58.3% of those responding) had done so in the last two years. More than 11 percent had mediated ten or more times. About one in five responding attorneys (21.1%) had participated in a mini-trial.

#### 3.1.3 Other non-binding processes

Other forms of dispute resolution were resorted to less frequently by respondents. Analysis of the results showed that less than a third (29.6%) had experienced non-binding arbitration. About one in ten (9.6%) had been involved in a summary jury trials. Relatively few attorneys had multiple experiences with such procedures.

---

Typically a court-annexed procedure, it is also aimed at facilitating settlement of disputes."

8. A limited budget prevented in-depth follow-up of the initial written response. Nevertheless, the sample is still large enough to provide useful observations and general conclusions on the use and abuse of dispute resolution techniques. Considering the length of the questionnaire, the collective response of attorneys represents an impressive aggregate of nonbillable time.

9. Although this is less than the 90 percent experience rate reflected in the earlier arbitration survey, it is still impressive.

### 3.1.4 Familiarity with processes

Not surprisingly, the construction bar's level of familiarity with alternative processes directly reflected the collective level of experience. Four in five respondents (81.2%) were "familiar" or "very familiar" with binding arbitration, while about 62 percent expressed familiarity with mediation. Despite frequent discussions of ADR in professional literature and education efforts by the ABA and other professional groups, more than one-fifth (21.0%) of those surveyed remained "unfamiliar" or "very unfamiliar" with mediation. Even more surprising, considering the long history of arbitration in the construction industry, was the fact that one in ten respondents admitted a lack of familiarity with binding arbitration!

Participants were less knowledgeable regarding other processes. Nearly two-thirds (63.0%) of the respondents indicated they were "unfamiliar" or "very unfamiliar" with summary jury trials; almost half (47.6%) were unfamiliar or very unfamiliar with mini-trial; and four of ten (40.3%) made the same statement with respect to non-binding arbitration.

## 3.2 Perceptions regarding mediation and mini-trial

Numerous survey questions elicited lawyer attitudes regarding when and under what circumstances mediation or mini-trial was appropriate. Responses indicated attorneys were much more familiar with mediation and generally favored the process over mini-trial.

### 3.2.1 Recommending processes

More than half (50.9%) of those responding would recommend the use of mediation to their clients in "most" or "all" construction-related disputes. Only a few (1.5%) said they would never recommend mediation. Eighty-six percent (85.8%) of those surveyed disagreed that proposing mediation was a sign of weakness in a party.

While just 12.3 percent would suggest a mini-trial in most or all cases, only six percent indicated they would not recommend a mini-trial under any circumstances. However, almost 90 percent (89.3%) of those polled did not believe that proposing a mini-trial indicated weakness.

### 3.2.2 Factors indicating use of process

Respondents regarded mediation as most appropriate where:

*the parties wished to maintain an ongoing business relationship;
*clients desired privacy and confidentiality;

*disputes needed to be resolved quickly; or
*an economical process for resolution of the dispute was essential.

Mediation was considered least appropriate where:

*the dispute turned on a novel question of law;
*the credibility of witnesses was at stake; or
*the opposing party or its counsel was considered
*untrustworthy or unlikely to compromise.

Responses concerning mini-trial were similar, although that process was regarded less favorably than mediation in most cases.

**3.2.3 Discovery**

Six of ten participants (60.8%) regarded mediation as "appropriate" or "highly appropriate" where no discovery had occurred. On the other hand, four out of five (81.7%) thought mediation appropriate where discovery had been completed and the case was ready to go to trial. Few attorneys (4.2%) believed discovery was never appropriate prior to mediation--more than half the survey group (56.2%) would prefer prior discovery in "most" or "all" cases.

The survey group considered discovery more important as a prerequisite to mini-trial. Less than forty percent (37.3%) generally regarded mini-trial as appropriate in the absence of prior discovery. Nearly four out of five attorneys (79.1%) believed discovery should precede mini-trial in most cases or all cases; only 2.6 percent regarded prior discovery as inappropriate in all cases.

**3.2.4 Advisors**

The survey group registered strong opinions on the role of the dispute resolution advisor in mediation and mini-trials. The majority (82.3%) thought mediators should be allowed to express their opinions to the parties regarding the issues in the dispute. The same result held for mini-trial (84.8%). Consistent with traditional practice (and the position of the American Arbitration Association), more than two-thirds (67%) of those responding believed that under no circumstances should mediators or mini-trial advisors serve as arbitrators in the same case.

Participants were asked to indicate the relative importance or unimportance of thirteen mediator attributes. Those attributes which the group regarded as almost always important were impartiality, managerial skills, personal discretion, listening ability, and the ability to understand complex issues. Patience and creativity were also important in most cases. The ability to explain complex issues, persuasiveness, design or construction experience, personal prestige and legal expertise were collectively regarded as important in some but not all cases. Familiarity to the parties was viewed as a relatively unimportant factor.

### 3.2.5 Contractual ADR provisions

Despite their generally positive attitudes toward mediation, less than half (42.7%) of those responding thought standardized contracts should require mediation prior to arbitration or litigation of disputes. On the other hand, over half (53.9%) thought standardized contracts should require mediation prior to arbitration or litigation of disputes involving large sums of money.[10] Respondents were less supportive of clauses requiring mini-trial prior to arbitration or litigation. For example, more than three-quarters (77.6%) of those responding warned against mini-trial provisions.

## 4 Actual Experiences with ADR

In addition to providing information on their perceptions, ABA members completing the survey provided detailed information on their actual experiences with mediation, mini-trial, summary jury trial, non-binding arbitration, and various other processes. Each respondent was permitted to describe three different experiences with these processes. A total of 548 separate experiences were reported by 320 respondents.

### 4.1 Range and Recency of Actual ADR Experiences

Of the 548 experiences, 459 (83.8%) involved mediation; 62 (11.3%) were mini-trial, and 20 (3.6%) were summary jury trial. The remainder included a minor assortment of alternatives such as technical advisory panels, expert negotiations, and "informal dispute settlement."

More than ninety percent of the reported experiences occurred between 1987 and 1990. Almost three-quarters (72.4%) took place in the two years preceding the survey. While this may result from the natural inclination of respondents with multiple experiences to report those of more recent memory, it seems clear that settlement-oriented procedures--particularly mediation--are gaining rapidly in popularity. The leading categories of disputes submitted to ADR processes were issues of defective work, payment, project delays, and changes. Less frequently, disputes involved jobsite administration, differing site conditions, and personal injury or property damage.

---

10. The questionnaire did not inquire how "large sums of money" should be defined, nor what implementing language might be included in a contractual mediation provision.

### 4.2 Experience with mediation

Statistics relating to two of the leading categories of reported ADR processes, mediation and mini-trial, reflect the diversity of experience and the enthusiasm of the construction bar for this process. Discussed together here are the 459 mediation and 62 mini-trial experiences.

#### 4.2.1 Role of respondent

Those participating in mediation acted as counsel for a party in nearly nine-tenths (87.4%) of the reported cases. In more than half of these (50.1%) the participant represented a contractor; another fifth (20.9%) involved representation of owners; 15.7 percent, design professionals or their insurers; and 10.5 percent, sureties. In nearly all of the remaining cases (12.4%) the respondents served as mediators.

With respect to mini-trials, 93.5 percent of respondents were attorneys for the parties. Over a fifth (12.5%) represented contractors, more than twenty-five percent (26.8) represented owners, and only a few (3.6%) represented sureties.

#### 4.2.2 Number of parties

In more than a third (36.9%) of the reported cases the mediated dispute involved two parties. Another 22 percent involved three parties; the remainder (58.9%) involved four or more parties. Of reported mini-trials, 54.8 percent included two parties, and 14.5 percent included three. One reported mini-trial involved 40 parties.

#### 4.2.3 Length of proceeding

The majority of mediations (82.9%) were concluded in three days, and almost half (49.4%) were completed, successfully or unsuccessfully, within one day. Typically, mini-trials were concluded within three days (64%), but slightly less than a quarter ended in only one day (23%).

#### 4.2.4 Amount in controversy

Amount in controversy ranged from $600 to $500,000,000. Across all types of ADR, the median amount in controversy was $1,000,000 with an average amount in controversy of $6,012,526. As expected, the average amount in controversy in mediation was $4,102,025, significantly lower than mini-trial ($9,542,016). [11] In the summary jury trial, the average amount in controversy

[11] *Cf.* Green, Marks and Olson, *supra* note 4, at 493 (mini-trial effective "settling *large* case litigation") (emphasis added); *see also* Edelman and Carr, *supra* note 4, at 11 (claims involving small sums of money will usually not be

was $15,160,500.

#### 4.2.5 Impetus for ADR

The decision to resort to mediation was a product of agreement between the parties nearly two-thirds of the time (65.1%). About a third of the time (29.4%) the process was court-initiated. Relatively few cases (3.7%) involved a process initiated pursuant to a mediation provision in a contract. In nearly nine out of ten cases (87.4%) a lawsuit or arbitration demand preceded initiation of mediation.

Similarly, most of the time (75.8%) the choice of mini-trial was the product of agreement by the parties. Under a fifth (19.4%) were undertaken as the result of court order, and one out of twenty (4.8%), slightly higher than mediation, were undertaken pursuant to a contractual provision. As was true for mediation, most mini-trials (90.3%) were preceded by an arbitration demand or prior suit.

#### 4.2.6 Status of discovery

Full discovery had been conducted in 43.3 percent of the cases sent to mediation. Document discovery was indicated in about one-fourth (26.5%) of the cases with depositions and interrogatories less frequently used (4.4% and 4.0%, respectively). In one-fifth of the cases (21.4%) no discovery was conducted prior to mediation. Where discovery occurred, the great majority of respondents (84.9%) found it "helpful" in the mediation.

The situation for mini-trial was usually full discovery (45.2% of the time), with no discovery only 14.5 percent of the time. Again, discovery in mini-trial was considered helpful by most respondents (92.5%).

---

attractive for the process).

The leading analysis of mini-trial to date, ABA Mini-Trial Evaluation, *supra* note 4, at 40 concludes mini-trial should be more attractive in cases involving larger dollar amounts because savings to parties are by comparison more significant in big cases. The report provides:

> Whether or not the mini-trial is suitable for disputes involving substantially smaller amounts [under $500,000], however, has yet to be shown, since almost all mini-trials have involved sims in excess of $100,000 (sic) and familiarity with the mini-trial format still remains largely the monopoly of a relatively small number of attorneys and corporate clients.

ABA Mini-Trial Evaluation, *supra* note 4, at 40.

#### 4.2.7 Source of procedures

The major sources of mediation procedures were: party-developed rules (34.1%); rules of the court or other judicially-imposed procedures (27.4%); and AAA mediation rules (20.1%). Other sources of procedures included the Center for Public Resources and various private services.

Quite different was the situation for mini-trial. For example, nearly two thirds of the time (62.3%) the parties developed their own mini-trial rules. Almost a quarter (24.6%) of the time court rules were used, and in only a few instances (3.3%) were specific mini-trial (e.g., AAA Mini-Trial rules) rules implemented by the parties.

#### 4.2.8 Selection of mediator

Mediators were appointed by independent organizations in about one-third (32.6%) of the reported cases; in a number of other situations (21.4%) appointment was by agreement of the parties. Selection was by some other method more than forty percent (44.9%) of the time. While most mediators (64.5%) were attorneys, one-fifth (21.1%) were retired judges. Design professionals, contractors, claims experts, and professors were employed far less frequently.

The same held true for mini-trial. Twenty-five percent of mini-trial advisers were appointed by the parties; an equal percent were appointed by independent organizations. The majority were appointed in some "other" way (48.1%).

#### 4.2.9 Presentations by parties

Nine out of ten mediations (90.2%) featured some form of oral presentation by each party before the mediator. Nearly two-thirds (64.7%) of the time this was supplemented by some form of written memorandum. In mini-trial, 87.1 percent of the experiences featured some presentation by each party before the mediator. A written memorandum was used less frequently (58.1%) in the mini-trial than in mediation.

#### 4.2.10 Advisor strategies

Mediators engaged in informal joint discussions with both parties in about half (52.5%) of the cases. Private caucuses with each party were employed in two of three cases (64.9%). In some cases mediators reviewed job records and other documents (26.4%). Less frequently they conducted discussions with third parties (11.1%), made jobsite visits (9.6%), or consulted with independent experts (6.5%). Only a few (3.3%) consulted technical reference works. Advisors expressed their views of the factual and legal issues in the dispute in most cases (72.2%).

A similar situation existed for mini-trial. Strategies implemented by mediators included: (1) job record review (35.5% of the time), (2) informal joint discussions with the parties (35.5%), (3) caucusing with the parties (19.7%), and (4) discussions with third parties (9.7%). As was true for mediation, the mini-trial advisers usually (88.9%) expressed their view on the law and facts of the dispute.

### 4.2.11 Potential drawbacks

The survey group was largely unconcerned with the most frequently mentioned potential drawbacks of mediation. A mere 13.5 percent of respondents were concerned with revelation of trial strategy in mediation; only 7 percent viewed revelation of confidential information as a drawback of mediation. Similarly, only one in twenty (5%) respondents saw delays or disruptions of litigation or arbitration processes as a problem. Even fewer attorneys (4.1%) were troubled by potential difficulties of addressing the rights of third parties not participating in the mediation.

Likewise, those responding generally (91.9%) indicated that sharing of confidential information was not a problem in a mini-trial setting. Only a few (3.3%) saw the mini-trial falling short of the mark because third party views were not included from the proceeding. Even with its resemblance to actual trial, only a quarter (27.4%) of the respondents indicated revelation of trial strategy was a serious disadvantage of the mini-trial.

## 4.3 Results

Across all types of ADR reported, full settlement occurred in 57.4 percent of the time and partial settlement occurred in 8.4 percent of the cases. As shown in Table 1, mediations accounted for more settlements, proportionally, than any of the other forms of ADR; mini-trial resulted in proportionally more partial settlements.

### 4.3.1 Mediation outcomes

Of the mediations shown in Table 1, full settlement occurred 59.1 percent of the time; a further 7.9 percent of the cases were settled in part. Of these, two-thirds (65.1%) of the time a monetary settlement was involved; an agreement to perform specific work tasks resulted in few cases (7.0%).

A few bivariate analyses underscore these results.[12] For example, analysis of variance showed no significant differences among the various

---

12. The results must be interpreted with caution, since only bivariate relationships are discussed. Several multivariate analyses are in progress.

mediation settlement outcomes (e.g., full settlement, partial settlement, no settlement) for amount in controversy ($p > .5514$). Without controlling for other influences, mediation thus seems to work as well (or as poorly) in cases

**Table 1: Type of ADR, By Settlement Outcome**

| | Mediation | Mini Trial | Summary Jury Trial | Row Total |
|---|---|---|---|---|
| Settlement | 269 | 33 | 7 | 309 |
| | (87.1)[a] | (10.7) | (2.3) | (57.6) |
| | [59.1][b] | [53.2] | [35.0] | |
| Partial Settlement | 36 | 8 | 1 | 45 |
| | (80.0) | (17.8) | (2.2) | (8.4) |
| | [7.9] | [12.9] | [5.0] | |
| No Settlement | 150 | 21 | 12 | 186 |
| | (82.) | (11.5) | (6.5) | (34.6) |
| | [33.0] | [33.9] | [60.0] | |
| Column Total | 455 | 62 | 20 | 537 |
| | (84.7) | (11.6) | (3.7) | (100.0) |

Notes: a Row percentages shown in parentheses
b Column percentages shown in brackets

where significant financial amounts are in dispute. Nor does mediation success (or lack of it) seem to depend on whether an arbitration demand or law suit had been filed ($p > .9139$). The "pressure to reach agreement" hypothesis thus receives little support in this preliminary analysis. However, mediation outcome did differ significantly by the number of parties involved ($p < .0008$), suggesting that mediation may be less effective where multiple parties are involved.

Chi-squared analyses showed settlement was more likely in cases where the parties agreed to ADR after a dispute arose than where they met pursuant to a court order, or an executory provision in the construction or design contract.

Where parties agreed to ADR, for example, settlement or partial settlement occurred most of the time (63.1% and 9.1%, respectively--or, 27.8% did not settle). However, when parties were required to use ADR--either by contract or by the court--a greater percentage (43%) failed to settle. Settlement rates were also significantly greater in cases where parties had developed their own rules of process and where they appointed their own neutral.

Other factors associated with higher settlement rates was where the mediators offered their views on factual and legal issues. Where advisers exposed their views, 62.6 percent of the disputes settled; where advisers did not, only 49.6 percent settled ($p < .0048$). Settlement rates also appear higher where some form of discovery occurred prior to mediation. Although difficult to pin down because of the lack of data, success rate also seemed to vary little depending on background of the advisor (attorney, retired judge, contractor, etc).

Where cases failed to settle, two-thirds (65.3%) of the time respondents indicated it was because of the unwillingness of a party or parties to compromise. Another fifth (20%) of unsettled cases were ascribed to ineffectiveness of the mediator. Lack of authority to settle the dispute stymied resolution in a further 7.5 percent of the cases.

### 4.3.2 Mini-trial outcomes

The smaller sample size for mini-trial makes investigation of outcome difficult at best. A few observatons can be made, however. A significant relationship appeared between the number of parties and the final mini-trial outcome ($p < .0005$). The more parties involved, the less likely the mini-trial would end in full settlement. Length of proceeding was not related to mini-trial outcome.

Although empirical relationships could not be established, in mini-trial party-developed rules seem to result in more full settlements, compared with situations where rules of the court were used. As was true of mediation, those responding indicated that where the mini-trial did not reach settlement, the primary reason was the parties' unwillingness to compromise.

## 5 Conclusion

The construction field has long served as a laboratory for experimentation with alternatives to the courtroom. In the past the primary focus has been on development of suitable arbitration processes. Today, however, increasing emphasis has been placed on procedures aimed at assisting parties to tailor their own solutions to disputes rather than referring the matter to adjudication.

The results here, to be expanded in the future, confirm the speed and success of non-binding dispute resolution in the construction industry. Few qualifications, practical or theoretical, were identified by those using these techniques. Moreover, the views of mediation and the mini-trial often portrayed in the literature do not square with the actual experiences of construction lawyers. Although settlement-oreinted processes such as mediation and mini-trial are less well understood than arbitration, the collective experience of the construction bar may encourage optimal use of such alternatives.

# THE DISPUTE RESOLUTION ADVISER IN THE CONSTRUCTION INDUSTRY

COLIN J WALL
Commercial, Mediation & Arbitration Services Limited, Hong Kong

Abstract
The evolutions of the concept of the DRA system, as implemented in Hong Kong, from the various hypothetical models promulgated in the UK and from the practical experience in the USA of Disputes Review Boards and Project Arbitration. Distillation of the role of the DRA from these sources, with particular reference to C Evans' 'independent intervenor', K Severn's Dispute Adviser, US Army Corps of Engineers Disputes Review Board, the research of the Center for Public Resources and the Foster-Conner Model.

Explanation of, and rationale behind, the consultancy procedure and the consensus building approach which was adopted prior to and during the detailed design of the system. Crystallization of the DRA role, as it would best serve the particular needs of the project and meet the concerns of the participants.

Expectations of the DRA system. Reduction in tender prices, development of closer cooperation between the parties, increase in joint problem solving at site level, enhanced budgetary control, preventative effects, greater sub-contractor involvement, general resolution by non-binding means but occasional and ultimate resolution by concurrent short-form arbitration.

Detailed exposition of how the DRA will operate in the Hong Kong project. Appointment, replacement, terms of reference, etc.

Potential applicability of the system to other projects: aspects which are directly applicable and those which should be tailor-made for individual circumstances of the project.
Keywords: Dispute Resolution Adviser, Project Mediation, Short-Form Arbitration

## 1 Introduction

On 23rd December 1991, the first Dispute Resolution Adviser (DRA) was jointly appointed by the Hong Kong Government's Architectural Services Department (ASD) and by the contractor awarded the refurbishment contract for the Queen Mary Hospital in Hong Kong.

The contract required the refurbishment of a 56 year old general hospital, including operating theatre block, the removal of asbestos and the maintenance of existing services and facilities. Hospital contracts are notorious for disputes as the considerable volume of hospital related English construction law indicates. The added complexities of this contract, which included a requirement to keep the hospital and operating theatres operational during refurbishment, meant that unless an innovative approach was adopted to the management and resolution of construction conflict, the contract would be beset by disputes. The innovative approach was provided by the DRA system.

## 2 The genesis of the DRA

### 2.1 A hybrid system

The DRA system has been designed to try and stem the tide of disputes which have now become the norm in the construction industry. The DRA system is one solution to the plea 'there has to be a better way'. Like the "Mini Trial", the DRA is a hybrid system combining elements from many ADR techniques. As a result, it forms a highly flexible set of procedures centered around the DRA for both dispute prevention and expeditious dispute resolution, preferably by non-binding means.

The DRA system is based on a belief that disagreements should be resolved within a very short time-frame and that the parties should, in the first instance, attempt to resolve these disagreements themselves at site level before they turn into full blown disputes which affect the parties working relationships. The DRA system borrows elements from the highly successful Disputes Review Boards used in the USA and Project Arbitration. It also draws on various hypothetical models promulgated in the UK, particularly the 'independent intervenor' and the 'Dispute Adviser'.

### 2.2 The independent intervenor

In 1986, Clifford Evans, in an address to the Wales Branch of the Chartered Institute of Arbitrators and the South Wales Association of the Institute of Civil Engineers, repeated a suggestion made four years earlier:

'Let there be appointed at the commencement of the contract a mediator or conciliator, call him an 'independent intervenor', paid a small retainer shared equally by the contractor and employer and called in to settle a dispute immediately it arises - without waiting until the end of the contract. Both employer and contractor would be bound by the decision, at least until completion, when either party would have the right to go to arbitration in the normal way.'

'The engineer (and here I must remind you that I'm talking about architects and their role in the JCT contract, as well as engineers) is not always right: occasionally he is late in issuing drawings or instructions or there are minor errors in drawings, all of which could cause claims to arise against the employer. How can one expect the engineer or architect to have an independent view when assessing these claims, knowing that acceptance of the claim would be an admission to the employer of his liability? If, however, the engineer or architect knew that there was an independent intervenor waiting in the wings who could, if a dispute arose over much matters, quite quickly attach the blame to him, would not this concentrate his mind even more and ensure that drawings and instructions were issued on time, and would not this help to make the contract run more smoothly?'

The concept of the independent intervenor has been adopted as adjudication and a number of standard forms of contract now provide for prompt adjudication of disputes. Sometimes the adjudicator is given a wide range of powers to resolve all manner of disputes whilst in other instances his or her powers are restricted to specific areas. Disputes relating to set-off are one such example.

Whilst the concept of 'the independent intervenor' has a number of advantages, including the prompt resolution of disputes, the likely reduction in posturing and an expected general increase in efficiency, the adjudication system does have several disadvantages. These include the fact that the adjudicator imposes an agreement upon the parties by providing an evaluation of the merits of the dispute based on law, legal precedent and perhaps his own expertise, the wider interests of the parties are not considered; the matter is not finally disposed of and there are problems with enforcement. An adjudicator's decision cannot be enforced as an arbitrator's decision. The law is clear on this point.

**2.3 The Dispute Adviser**

Kenneth Severn presented a paper in July 1989 entitled 'New Concepts in the Resolution of Disputes in International Construction Contracts'. The paper was the product of ideas from a small working party of the Chartered Institute of Arbitrators, including Clifford Evans, and contained a number of interesting ideas. One of these was the concept ofusing a Dispute Adviser much along the lines of the independent intervenor but with an important difference. The Dispute Adviser was to advise on the means of settling the dispute. The paper described the concept in the following terms:

'The parties, at the time of agreeing to appoint a Dispute Adviser, select one with no connection to either party. His task is to advise on the means of settling disputes and should be selected for that purpose. He may, in some circumstances, assist in their resolution but it is anticipated that his primary duty as an adviser will not undermine the authority of the Engineer under the contract.'

The paper explained that the Dispute Adviser would enter immediately upon his duties should a dispute arise during the progress of the works. It continued

'If the dispute be of limited extent - for convenience identified as a 'minor dispute' - it may be within the competence of the Dispute Adviser to resolve, depending upon his qualifications. Alternatively it may be a case for calling in an expert to hear the parties and resolve the matter. The choice is with the parties with the advice of the Dispute Adviser.

In the event of a failure to reach a settlement or of the reluctance of the parties to accept the 'minor dispute' procedure, the issue becomes a 'major dispute' for resolution either immediately or at a later date. If later, the Dispute Adviser would investigate the matter and record the facts for use in the determination of the dispute at some future date.'

A 'major dispute' can either be conciliated or mediated or the Dispute Adviser can make a recommendation. Conciliation is defined as the 'intervention of a neutral third party for the purposes of bringing the parties closer together', whilst mediation is described as the appointment of a technical person who, if he cannot reach a settlement with the parties, provides his or her written reasoned opinion. This effectively becomes an adjudicator's award and is binding until overturned in arbitration. The parties also have the choice of arbitration and the Dispute Adviser will help them with the choice of procedure and to choose a technical arbitrator.

Flexibility is the main advantage of this proposed system as the Dispute Adviser does not necessarily resolve the dispute himself. The system also overcomes the criticism levelled at project arbitration whereby the arbitrator is chosen before the nature of the dispute is known and it also allows the parties a choice of procedure based upon neutral advice.

There are some disadvantages too. For example, if the conciliation process fails, the dispute is not resolved until it is finally arbitrated upon. The mediation process may not dispose of the dispute. If the mediation fails and the matter is adjudicated, then there is the possibility of one side ignoring the adjudicator's decision, which would lead to further disputes and, like a failed conciliation, the dispute cannot finally be resolved until it is arbitrated. Often this is at project completion. Finally, there is a concern that if the mediator is required to evaluate the dispute, should the parties be unable to reach a settlement, the parties may not be as candid as they might otherwise be during the mediation process. Med/Arb as employed in the USA, has not proved popular for this reason.

The Dispute Adviser has not, as yet, been used in the form described in Mr Severn's paper so there is no practical experience of how successful it would be.

### 2.4 Disputes Review Boards

In the USA the Disputes Review Boards have been an outstanding success. The Disputes Review Board process is an expedited non-binding ADR procedure whereby an independent board, usually of three persons, is established to evaluate disputes and make settlement recommendations to the parties. The Board members become knowledgeable about the project by periodically visiting the site. The Board meets at regular intervals and hears presentation on the disputes which have arisen since last time it met.

The goal of the Disputes Review Board process is not only to resolve disputes, but to prevent them. Because the same Board will sit on all disputes occurring during the project, the parties will become familiar with how the Board members look at particular types of issues. The parties will be able to predict how the Board will react and will take that reaction into account when negotiating between themselves. Many disputes will consequently settle before reaching Board level.

One of the major advantages of the Disputes Review Board is the preventative effects described above. Obvious disadvantages include the cost of maintaining a three person tribunal, the non-binding nature of the recommendations means disputes are not necessarily finally disposed of and may be litigated or arbitrated. Because all disputes are evaluated, there is a danger of the Board losing credibility in the eyes of one or both parties should a situation arise where the Board makes decisions which appear to favour one party at the expense of the other, or where the Board makes decisions which neither party finds acceptable. Once this lack of credibility occurs, the advantages of the system are compromised.

**2.5 Project arbitration**

Project arbitration offers an answer to the criticism that matters are not necessarily disposed of in the ADR techniques outlined above and also overcomes the enforceability problems associated with adjudication. Project arbitration is, by its very nature, evaluative - the wider interests of parties are not considered and there is the obvious drawback mentioned previously that the arbitrator is appointed before the nature and subject matter of the dispute is known. This drawback has been mitigated in the USA by adopting the Foster-Conner Model. This is a project arbitration system whereby several arbitrators, embodying a wide range of legal and technical skills, are appointed at project commencement. The arbitrators are required to give final and binding decisions by short-form arbitration, within a short time of the dispute first arising.

The Foster-Conner model was first used on the Ocean Creek Project - there were no arbitrations. Preliminary research by New York's Center for Public Resources has indicated that this pattern has been repeated. In a survey, those US lawyers who had direct knowledge of the use of project arbitrators, reported that in their experience, no disputes ever arose which were presented to the arbitrator. The use of a project arbitrator is a highly effective preventative measure. The parties become more reasonable and negotiate settlements to meet their interests rather than having a final and binding decision imposed upon them. The major disadvantage of this particular system is the cost of maintaining a large panel of project arbitrators all of whom visit the project on a regular basis to maintain familiarity with the progress.

## 2.6 Other ADR techniques

The research by the Center for Public Resources has resulted in a paper 'Preventing and Resolving Construction Disputes'. It is a collection of articles on both binding and non-binding ADR and dispute prevention techniques. One of the most successful techniques, pioneered by the US Army Corps of Engineers, is partnering. Partnering is essentially a consensus building process that re-orientates the parties from a 'them and us' mentality to a 'we' mentality. Another useful technique noted by the Center is step negotiation. This is a process whereby the representatives of the parties to whom the dispute is referred will be required, in the event they cannot resolve the problem, to refer the matter to their superiors, in both organizations. This provides an additional incentive to the representatives to resolve the problem without having to bother their superiors.

# 3 The crystallization of the DRA

## 3.1 The elements of the system

Having considered all of the above techniques and looked at the strengths and weaknesses of each one, the challenge was to devise a system which would gain acceptance in the Hong Kong Construction Industry, would utilize the positive attributes of these other ADR techniques, but above all, would act as an effective, economical and expeditious system of dispute prevention and resolution.

The DRA system draws upon the independent intervenor concept as modified by the Dispute Adviser but provides a far more flexible approach. It embodies the dispute prevention attributes of the Disputes Review Board and Project Arbitration, it uses partnering techniques to re-orient the parties' thinking and encourages negotiation by using a tiered dispute resolution process. It is based on giving the parties maximum control through the use of mediation techniques but also includes prompt short-form arbitration which encourages voluntary settlement and, if necessary, provides a final and binding resolution to the dispute.

## 3.2 Consultancy procedure

Commercial, Mediation & Arbitration Services Ltd (CMA), together with Endispute Incorporated (Endispute) of the United States, were appointed by ASD to design a suitable DRA system for the Queen Mary Hospital refurbishment in Hong Kong. CMA/Endispute, although appointed by ASD, acted as neutral consultants and obtained candid opinions on a confidential basis. The first stage of the consultancy was to understand the nature of the project, to discuss with the consultants and with the potentially pre-qualified contractors the likely areas of dispute and to determine their objectives and concerns about the possible dispute resolution models, including the DRA system.

The next stage of the consultancy was to produce a report setting forth the likely areas of dispute and in an anonymous manner the concerns raised by the parties regarding the use of a DRA system. The report developed specific recommendations for a dispute resolution system including contract language. Following analysis of this report by ASD, there was another round of discussions and consensus building leading to modifications and the production of the final system. In the event 86% of the potential pre-qualified contractors' suggestions for improving contractual arrangements to minimize potential areas of dispute and to clarify areas of responsibility, were accepted by ASD for final incorporation. The most significant of which was that there be no Nominated Sub-Contractors.

During the course of the consultancy it also became necessary to modify the Hong Kong Government's standard Conditions of Contract by inserting preliminary time frames for the making of decisions, giving of certificates, evaluation of variations etc... The requirement to impose strict time frames applied equally to ASD/consultant and contractor. ASD had indicated that they must have strict budget control and did not want to be faced with a massive claim at the end of the contract. CMA/Endispute wished to ensure that all potential decisions which may give rise to disputes were made within a short time of the event giving rise to the decision, in order that the matter could be decided whilst the facts were fresh in everyone's mind. If there was a dispute it could be promptly resolved without having to rediscover the key facts. Adherence to strict timeframes achieved these objectives.

The DRA system, in its final form, incorporated a 'final offer' model for the short-form arbitration of quantum disputes. CMA/Endispute considered disclosure of each parties' figures would help discourage exaggeration and encourage settlement.

## 4 Expectations of the DRA system

### 4.1 Reduced tender price

Research by the Technical Committee on Contracting Practices of the Underground Technology Research Council, has indicated that significant tender price reductions can be achieved when Disputes Review Boards and other techniques are adopted. There were similar expectations of the DRA system. The consensus building stage had convinced the contractors that the DRA system would not work against their interests, would improve their cash-flow and would avoid long and costly arbitrations post project completion. In the event the successful tender was considerably below the budgeted price.

### 4.2 Initial familiarization period

At the commencement of his or her functions, the chosen DRA would engage in a series of familiarization meetings which would build support for the system and encourage party cooperation. The participants, especially those working on site and having daily contact with each other were told to be candid, to trust each other and if something unexpected did happen to discuss their concerns openly and orally rather than fire the first shot in a salvo of contractual correspondence. Particular attention was given to emphasizing the need for clear communication, preferably oral, to try and reduce the volume of written correspondence. The participants were encouraged to think of problems as problems which affected the project rather than 'their' problems or 'our' problems.

### 4.3 Budgetary control

Provision was made for ASD's need for close budgetary control. The valuation of variations and claims is made within a very strict timetable but provision is made in exceptional cases for extending these time frames if circumstances so dictate. If the parties are unable to agree on a suitable extension, the DRA will decide the matter.

### 4.4 Sub-contractor participation

One important feature of the DRA system is the recognition that it is often the specialist sub-contractor and not the contractor who is the real party with an interest in the settlement of a dispute. The chain of contractual responsibility, especially where all sub-contractors are domestic, often makes communications unnecessarily long and can give rise to misunderstandings. Three way communication is encouraged, especially on specialist technical issues, but the contractor is kept fully informed at all times. This theme of sub-contractor involvement is continued right through the DRA system, so, for example if a dispute reaches short-form arbitration and it is the sub-contractor who has the real interest in the dispute, it is he, together with ASD who chooses the arbitrator in the consolidated arbitration.

### 4.5 Early settlement of disputes

It is anticipated that few, if any, disputes will reach arbitration; the parties to the dispute will generally prefer to settle the matter themselves rather than have a result imposed upon them by an arbitrator. The parties are encouraged to resolve disputes by non-binding means with, if necessary, the assistance of the DRA. How the DRA system will operate on this project is described below.

## 5 The details of the DRA system

### 5.1 DRA appointment

The DRA is jointly chosen and appointed by the parties at contract commencement. The fees of the DRA are shared equally between the parties. If the parties cannot agree on a suitable individual, the DRA is chosen by a ranking system. If this also fails, then the DRA is chosen by the Hong Kong International Arbitration Centre, which maintains panels of mediators and arbitrators. The DRA is an individual, familiar with construction, who possesses dispute resolution skills and preferably has some knowledge of arbitration.

**5.2 Familiarization and monthly site visits**

The DRA conducts a series of familiarization meetings as described above and becomes familiar with the project, the construction programme and the participants, including representatives from the specialist sub-contractors and hospital management. The DRA visits the site on a monthly basis and assists the site level representatives to facilitate settlement of any disagreements or disputes which have arisen during the previous month. This assistance can either be formal or informal. The DRA operates on an informal level where there is a disagreement but it has yet to become a full blown dispute.

**5.3 Timeframes**

The parties have 28 days in which to challenge any decision, certificate or evaluation made under the contract; failure to do so renders the decision, certificate or evaluation final and binding. Preliminary timeframes have been inserted into the contract for the making of certain decisions and evaluations.

If the decision, certificate or evaluation is challenged, the site level representatives of the parties have 28 days in which to attempt to resolve the matter by good faith negotiation. If desired, the DRA may assist the representatives with their negotiation during one of the monthly visits. If the matter has not been resolved before the expiry of the 28 days, then the aggrieved party is required to give a formal, written Notice of Dispute. If no notice is given, the right to dispute is deemed waived. The DRA and the site level representatives then have 14 days in which to attempt to resolve the dispute. This is the formal stage of dispute resolution involving the DRA. The DRA is free to choose the most appropriate ADR technique to help the parties resolve the dispute. This may be by formal mediation, mini trial, expert fact finding, expert opinion etc. If there is to be a formal mediation which is likely to be evaluative or involve specialist knowledge, then the mediation will most likely be carried out by another neutral person rather than the DRA. If necessary, the 14 day time period may be extended to accommodate the availability of the chosen neutral person. The DRA will carry out facilitative mediations.

The process ends if the dispute is resolved.

**5.4 Involvement of senior staff members**

If the DRA and the site level representatives have been unable to settle the dispute, the DRA produces a report which contains an analysis of the dispute, the key issues and the DRA's perception as to the barriers of settlement. This report is given to the senior staff members of the parties, in order that they may appreciate the true nature of the dispute. If both senior staff members request it, the DRA report will also contain either a non-binding recommendation for resolution or a non-binding evaluation of the dispute. The senior staff members, who are the individuals in overall charge of the project but who are not involved in the day to day management of the contract and are not involved in the decisions that are being disputed, can bring a non-emotional, broader perspective to the dispute. Hopefully, the senior staff members can meet and resolve the matter. If desired, the DRA will attend these meetings with the senior staff members.

**5.5 Short-form arbitration**

If the matter is still not resolved within 14 days of the DRA's report, then the DRA will convene a short-form arbitration, unless the parties have accepted a recommendation from the DRA to resolve the matter by some other means.

The arbitration is governed by special rules, which are written in to the contract, and will take place within 28 days of the date that the senior staff officer settlement efforts have terminated. The arbitrator will be selected by the parties in the dispute, including, where appropriate, those sub-contractors who may be required, by the terms of their sub-contracts, to participate in the arbitration. The arbitrator will be appointed by an exchange of lists, but if this proves unsuccessful, will be selected by the DRA and appointed by the parties.

**5.6 Short-form arbitration - key elements**

The short-form arbitration has the following characteristics:

(a) It should involve one claim or issue or with the written agreement of the parties and the DRA, a limited number of distinct claims or issues.
(b) If it involves one claim or issue the arbitration will be conducted and concluded in one day. If it involves more than one distinct claim or issue, the parties and the DRA will agree on a revised timescale, taking into account whether or not the arbitration is a consolidated one.
(c) Each party will have the opportunity to present its case to the arbitrator, either through a written presentation, oral evidence, or the use of affidavits and documents only.
(d) The arbitrator will fairly allocate the amount of time within the day for each presentation as well as allowing time for questions and answers.
(e) The arbitrator will have seven days to make a written award, which will contain concise, reasoned decisions. The award will enable the parties to appreciate the outcome, but will not contain sufficient detail to enable the parties to mount an appeal.
(f) The decision of the arbitrator will be final and binding, subject only to the limited rights provided to the parties under the Hong Kong Arbitration Ordinance.

(g) If the arbitration concerns a quantum (time or money) dispute, then this will be resolved in accordance with a final offer arbitration. The arbitrator will have limited authority to render an award selecting one or other figure as the more reasonable. The arbitrator will not be permitted to make his or her own award.

## 6 DRA appointment, replacement, terms of reference

### 6.1 Appointment, replacement and tenure

As noted above, the DRA is a joint appointment. DRA costs are split 50%-50% irrespective of the nature of the DRA activity.

There is a presumption that the tenure of the DRA will be for the life of the project but provisions exist for replacement in the event of death, illness or resignation. The parties may also terminate the DRA's employment if they are dissatisfied with his or her performance. No one party is allowed to discharge the party unilaterally during the first six months of the project or during the first six months of the tenure of a subsequently appointed DRA. If one party wishes to discharge the DRA after six months, he is first required to confer in good faith with the other party.

Without this restraint on unilateral discharge, the DRA might be inhibited from taking such steps as encouraging one party to compromise or fairly evaluate a disagreement when, for example, the parties are unable to agree on an extension to any of the time frames. By precluding unilateral discharge until after six months, a precipitous response by a party to a DRA action is discouraged. After six months, if one party is dissatisfied with the DRA, that party can effect discharge, otherwise they would have no confidence in the DRA and this would undermine the effectiveness of the whole system.

### 6.2 Terms of reference

The terms of reference for the DRA are broad and flexible and consist of using professional expertise to help the parties foster and maintain a good working relationship, engaging in dispute prevention activities, including using the monthly site meetings to look ahead for potential problems, helping the parties both informally and formally to resolve disagreements and disputes, providing neutral advice on the most appropriate means of settling a dispute and providing advice and assistance in appointing other neutral persons including arbitrators.

## 7 Applicability to other projects

There is no reason why the DRA system should not be used on other building and engineering projects. It is relatively inexpensive to use and promises to create an atmosphere which will maximize the chances of the project being completed successfully. It will certainly avoid long, costly and destructive arbitrations.

Most aspects of the DRA system can be readily applied to other projects. The initial stage of establishing the likely areas of dispute will be project specific and if honest opinions are sought, is best carried out by a neutral consultant. The preliminary timeframes for the giving of decisions and raising of claims will also be project specific. Timeframes for the various stages of the dispute resolution process may also need changing to reflect the duration and complexity of the project and the local availability of mediators and arbitrators.

## 8 Conclusion

The Center for Public Resources paper, 'Preventing and Resolving Construction Disputes', concludes that the appointment of a standing neutral before any disputes have arisen appears to have great promise as the most effective approach which has yet been developed for early resolution of construction disputes. The DRA system, combining the best elements from various ADR techniques fulfills this promise.

## 9 References


Center for Public Resources, Inc (1991) **Preventing and Resolving Construction Disputes.**

Evans, C.J. (1987) 'Construction contracts - time for a change'. **53, JCI Arb. 3**,172-175 & 216.

Severn, K. (1989) **New Concepts in the Resolution of Disputes in International Construction Contracts**

Underground Technology Research Council (1989) **Avoiding and Resolving Disputes in Underground Construction** (Sponsored by the American Society of Civil Engineers and the American Institute of Mining, Metallurgical and Petroleum Engineers)

# WHITHER SMALL VALUE RESIDENTIAL DISPUTE SETTLEMENT IN AUSTRALIA?

I. EILENBERG, Royal Melbourne Institute of Technology, Melbourne, Australia.

Abstract
The costs of resolving small value disputes in the residential arena are often as large as for a major dispute. The causes for this are numerous but relate in many respects to the underlying legal system and that in the past most non court resolution systems have involved ex-builders as the mediator or arbitrator. Today the consumers are far more aware of their rights and are insisting on impartial hearings. Australian governments are hearing this call and are now proposing systems where the builder is beginning to consider that he is being disadvantaged. A system is required that will provide both sides with a quick, cheap but effective resolution of the small value claims.
Keywords: Alternative Dispute Resolution, Residential Construction Disputes

## 1 The Australian legal system

### 1.1 Early introduction

The mechanisms for the resolution of building disputes in Australia has been in existence from the time of the first settlement in Sydney in 1788. One of the earliest accounts of arbitration is recorded by Turner in his History of the Colony of Victoria(1). This is hardly surprising when considering that building was a major undertaking of that first colony, with only 12 carpenters amongst the convicts and ten more ships carpenters. The need to house the convicts and soldiers as well as stores and administration took much of the time of the first arrivals.

Famous Australians have been involved in dispute resolution from early in our history. In Australia, as in the United States of America, the period of 1870-1890's was one of great railway construction with the expected number of disputes. Sir Edmund Barton (later Australia's first Prime Minister) is recorded as acting as an arbitrator in a railway dispute in this period(2).

## 1.2 Basis for Australian law

The Australian legal system is based on the English law. The enabling Act to extend British laws to the new Colony passed through the British Parliament in January of 1787[3]. The Act gave the Governor wide powers to ensure that the early problems that had been faced in the Americas would not be repeated in Australia. From the day of landing at Sydney Cove on January 26th 1788 the colony was thus under British law and it was so proclaimed officially on the 7th of February 1788.

## 1.3 The court system

The Australian court system is based on the British system comprising as it does of three levels - the lower or Magistrates Court (so called in Victoria and also known as Court of Petty Sessions, Local Courts etc.), and hears disputes up to $25,000 (£10,500). These courts are widely spread throughout Melbourne and the towns throughout the State with similar situations in each of the other states. These courts deal with a wide range of matters both civil and criminal. The majority of the residential building disputes which are brought before the courts are heard at this level of the court system. The majority of its cases go unreported in building law related journals. The County Court and the Supreme Court deal with larger disputes and the more serious criminal matters, outside of our consideration. Both these courts have their own building lists, where building related matters (after a brief pre-hearing) are normally sent to a mediator prior to a full court hearing of the dispute. Disputes less than $5000 are heard in the Magistrates court under arbitration rules and thus full rules of evidence can be waived. There is also separate Federal Family Court system and a State Residential Claim Tribunal for dealing with rent related matters.

Appeals from the State Supreme courts go to the Federal High Court.

# 2.0 Arbitration

A large proportion of small value residential building disputes are taken to the Magistrates Court or to one of the other resolution systems - arbitration or a series of tribunals. The majority of building contracts include a reference of disputes to arbitration. This includes the Victorian "Uniform Housing Contract" which is the contract drafted for the guarantee body (the Housing Guarantee Fund Limited (HGFL) and the various builder bodies [4] which is used for the vast majority of new houses.

Since the mid 1980's each of the Australian states has introduced a Commercial Arbitration Act (based on the British Act) which are fundamentally uniform with each other, with the differences not being fundamental to the whole. In Victoria the act is the Commercial Arbitration Act of 1984". One important inclusion has been Section 55 which specifically over rides Scott and Avery clauses, permitting legal proceedings to be brought, even though an arbitration clause is inserted in the contract.

Arbitration proceedings can be initiated by either party to the dispute and the arbitrator may be appointed by mutual agreement at the start of the contract, by mutual agreement when the dispute arises, or if the parties cannot agree upon an arbitrator, by a third party. In most cases today, this is the Institute of Arbitrators Australia.

The system works very well, with the reservation being the scope of the matter under dispute. For most residential disputes there will be one arbitrator appointed. The parties are usually represented by a barrister and at least one expert technical witness will be called for each side. The solicitors may or may not be present at the main hearing. The average cost to the parties in arbitration who have a pre hearing conference and a one day sitting, with the parties represented as above will be in the region of $12,000 (£5000). This will apply equally for a $5000 (£2000) as it will for a $20,000 (£8,500). Thus the cost of running a dispute of this size can outweigh the award.

## 3 Tribunals

### 3.1 Small claims tribunal

The other system available to the consumer (building owner) in Victoria (with similar bodies in most states) is the "Small Claims Tribunal". This is a tribunal attached to the Ministry of Consumer Affairs. At this time, only the consumer can bring a claim. However legislation is planned to go before the Victorian government to permit the supplier (read also trader or builder) to bring an action. The limit of the value of the dispute is $5000 (£2,100). The hearing is in front of a one person tribunal, who has a legal background. No legal representation is permitted but expert technical witnesses are permitted to assist the parties.

### 3.2 Housing guarantee fund

The Housing Guarantee Fund Limited (HGFL) in Victoria has not been satisfied with any of these systems and has been trialing a system of simple or expedited arbitration. The rules are generally that the hearing should have only one (or more by agreement) clearly defined matter in dispute, the arbitrator does not provide any reasons for his award, the hearing is usually only for one day, no legal or expert representation is permitted at the hearing and the costs are minimal, with the parties each paying only $200 (£85) per day with any additional expenses being picked up by the HGFL. This system has been running for twelve months and so far only 10 or so cases have been heard. The initial reaction appears to be favourable. A similar system is being trialed by the Housing Industry Association.

### 3.3 Proposed residential dispute tribunal

The present socialist government in Victoria has come under considerable pressure to provide a cheaper and speedier resolution to residential disputes. Many consumers see the present system as being dominated by builders or at best ex-builders and perceive a bias against themselves [(5)]. Both the states of Queensland and New South Wales have recently introduced a new (and different in each state) tribunal system for the settlement of this type of dispute. The Victorian Government has drafted its own act, with the assistance of the Ministry of Consumer Affairs, to introduce a tribunal system, different to both of the other states. Thus, the East coast of Australia could have three different compulsory settlement tribunals.

The Victorian Tribunal would be divided into two divisions; disputes between $0-10,000 and those between $10-40,000. The former would be heard before a one person tribunal - drawn from the legal profession or an architect or engineer - but no builders, as the registrar considers appropriate. In most cases there would be no legal representation or expert technical support permitted. For the larger dispute two persons would sit - one legal and one technical member - again not a builder.

Legal representation will be permitted but as the explanation of the proposed legislation reads, "it is intended that the prescribed scale of costs will be structured to discourage the use of expert witnesses".[(7)]

The proposed legislation has been warmly welcomed by the consumer related organisations but with severe reservations by the builder related organisations. They claim that the system will deny them 'natural justice', as the Tribunal will simply be an arm of the Ministry. The Minister has denied this claim, yet in his reply to a number of questions put to him, the Minister states that the appointment of referees will be the same as for the Small Claims Tribunal and the Residential Tenancies Tribunal[(8)]. Both of these Tribunal suffer from the very problem of being perceived as consumer oriented.

A further objection has been the intention to stop the use of arbitration in the resolution of this level of residential dispute by removing or over-riding any arbitration clause in existing or future residential construction contracts. The Minister has agreed to modify this if at the time of the dispute arising BOTH parties agree to arbitration. It cannot be part of the general contract conditions as at present[(9)]. The cost of running this system in its first year is put as high as $718,573 (£301,800) with income of $395,000 (£165,9000) thus leaving a shortfall of $323,573 (£135,900). It is planned to recover this shortfall by an additional levy on all project registrations for insurance with the HGFL, higher application fees and higher fees payable by the respondents to the hearings. A lower costing is also provided.[(10)]

## 4 Problems with the current systems

The building industry is in general agreement that some change to the resolution of small value disputes is overdue. The need for speedy, cheap yet reliable resolution procedures are overdue. The manner of this change is not so easily agreed. The attitude of the present governments in Australia appear to be socially driven with an economic overlay. The industry is not willing to pay more (or even as much) for a system which they perceive as denying them full equity.

### 4.1 Arbitration

It may be that the current problems are, at least to some extent, one of the participants own making. Many arbitrators are aware of their responsibility and so tend to reproduce the court system of operation. To a degree this is reinforced by the advocates, who practice in both the court and the arbitration arena. They thus bring the same systems with them and are most comfortable with them - pleadings, discovery etc., the formal opening and closing addresses, examination in chief and cross examination of numerous witnesses and strict rules of evidence. Many arbitrators would prefer the simpler "Expedited Commercial Arbitration Rules", but these must be agreed to at the preliminary hearing and are seldom accepted by both advocates. (11) The result is that what was originally perceived as a relatively quick and cheap settlement of a dispute by a person expert in that particular matter, has in many ways become yet another arm of the court system.

One major draw back to arbitration is the scale of costs. Arbitration costs are based on the Supreme Court scale of costs. This applies to the smallest claim of a few hundred dollars all the way to the largest dispute. Thus the taxing master has little scope to minimise the costs.

## 5 Alternative Dispute Resolution

Alternative Dispute Resolution (ADR) is regarded as including arbitration, but is moving into other forms of resolution. The Institute of Arbitrators has been running courses for prospective arbitrators for many years. There are now formal mediation courses being run to train people in the art of mediation. The organisation Lawyers Engaged in Alternative Dispute Resolution (LEADR) run annual courses in mediation techniques. In each state, groups of lawyers within the formal Law Societies are establishing sub-committees whose chief interest is in mediation. In Victoria this group is considering criteria for accreditation of lawyers involved in ADR.

Other areas under consideration in relation to dispute resolution are the community legal centres. The idea is to bring the disputes to local levels and to resolve matters without recourse to the formal legal system.

The question of neighbours' trees being but one example. Some work has been done in this area in Victoria where legislation for the Consumer Credit Legal Service has been trialing with a view to extending the system. The neighbourhood centres are widely used in NSW with some apparent success.

The research by the author will consider the existing and proposed Australian situation as well as the method of resolving similar disputes around the world. It is hoped that a body of knowledge can be accumulated which will permit the drafting of a set of criteria which can be adopted into a resolution process which might provide a cheaper, quicker and more satisfactory way of resolving small value residential disputes.

## 6 References


1 Turner H.G. **A History of the Colony of Victoria,** Heritage Publications, Melbourne 1973, 147-148.

2 Reynolds John, Edmund Barton, Angus & Robertson Publishers, Melbourne, 1948, 44-46.

3 Windeyer J.V. A birthright and Inheritance in **Tasmanian Law Review**,I 5, November 1962, 636-8.

4 Master Builders Association of Victoria and the Housing Industry Association.

5 In the opinion of the writer this is not the case and most ex-builder arbitrators may in fact tend too far towards the consumer in an effort to avoid this situation.

6 Proposal for Establishment of Building Disputes Tribunal - Victorian Ministry of Consumer Affairs - 1992.

7 Ibid - clause 18.0

8 Letter from the Minister of Consumer Affairs, Mr. Theo Theophanous dated 24 March 1992.

9 Ibid page 4.

10 **Estimated Costing for Building Disputes Tribunal** (1992) Ministry of Consumer Affairs.

11 **Expedited Commercial Arbitration Rules** (1988) - The Institute of Arbitrators Australia. - August.

12 ADR - On The Move (1992) 66 **L.I.J.** 409.

# PEACE, LOVE AND HARMONY

M. PAUL NICHOLSON
School of Architecture, University of Nottingham, England

Abstract
It has frequently been said that to meet trouble halfway, or to anticipate problems in advance, helps to solve the problem. The construction industry in the UK has an unfortunate history of disputes and litigation which has developed alongside the separation of roles and responsibilities of construction professionals.

This paper attempts to identify areas of conflict and to offer alternatives of contract methods and procedures. It will advocate that the merging of professional disciplines will assist in reducing conflict. It will also show how such systems as 'Partnering', 'Design and Build' and 'Build-Operate-Transfer contracts (B.O.T.)', offer fresh approaches to construction developments.
Keywords: Reconciliation, Partnering, Alternative Dispute Resolution (A.D.R.), Mediation.

It seems that since the construction industry became institutionalised in the nineteenth century, the modus operandi of its constituent members has become dehumanised and to a large extent litigation-driven. This has developed alongside other branches of law which in many major regards have become more humane in dealing with conflicts between two people, especially within the area of family law.

S. M. Cretney refers to the situation of people suffering severe financial and emotional difficulties being brought before the UK magistrates' matrimonial jurisdiction. This led to a widespread feeling that courts should move away from a concern with exclusively legal issues to try instead to bring about reconciliation.

The Magistrates' Courts Act 1952 gave the court power to request a probation officer to effect a conciliation between two parties. But it was widely recognised that once proceedings had started the prospects of success were not high. In some areas the practice developed of holding a preliminary meeting (or 'applications court') between the applicant, the justice's clerk and a justice, often with the court probation officer in attendance. At this meeting reconciliation was mentioned and where there was a prospect of this being brought about the probation service would be asked to look into the case. Furthermore, even if such discussion did not result in 'reconciliation' (i.e. persuading the parties to resume cohabitation) 'conciliation' might help the parties to resolve their conflicts, make decisions affecting their relationship or their children, "and come to terms emotionally with the inevitable personal problems accompanying marital distress." The Law Commission favoured the establishment of 'applications courts', but thought this was a matter best left to statutory procedure. The Commission also considered that magistrates should be "alert at all times to the possibilities" of reconciliation and conciliation, but stressed that the primary function of any court is adjudication; courts should not become too closely involved in the processes by which conciliation work is carried out.

It is now provided that where an application is made for a financial provision order the court has a duty before deciding whether to make an order to consider whether there is any possibility of reconciliation between the parties to the marriage in question. If it thinks there "is a reasonable possibility" it may adjourn the proceedings to enable attempts to be made. This power of adjournment may be exercised at any stage in the proceedings.

In the United States there are at least four different types of organisation offering mediation services:

(a) The non-profit making Centre for Public Resources. This encourages corporations and lawyers to think of and use A.D.R., provides suitable people to mediate and to preside over mini-trials and administers those procedures. The panel from which their names are drawn is described by others as a 'Famous People Panel'. Its members are lawyers and retired judges distinguished by their position and status and have no training in mediation beyond their experience. The C.P.R. is now in the process of creating local panels of 'slightly less famous people' in response to its growing success. The New York panel was published early in 1990.

(b) The non-profit making American Arbitration Association (A.A.A.). The A.A.A. provides mediators and will arrange mini-trials. It will also undertake the administration of A.D.R. procedures. Mediation under the auspices of the A.A.A. is a pre-trial requirement of the courts in some states.

(c) The 'Rent-a-Judge' organisations. These include Judicate Incorporated, which is a publicly quoted corporation and 'J.A.M.S.' (Judicial Arbitration and Mediation Service). These companies provide retired judges to undertake non-binding arbitration or mediation. They are popular. The judge will normally be expected to give his opinion of the likely result of a trial. So it can be a 'dry run' for the parties.

(d) The professional mediation companies. These companies offer the services of professional or semi-professional trained mediators who may be employed by the company. Not all these mediators are qualified lawyers but they will have a case load of up to 100 mediations and may actually mediate in 150 or more disputes in a year. So, in time, they acquire enormous experience.

A healthy development of A.D.R. has taken place in the United States in that it has grown mainly within the courts rather than 'private' intervention. This may have been prompted originally by the courts wishing to clear up a backlog of work but nevertheless the resultant situation has brought A.D.R. onto the legitimate stage. It is clearly in the interests of both the courts and the litigant to intervene early in a dispute to help the parties to recognise the strengths and weaknesses of their case and to be realistic about the true benefits of the ultimate outcome.

The adversarial relationship established by the traditional contractual framework has become so much part of the construction industry fabric that innovation in this area has been stifled. Claims and counter-claims between clients and their architects; contractors and their employees; contractors and their sub-contractors are legion. They often continue for years after the completion of a contract, exhausting the industry from energy and resources. The construction industry in fact spends more on contesting claims than it spends on research and development.

Whilst the motive for advancing A.D.R. in the United States was most probably the need to reduce court lists, the U.K. has welcomed A.D.R. with a loud sigh of relief. One of the prime movers of A.D.R. in the U.K. is the Centre for Dispute Resolution (C.E.D.R.) which involves itself with private resolutions of disputes without resort to the courts. It seems that the legal profession does not object to non-lawyers acting as mediators except in as much as those people outside the profession are not bound by any regulatory body to maintain standards or ethics. Philip Norton offers the belief that good mediators normally are lawyers, subordinate to the ability to bring parties, between whom there may be little trust, to a common position.

There appears to be a distinct and marked difference between a judge or magistrate, and a mediator. Whereas the judge is trained to reach decisions on the merits of the case (separating facts from untruths) the mediator's task is to bring the parties together without expressing his or her own view. The mediator does not drive or force the discussions but he does keep the parties talking and does insist that each side addresses the points put against them. The mediator's greatest skill is in persuading the parties to aim towards a settlement and that the case will settle.

Who is my neighbour?

The one paramount and key quality which separates the Japanese from the Western industrial culture is that of trust; trust between client and advisers and so on. The Japanese form of building contract may be contained on one sheet of paper - the party of the first part (the contractor) promises to build the 'hotel' in return for which the party of the second part (the employer) promises to pay one-third of the agreed total sum on the signing of this document, one-third when the roof is completed and the final third on completion. There is no mention of further conditions, quality standards nor 'what if' clauses - they are implicit in the culture and business attitudes of the parties.

The Japanese contractor's approach is that the client is always right and that one happy customer will seek further contracts; a business relationship based on mutual trust will develop. The Japanese cultural heritage of non-argument is probably a contributory factor in preventing potential conflicts involving Japanese firms. De Bono goes further by suggesting that 'we (the West) must replace the dialectic argument system of conflict resolution with a new idiom; the parties to the dispute are incapable of this and the intervention of a third party is essential - thus bringing the argument back in a full circle'.

Who then is my partner?

According to the national Economic Development Council's publication 'Partnering: Contracting without conflict', partnering is:

> "The relationship based upon trust, dedication to common goals and an understanding of each other's individual expectations and values. Expected benefits include imposed efficiency and cost-effectiveness, increased opportunity for innovation, and the continuous improvement of quality products and services."

In other words, it is a relationship wherein:

- all seek win-win solutions
- value is placed on long-term relationships
- trust and openness are norms
- all are encouraged to openly address any problems
- all understand that neither benefits from exploitation of others
- innovation is encouraged
- each partner is aware of the other's needs, concerns, objectives, and is interested in helping their partner to achieve this

The main objective of partnering is to meet the client's requirements in the most cost-effective way. Due to many clients' need to reduce certain overheads, such as maintenance contractors being in-house or employed on an ad-hoc basis, a system of continuous or regular partnering offers advantages. The benefits to the contractor are clearly continuity of work, the ability to plan resources, better control of quality and the ability to train which are all realised over time. The ability to provide a significant core work programme and then to retain a core team is essential to the maintenance of any ongoing partnering arrangement.

The full potential of partnering can only be achieved if both parties are prepared to become totally integrated into a single team and to be receptive to each other's requirements. The selection of suitable individuals, intensive induction and training in the requirements of the new culture are therefore necessary within the Western world whereas it appears that the Japanese, by contrast, approach all business matters on the basis of trust and from that follows co-operation.

According to Adams the U.S. Army Corps of Engineers is probably the major proponent of partnering today, its Mobile District published a guide for partnering implementation in January 1990. The Associated General Contractors of America has also recently published its own document, Partnering: A Concept for Success and has made the implementation of the process a major focus in 1992. Both of these approaches are complementary in spirit and implication.

The Associated General Contractors of America partnering process model consists of seven progressive steps:

a) organisational education
b) explicit statements of partnering intentions
c) initial and continuing top management commitment
d) partnering workshop
e) periodic evaluation
f) occasional issue escalation
g) final evaluation and 'celebration'

It appears that the critical element in this model is the partnering workshop (d) which may discuss areas around three main agenda items:

(i) barriers, problems and opportunities
(ii) interests, goals and objectives
(iii) issue resolution and team evaluation

This system of partnering which is now becoming established in the U.S.A. has yet to get a foothold in the U.K.

**Is this the way?**

Although the French have used this system for some time, the use of Build Operate and Transfer (B.O.T.) appears to have been introduced only recently into the U.K. A team from U.M.I.S.T. recently wrote a paper extolling the virtues of this method of procurement. In essence the system provides for the gathering of funds, the purchase of the land (if necessary), the design and construction of the project. All of these areas are familiar to the developer. B.O.T. extends this role further into operating the building (or bridge or toll road etc.) until the funds have been repaid and profits received; then transferring the project over to the custody of its 'owner' - the local authority or government for a nominal sum of money.

All parties to such a project are clearly focused to the one aim - the completion of the project and successful transfer to its ultimate owner/user. In order to achieve these goals the essential element of 'trust' must be implicit in all of the financial and business transactions. Such teams of firms may be pooling their resources on a one-off or regular ongoing basis.

Whilst price, value and money are involved throughout this process, human values appear to be the overriding qualities of success. Firms are selected on the basis of the staff they employ, the personal qualities of the project team and, (dare one say it?) Merit!

The time has surely come when the traditional arbitration clause in forms of contract should be replaced by one expressing the opportunities for mediation so that the whole ethos of participation and partnering, the development of team activity and joint ventures should become the norm. Even in areas where architects may be stepping into unknown territory and experimenting with little-tried materials or specifications they should harken to the warnings of Judge Newey that it needs only to be added that the warning of potential problems and approval or acceptance of the risk should be written as in an exchange of correspondence. This will remind both parties of what was said and help to avoid a dispute arising from what has been called by Lavers et al, selective amnesia. The main point of emphasis is that disputes, like accidents, can be avoided and it is possible the eliminate the traditional adversarial system which is so ingrained into the construction industry, and to have Peace, Love and Harmony as a real alternative to the gladiatorial fights in legal areas.

**References**


Adams, Michael S. **Value Management and Construction Dispute Prevention**, (1992) Architectural Management International Symposium Proceedings. Ed. Nicholson, M.P., Nottingham.

Alternative Forms of Dispute Resolution (A.D.R.). See also **A New Concept** by Kenneth Severn, given at the conference, 'New Concepts in the Resolution of Disputes in International Construction Contracts', June 15-16, 1989.

Associated General Contractors of America A.G.G. (1991), **Partnering: A Concept for Success**, Washington, pp.1-19.

Ibid. pp.14-15.

Cretney, S.M. (1979) **Principles of Family Law**, London.

De Bono, E. **Conflicts**, (1985) Harrop, London.

Fenn, Peter **Settling disputes on construction projects**, (1990) CIB.W92 Procurement Systems, Zagreb.

For a detailed description see **Court Annexed Arbitration in the United States**, by the Hon. Richard Enslen, as a paper in the series 'Current problems in Arbitration and Litigation' in 1988.

Op.Cit. ref.3.

Lavers, Anthony. **Communication and Clarification between Designer and Client: Good Practice & Legal Obligation**, (1992) Architectural Management International Symposium. Ed. Nicholson M.P., Nottingham.

Merna, A. and Smith, M.J., UMIST. **The allocation of procurement risks in 'Build, Operate and Transfer' contracts.** (1992) Las Palmas, CIB Proceedings W92.

N.E.D.C. Construction Industry Sector Group (June 1991) **Partnering: Contracting without Conflict**, N.E.D.O. London.

Naughton, Philip **Alternative forms of dispute resolution - Their Strengths and Weaknesses.** (1980) Construction Law, Vol.6, No.3, pp.195-206.

Sec. e.g. Claude Mullins: **Wife and Husband in the Courts** Lord Listowel's Summary Jurisdiction (Domestic Procedure) Bill introduced in 1935, providing for a 'conciliation summons' procedure.

U.S. Army Corps of Engineers. **A Guide to Partnering: A Process for Implementation** (1990), Mobile, Alabama: U.S. Corps of Engineers, 1-6 Attachment B.

Victoria University of Manchester v. Hugh Wilson, quoted (1985). Two Construction Law Reports 43.

# THE USE OF MINI-TRIALS TO RESOLVE CONSTRUCTION DISPUTES

G.J. SIEDEL
The University of Michigan, Ann Arbor, Michigan, U.S.A.

Abstract
The mini-trial is a structured settlement procedure that is considered by many experts to be the most successful of the new methods of alternative dispute resolution (ADR). This paper opens with an overview of three alternatives to litigation -- dispute prevention, dispute management, and dispute resolution -- and the relationship between the mini-trial and these processes. Mini-trial procedures that have been used to resolve construction disputes are then analyzed on the basis of a variety of factors, including the timing of the mini-trial in relation to litigation, the number of parties involved in the dispute, the participant representatives, the role of neutral parties, the mini-trial schedule, the amount in dispute and savings in legal costs. The paper concludes with a summary of construction mini-trial variations and recommendations for improving the mini-trial process. Especially recommended is the ADR pledge, which has not yet been utilized in construction mini-trials.
Keywords: Alternative Dispute Resolution, Arbitration, Dispute Management, Dispute Prevention, Litigation, Mediation.

## 1 Introduction

### 1.1 Alternatives to litigation

In recent years the high cost of litigation and concern about potential liability have caused litigants to search for alternatives to traditional methods of resolving disputes. These alternatives fall within three broad categories: (1) dispute prevention, (2) dispute management, and (3) dispute resolution.* The primary focus of dispute prevention is on the causes of litigation rather than on outcomes in court. Examples of dispute prevention techniques include consensus building, legal

*For more detail regarding these processes, see Siedel (1988) and the references cited therein.

audits, and dispute analyses. Dispute management involves the use of traditional business techniques such as budgetary planning to control legal costs and manage the law function. Dispute resolution, also known as alternative dispute resolution (ADR), focuses on the development of new processes to resolve disputes.

Of these three areas, ADR has received the greatest attention. Most ADR processes are based on two models that involve the use of neutral third parties: arbitration and mediation. With arbitration (like litigation) the disputing parties control the process and the neutral party renders a decision, while the mediation process (like a negotiation that does not include a third party) allows the disputing parties to control both the process and the decision (Thibaut and Walker, 1978). Of the processes based on these models, the most promising is the mini-trial, which is considered the "Cadillac of the corporate dispute resolution system" (Berreby, 1986). A structured settlement procedure that combines negotiation, mediation and adjudication (Goldberg et al. 1985), the mini-trial will be described in greater detail below.

## 1.2 Alternative dispute resolution and the construction industry

The construction industry has been especially receptive to alternative dispute resolution. The number of annual construction arbitration cases filed with the American Arbitration Association ("AAA") has grown from 2,683 in 1982 to 5,189 in 1991, with the 1991 claims and counterclaims totalling $1,425,987,801. During the three and one-half year period from January 1, 1988 to July 1, 1991, 545 construction cases were submitted to AAA mediation, with the claims ranging from $1000 to $56 million.*

The construction industry has also been willing to experiment with new forms of ADR like the mini-trial. In fact, out of sixty-four publicly-reported mini-trials from 1977 (when the mini-trial was first used) through 1987, twenty involved construction claims. This paper analyzes these twenty cases in light of the 1977 prototype, Telecredit v. TRW, which is summarized in the next section in order to illustrate the mini-trial process. In Section 3, the twenty cases are examined on the basis of a variety of factors, including the timing of the mini-trial in relation to litigation, the number of parties involved in the dispute, the participant representatives, the role of neutral parties, the mini-trial schedule, the amount in dispute and savings in legal costs. The paper concludes with a summary of construction mini-trial variations and recommendations for improving the mini-

*These figures were provided by Laurie J. Kaufman, AAA Director of Marketing.

trial process. Especially recommended is the ADR pledge, which has not yet been utilized in construction mini-trials.

## 2 Telecredit vs. TRW

The Telecredit vs. TRW mini-trial ("TRW mini-trial") originated in 1974 when Telecredit brought suit against TRW, claiming that TRW had infringed on its patent rights.* Telecredit sought damages of $6 million and an injunction prohibiting further infringement. The lawsuit proceeded over the next two and one-half years like many others, with the parties exchanging approximately 100,000 documents as part of the discovery process and spending around $500,000 to cover legal costs. Finally, with no date for trial having been set, the parties realized that the legal costs would continue to mount over the next several years and began to discuss an alternative process to resolve their dispute.

This process, called an "information exchange" by the parties but which later came to be known as a mini-trial, allowed lawyers for each side a limited time period (four hours each) to present their case to senior executives who had authority to settle the litigation. Telecredit was represented by its president and TRW by its vice-president. Following each side's presentation, a two-hour time period was scheduled for the other side's reply and for a rebuttal to the reply. The entire process lasted two days and was moderated by a neutral party (a former judge with patent law expertise), who was given authority to provide a nonbinding opinion in the event the executives did not settle the case following the attorneys' presentations. This opinion was unnecessary, however, because the executives were able to resolve the dispute in a thirty-minute private meeting without attorneys. The settlement saved an estimated $1 million in legal fees.

In addition to the obvious savings in time and money, the mini-trial provides two major benefits that distinguish it from traditional litigation. First, the process allows each company's representative the opportunity to hear the case presented by the other side's attorney before trial, a presentation that often yields new perspectives on the issues in dispute. And, second, the executives who meet without attorneys to discuss settlement are not constrained by legal remedies that are based on the assumption that litigation is a zero-sum game but, instead, can develop creative solutions that benefit

*For a more detailed account of the TRW mini-trial, see Henry and Lieberman (1985), from which this summary is derived.

both sides.

## 3 Construction mini-trials

Although mini-trials are often confidential, sixty-four mini-trials were publicly reported in the decade following the TRW mini-trial (1978-87).* When these mini-trials are analyzed in terms of type of dispute, construction claims (twenty mini-trials) represent the largest category. Seventeen of the twenty construction mini-trials were successful, one resulted in a partial settlement, and two were not settled through the mini-trial process. These construction mini-trials were not limited to the United States; one case involved the construction of a $20 million hospital in the Mariana Islands while another case dealt with a dispute over construction of a pipeline in Great Britain. In the following sections various features of the twenty construction mini-trials will be compared and contrasted with the TRW mini-trial.

### 3.1 Commencement of litigation

The TRW mini-trial took place two and one-half years following commencement of litigation. This timing may be advantageous in that the parties, having completed a significant amount of discovery, may be in a better position to discuss settlement. On the other hand, discovery represents a major litigation expense that might be reduced if expedited through a mini-trial process.

The construction mini-trials show that the commencement of litigation is a common, although not essential, prerequisite to the mini-trial procedure. Of the fifteen construction cases in which information regarding litigation was available, litigation (or a comparable administrative proceeding) had been commenced in thirteen.

### 3.2 Number of parties

The TRW mini-trial was fairly simple procedurally in that only two parties were involved. Construction disputes are frequently more complex because they may involve a wide variety of parties including architects, contractors, several tiers of subcontractors and the client.

---

*The gathering of data regarding these mini-trials was conducted by the author under a grant from the Peat Marwick Foundation. A primary source of information about the mini-trials described in this paper is **Alternatives to the High Cost of Litigation**, which is published monthly by the Center for Public Resources ("CPR") and includes occasional mini-trial reports. Several other CPR publications also contain useful data. As noted later in this paper, however, complete information is not available for every reported mini-trial.

Experience in the twenty construction mini-trials, however, indicates that a multiplicity of parties does not impede the success of the process. Fifteen of the twenty cases directly involved two parties, four cases were three-party disputes, and one case was a nine-party dispute. Furthermore, the impact of the mini-trial occasionally reaches beyond the participants. Three of the construction mini-trials that directly involved only two or three parties resulted in the settlement of multiple subcontractor claims. In one case, for example, the resolution of a dispute between a contractor and client led to the settlement of forty-nine claims by subcontractors.

### 3.3 Representatives of the participants

In the TRW mini-trial, the disputants were represented by company executives who had authority to settle. Information regarding representatives was unavailable in six of the twenty construction mini-trials. In twelve of the remaining cases, company executives (or government officials, where the government was a party) served as representatives for at least one of the parties. Other representatives included engineers or project managers (three cases), a contracting officer (one case) and in-house lawyers (one case).

The representatives were given authority to settle in nine of the ten cases in which such information was available. Information provided for four of the cases also indicated that the representatives had no personal involvement in the dispute. In contrast, one representative (a contracting officer) was selected specifically because he had personally denied the claim, as a test to determine whether mini-trials were feasible earlier in construction disputes.

### 3.4 The role of neutral parties

In the TRW mini-trial, a retired judge moderated the proceedings and was prepared to provide an advisory opinion if the parties had been unable to settle the case. In seven of the construction mini-trials, a neutral party was not used (although in one of these cases a consulting engineer was present). The failure to name a neutral has occasionally been regretted by participants in the process. In the British mini-trial, for instance, a barrister concluded that "the mini-trial would have been more useful if we had used a neutral.... [A neutral] may have provided each side with an objective assessment of the case upon which they were more willing to rely."*

In eight construction cases a retired judge served as a neutral and in the remaining cases the neutral was a law

---

*The barrister noted, however, that the general structure of the mini-trial was an improvement over the unstructured

professor (two cases), an engineer (one case), a construction lawyer (one case) or unspecified (one case). The role of these neutral parties was similar to the role played by the retired judge in the TRW mini-trial.

### 3.5 The mini-trial schedule

The TRW mini-trial took place over a two-day period and allowed each side's attorneys to present case summaries to the participating executives. The construction mini-trials followed a similar pattern. Information regarding the length of the mini-trial was unavailable in two cases; in the remaining cases the process lasted from one day to one week with an average of two days each. One of the cases varied from the TRW mini-trial in that lawyers were not present.

One interesting, and appropriate, variation from the TRW mini-trial is that at least two of the construction mini-trials were held in the buildings that were the subject of the dispute. In one of these cases, which involved a dispute over massive leaks in the glass wall of a fourteen-story building, the attorney for the client expressed the hope that it would rain during the proceeding.

### 3.6 Amount in dispute

The TRW mini-trial resulted in a settlement of a $6 million dispute, with an estimated $1 million savings in legal costs. The construction mini-trials indicate that the process is equally successful when amounts much smaller (or larger) are in dispute. In the twelve construction cases in which information was available, the claims ranged from $120,000 to $66 million and averaged a little over $12 million. In seven of these cases, in which claims averaged $10,454,510, settlement figures were also available. The average settlement in these cases was $3,880,696.

Only three of the cases contained information regarding savings in legal costs. The average savings in these cases was $279,000.

## 4 Conclusion

The examination of construction mini-trials in section 3 indicates that, for the most part, they are very similar to the TRW prototype. The analysis also reveals, however, a number of variations on the TRW theme. In some of these

---

(continued from prior page) construction settlement negotiations in which he had participated over the years. For further information, see Mini-trial Premieres in Britain (1989) **Alternatives to the High Cost of Litigation**, 7, 125.

variations, the mini-trial (1) is used before the commencement of litigation, (2) has resolved disputes among more than two disputants, (3) includes company representatives other than executives, such as engineers, who sometimes are selected because they have no personal involvement in the dispute, (4) either does not include a neutral party or uses a neutral party with engineering rather than judicial expertise, (5) may take place in the building that is the subject of the dispute, and (6) involves both small and large claims.

As noted in section 3, two of the mini-trials were failures. One of these cases was unsuccessful because the parties failed to establish ground rules for the proceeding. For instance, an adversarial atmosphere was created when each company representative arrived with several subordinates who had been personally involved in the dispute. In the other case, the failure was ascribed to the fact that the process did not include a neutral advisor. However, as noted earlier in this paper, a number of successful mini-trials were conducted without a neutral party.

In addition to addressing these possible shortcomings, parties facing potential construction disputes should consider using ADR pledges, which are conspicuously missing from even the successful construction mini-trials. Developed by the Center for Public Resources in 1984, the pledge provides in pertinent part: "In the event of a business dispute between our company and another company which has made or will then make a similar statement, we are prepared to explore with that other party resolution of the dispute through negotiation or ADR techniques before pursuing full-scale litigation."

This pledge has been adopted as a matter of corporate policy by close to 500 of the largest U.S. and multinational corporations.* The pledge is especially useful because it allows a company to propose an ADR procedure such as the mini-trial as a matter of corporate policy, thus mitigating any possible perception that it is making the proposal because its case is weak. The adoption of this pledge by construction firms and their clients will in all likelihood accelerate the use of the mini-trial, a procedure that, on the basis of experience to date, has the potential to become the premier vehicle for the resolution of construction disputes.

*For a listing of pledge signers, see 'Pledge' Registry Nearing 500 (1990) **Alternatives to the High Cost of Litigation**, 8, 9.

## 5 References


Berreby, D. (1986) Thoughts on ADR: An Interview with a Veteran Neutral. **Alternatives to the High Cost of Litigation**, 4, 3.

Goldberg, S.B., Green, E.D. and Sander, F.E. (1985) **Dispute Resolution**. Little, Brown and Company, Boston, MA.

Henry, J.F. and Lieberman, J.K. (1985) **The Managers Guide to Resolving Legal Disputes**. Harper & Row, New York, NY.

Siedel, G.J. (1988) Intersections of Business and Legal Dispute Resolution. **J. Dis. Res.**, 1988, 107-141.

Thibaut, J. and Walker, L. (1978) A Theory of Procedure. **Cal. L.R.**, 66, 541-566.

# FIDIC STUDY ON AMICABLE SETTLEMENT OF CONSTRUCTION DISPUTES

DAVID E HOLLANDS
FICE, FCIArb, FIPENZ, FIEAust, FIArbA, FArbINZ, MSPIDR, MASCE
Arbitrator & Mediator, Auckland, New Zealand
Task Committee Chairman

Abstract
This paper is a progress report on work done by FIDIC's ADR Task Committee. FIDIC is the international body of consulting engineers' associations. The committee has studied amicable settlement processes (i.e. party-controlled pre-arbitration methods of dispute resolution) in use within the construction industry of member association countries, including local standard clauses, rules, and the perceived or actual success or failure of current amicable settlement practices.
<u>Keywords:</u> FIDIC, Construction, Amicable Settlement, ADR, Conciliation, Mediation, Dispute Resolution.

## 1 FIDIC's Clauses

The FIDIC Conditions of Contract for Works of Civil Engineering Construction (Fourth Edition, 1987, known as the Red Book) deals with the settlement of disputes in Clause 67. A new provision in Sub-Clause 67.2 states:

> "Amicable Settlement - Where notice of intention to commence arbitration as to a dispute has been given in accordance with Sub-Clause 67.1, arbitration of such dispute shall not be commenced unless an attempt has first been made by the parties to settle such dispute amicably. Provided that, unless the parties otherwise agree, arbitration may be commenced on or after the fifty-sixth day after the day on which notice of intention to commence arbitration of such dispute was given, whether or not any attempt at amicable settlement thereof has been made."

Amicable settlement is also referred to in the final paragraph of Clause 2.7 of FIDIC's Conditions of

Contract for Electrical & Mechanical Works (Third Edition, 1987, known as the Yellow Book), as follows:

> "If either party disagrees with the action taken by the Engineer, or if the Engineer fails to reply to the Contractor's notice within the stipulated 28 days, and the matter cannot be settled amicably that party shall be at liberty, subject to Sub-Clause 50.1, to refer the matter to arbitration in accordance with the Contract."

## 2 ADR Task Committee

FIDIC's Alternative Dispute Resolution Task Committee (ADRTC) was asked to gather information on the use being made within the construction industries of Member Association countries of amicable settlement methods, and to make suggestions for further FIDIC action.

To carry out the task, corresponding members nominated by Member Associations reported on the situation in their countries. A number of "Friends of FIDIC" in the legal profession also assisted.

The results are summarised in a report, which at this time (April 1992) is in near-final draft form and about to be considered by FIDIC's Executive Committee. The final report is scheduled for publication in late 1992 or early 1993.

- Part I of the report contains general information and a discussion of amicable settlement processes.
- Part II of the report contains status reports from 17 countries.
- Part III of the report contains examples of rules and guidelines, used world-wide for conciliation and mediation.

The following discussion arises from the committee's work.

## 3 Amicable Settlement

Amicable settlement involves processes in which the parties retain the right to decide the outcome of their dispute, e.g. negotiation, conciliation and mediation, rather than leaving it to the imposed decision of an arbitrator or judge. Any opinion or assessment of a conciliator or mediator is not binding on the parties, except to the extent they agree with it.

Unresolved disputes undermine construction projects and hamper their successful completion. All parties stand to benefit from minimising the time and cost associated with fighting claims and, instead, working together as joint problem-solvers and enhancing their on-going relationships.

Only amicable settlement has the potential to deal properly with the technical and legal uncertainties involved and any other concerns of the parties. It offers disputants the opportunity to participate in the process and empowers them to be creative in resolving their own problems. There may also be cultural reasons for preferring amicable settlement. Furthermore, it can be particularly helpful in resolving multi-party disputes.

Amicable settlement involves bringing the parties to the negotiating table, identifying problems, establishing facts, clarifying issues, developing settlement options and reaching agreement. It works because it can control losses, contain damages, preserve working relationships, clarify issues, and eventually secure agreements. It is voluntary and always within the control of the parties to a dispute.

There may however be reasons why amicable settlement is not favoured. For example, one party may be owed money and simply be looking for the final and enforceable decision which can be obtained by resorting directly to arbitration or litigation. On the other hand, a party may owe money and seek to use amicable settlement as a delay and discovery mechanism - the other party may therefore be concerned about the delay, incurring extra costs and being disadvantaged in subsequent arbitration or litigation. Furthermore, adjudicative methods may be most appropriate for resolving some situations; such as frivolous claims, outrageous claims, claims which compromise a particular principle, cases which involve bodily injury or alleged criminality, claims to which there is an adequate legal defence, and claims which one party has no intention of settling. Other factors, such as already sour relationships, the determination to never work together again, etc, may also militate against amicable settlement.

## 4 Relationship to Traditional Legal Processes

Courts have generally been unable to provide either the necessary speed or expertise, so the industry long ago embraced arbitration as the forum of choice for construction disputes.

Arbitration can be as simple, speedy and inexpensive as the parties in dispute wish and allow it to be.

However, users have often adopted a "hands-off" approach, leaving it to the lawyers to resolve in their own way and in their own time. Perhaps because it results in an imposed decision, which is generally final and unable to be appealed, lawyers have in turn often been cautious and tended to follow more familiar litigation procedures. The result is that arbitration has sometimes become expensive, in both time and money. Arbitration is also limited, in that it can only deal with differences between the parties directly involved, in terms of their contract, the law, and remedies able to be enforced through the courts.

If negotiations stall, many parties do not consider other alternatives to settle or restructure the problem. Their immediate reaction is that there is no choice but to initiate arbitration. However, the best time for amicable settlement is when the parties are still cordial and on good speaking terms. This is generally before either party embarks on a legal route for dispute resolution. Hence it is desirable to be able to initiate amicable settlement before having to file for arbitration.

Amicable settlement still depends on consideration of the law, input from lawyers on legal issues, legal advice or representation where warranted, the fall-back position of arbitration or litigation to enforce legal rights (albeit that these may be uncertain), and court action if necessary to enforce any contractually binding agreement reached through the process. The emphasis however is on minimising the need for formal legal procedures, especially if adversarial, by involving the parties and their advisers as joint problem solvers to develop acceptable outcomes and enhance their long-term relationships.

## 5 Conciliation or Mediation

Conciliation and mediation both involve a neutral adviser in seeking to encourage the parties to communicate, understand and evaluate the other's viewpoint, and negotiate an agreed settlement.

The report analyses the different ways the words 'conciliation' and 'mediation' are used in different parts of the world. Many corresponding members perceive a difference between conciliation and mediation, but their perceptions differ. It may be possible in the future to unify these perceptions and agree on definitions for two types of processes. In the meantime it is sufficient to note that a clear distinction is sometimes made.

However, it seems that whether the process is called conciliation or mediation, existing rules and guidelines used for construction disputes generally provide for a mediator or conciliator to be able to offer an independent assessment of the merit of claims.

Independent assessment of the evidence and the merits of the claims provides legitimacy, allowing each party to review the claim (possibly seeing things in a new light) and to retreat from its previously stated position. It is therefore desirable for a conciliator or mediator to have knowledge of construction law and practice.

However, it is not just a matter of a conciliator or mediator forming an opinion and trying to convince the parties to agree with it. The success of conciliation or mediation depends on process skills and a strategy which enables the parties to reconcile their interests and concerns and develop an agreed settlement. An assessment should not be offered at too early a stage.

## 6 General

It is expected that as more engineers and lawyers become familiar with amicable settlement processes, the number of conciliations and mediations will increase.

In summary, the parties will benefit from amicable settlement processes, because of:

- savings of time and cost
- better relationships
- less job disruption
- the inclusion of all parties (e.g. subcontractors)
- an outcome which may be different, and better, to that of arbitration
- participation of the parties, to control the outcome

Amicable settlement should therefore be encouraged, but at any time and not just after referral to arbitration. The involvement of a neutral conciliator or mediator will get the best out of the process. Credible and competent dispute resolution practitioners will therefore be needed.

# Part Five

## Education

This part focuses on educational issues, relating to methods and techniques which may be utilised to prepare professionals to deal with conflict more effectively.

'Managing conflict in organizations' (Rahim) discusses the nature of conflict, styles of handling interpersonal conflict, its diagnosis, and the use of intervention to manage organizational conflict.

'Planning for disputes - educating construction management' (Bishop) argues that insufficient emphasis is placed upon problem solving in academic and professional training, and suggests areas where improvements could be made.

'Conflict in the context of education in building ethics' (Powell) explores the ethical aspects of the relation between conflict and trust, and conflict and cooperation, and puts forward ideas on student study of conflict and ethics.

'Educating construction professionals to improve the built environment' (Hancock) contends that the root of conflict in the construction industry comes from the education which construction professionals receive, and discusses issues relating to this such as; willpower, competition and specialisation.

'Construction conflict management - the role of education and training' (Franks) asserts that, by education and training, confrontational attitudes among professionals can be reduced leading to more collaboration between professionals and their clients.

'The construction industry's male culture must feminize if conflict is to be reduced: the role of education as gatekeeper to a male construction industry' (Gale) discusses the role of education in supporting the management of change to produce a fundamental shift in the industry's traditionally male culture.

# MANAGING CONFLICT IN ORGANIZATIONS

M. AFZALUR RAHIM
Western Kentucky University, Bowling Green, KY, USA

Abstract
The management of organizational conflict involves the (a) maintenance of a moderate amount of conflict at intrapersonal, interpersonal, intragroup, and intergroup levels, and (b) enabling the organizational members to select and use the styles of handling interpersonal conflict so that various situations can be effectively dealt with. The management of organizational conflict involves the diagnosis of and intervention in conflict at various levels. A diagnosis should indicate whether there is need for intervention and the type of intervention needed.
Keywords: Handling Conflict, Organizational Conflict, Diagnosis, Intervention.

## 1 Introduction

Even though conflict is often said to be functional for organizations, most recommendations relating to organizational conflict still fall within the realm of conflict resolution, reduction, or minimization. Action recommendations from the current organizational conflict literature show a disturbing lag with the functional set of background assumptions which are endorsed.

## 2 Nature of conflict

Conflict is defined as an interactive process manifested in incompatibility, disagreement, or difference within or between social entities (i.e., individual, group, organization, etc.). Calling conflict an interactive state does not preclude the possibilities of intraindividual conflict, for it is known that a person often interacts with oneself. Obviously, one also interacts with others. Conflict occurs when a social entity (a) is required to engage in an activity which is incongruent with his or her needs or interests, (b) holds behavioral preferences, the satisfaction of which is incompatible with another person's implementation of his or her preferences, (c) wants some mutually desirable resource which is in short supply, such that the wants of everyone may not be satisfied fully, and (d) possesses attitudes, values, skills, and goals which are salient in directing one's behavior, but which are perceived to be exclusive of the attitudes, values, skills, and goals held by the other party. Conflict also occurs when two or more social entities (e) have partially exclusive behavioral preferences regarding their joint action, and (f) are interdependent in the performance of their functions or activities.

## 3 Styles of handling interpersonal conflict

There are various styles of behavior by which interpersonal conflict may be handled. The styles of handling interpersonal conflict in organizations was first conceptualized by Follett (1940). She found three main ways of dealing with conflict: domination, compromise, and integration. She also found other ways of handling conflict in organizations, such as avoidance and suppression. Blake and Mouton (1964) first presented a conceptual scheme for classifying the modes (styles) for handling interpersonal conflicts into five types: forcing, withdrawing, smoothing, compromising, and problem solving. They described the five modes of handling conflict on the basis of the attitudes of the manager: concern for production and for people. Their scheme was reinterpreted by Thomas (1976). He considered the intentions of a party (cooperativeness, i.e., attempting to satisfy the other party's concerns; and assertiveness, i.e., attempting to satisfy one's own concerns) in classifying the modes of handling conflict into five types.

Using a conceptualization similar to the above theorists, the styles of handling conflict were differentiated on two basic dimensions, concern for self and for others (Rahim & Bonoma, 1979; Rahim, 1992). The first dimension explains the degree (high or low) to which a person attempts to satisfy his own concern. The second dimension explains the degree (high or low) to which a person wants to satisfy the concern of others. It should be pointed out that these dimensions portray the motivational orientations of a given individual during conflict. A study by Vliert and Kabanoff (1990) yielded support for these dimensions. Combination of the two dimensions results in five specific styles of handling interpersonal conflict (Rahim & Bonoma, 1979, p. 1327). These styles are described as follows:

### 3.1 Integrating: high concern for self and others

This involves collaboration between the parties, i.e., openness, exchange of information, and examination of differences to reach a solution acceptable to both parties. "The first rule . . . for obtaining integration is to put your cards on the table, face the real issue, uncover the conflict, bring the whole thing into the open" (Follett, 1940, p. 38).

Prein (1976) suggested that this style has two distinctive elements: confrontation and problem solving. Confrontation involves open and direct communication which should make way for problem solving. As a result, it may lead to creative solutions to problems.

**3.2 Obliging: low concern for self and high concern for others**
This style is associated with attempting to play down the differences and emphasizing commonalities to satisfy the concern of the other party. There is an element of selfsacrifice in this style. It may take the form of selfless generosity, charity, or obedience to another person's order.

An obliging person neglects his or her own concern to satisfy the concern of the other party. Such an individual is like a "conflict absorber," i.e., a "person whose reaction to a perceived hostile act on the part of another has low hostility or even positive friendliness" (Boulding, 1962, p. 171).

**3.3 Dominating: high concern for self and low concern for others**
This style has been identified with win-lose orientation or with forcing behavior to win one's position. A dominating or competing person goes all out to win his or her objective and, as a result, often ignores the needs and expectations of the other party. Dominating may mean standing up for one's rights and/or defending a position which the party believes to be correct.

Sometimes a dominating person wants to win at any cost. A dominating supervisor is likely to use his position power to impose his will on the subordinates and command their obedience.

**3.4 Avoiding: low concern for self and others**
It has been associated with withdrawal, buckpassing, sidestepping, or "see no evil, hear no evil, speak no evil" situations. It may take the form of postponing an issue until a better time, or simply withdrawing from a threatening situation. An avoiding person fails to satisfy his or her own concern as well as the concern of the other party.

This style is often characterized as an unconcerned attitude toward the issues or parties involved in conflict. Such a person may refuse to acknowledge in public that there is a conflict which should be dealt with.

**3.5 Compromising: intermediate in concern for self and others**
It involves give-and-take or sharing whereby both parties give up something to make a mutually acceptable decision. It may mean splitting the difference, exchanging concession, or seeking a quick middle-ground position.

A compromising party gives up more than a dominating but less than an obliging party. Likewise, such a party addresses an issue more directly than an avoiding party, but does not explore it in as much depth as an integrating party.

Additional insights may be gained by reclassifying the five styles of handling interpersonal conflict according to the terminologies of the game theory. Integrating style can be reclassified to positive-sum (win-win) style, compromising to mixed (no-win/no-lose) style, and obliging, dominating, and avoiding to zero-sum or negative-sum (lose-win, win-lose, and lose-lose, respectively) style.

Further insights into the five styles of handling interpersonal conflict may be obtained by organizing them according to the integrative and distributive dimensions of labor-management bargaining suggested by Walton and McKersie (1965). The integrative dimension (integrating-avoiding) represents the degree (high or low) of satisfaction of concerns received by self and others. The distributive dimension (dominating-obliging) represents the proportion of the satisfaction of concerns received by self and others. In the integrative dimension, integrating attempts to increase the satisfaction of the concerns of both parties by finding unique solutions to the problems acceptable to them. Avoiding leads to the reduction of satisfaction of the concerns of both parties as a result of their failure to confront and solve their problems. In the distributive dimension, whereas dominating attempts to obtain high satisfaction of concerns for self (and provide low satisfaction of concerns for others), obliging attempts to obtain low satisfaction of concerns for self (and provide high satisfaction of concerns for others). Compromising represents the point of intersection of the two dimensions, i.e., a middle-ground position where each party receives an intermediate level of satisfaction of their concerns from the resolution of their conflicts.

It is generally agreed that the above design for conceptualizing the styles of handling interpersonal conflict is a noteworthy improvement over the simple cooperative-competitive dichotomy suggested by earlier researchers.

The previous discussion presented the five styles of handling interpersonal conflict, such as integrating, obliging, dominating, avoiding, and compromising. Although some behavioral scientists suggest that integrating or problem-solving style is most appropriate for managing conflict (e.g., Blake & Mouton, 1964; Burke, 1970; Likert & Likert, 1976), it has been indicated by others that, for conflicts to be managed functionally, one style may be more appropriate than another depending upon the situation (Hart, 1991; Rahim & Bonoma, 1979; Thomas, 1977). Following is a list of styles of handling interpersonal conflict and the situations where they are believed to be appropriate (Rahim, 1992).

## 4 Situations where each style is appropriate

### 4.1 Integrating

a. Issues are complex.
b. Synthesis of ideas is needed to come up with better solutions.
c. Commitment is needed from other parties for successful implementation.
d. Time is available for problem-solving.

e. One party alone cannot solve the problem.
f. Resources possessed by different parties are needed to solve their common problems.

### 4.2 Obliging

a. You believe that you may be wrong.
b. Issue is more important to the other party.
c. You are willing to give up something in exchange for something from the other party.
d. You are dealing from a position of weakness.
e. Preserving relationship is important.

### 4.3 Dominating

a. Issue is trivial.
b. Speedy decision is needed.
c. Unpopular course of action is implemented.
d. Necessary to overcome assertive subordinates.
e. Unfavorable decision by the other party may be costly to you.
f. Subordinates lack expertise to make technical decisions.
g. Issue is important to you.

### 4.4 Avoiding

a. Issue is trivial.
b. Potential dysfunctional effect of confronting the other party outweighs benefits of resolution.
c. Cooling off period is needed.

### 4.5 Compromising

a. Goals of parties are mutually exclusive.
b. Parties are equally powerful.
c. Consensus cannot be reached.
d. Integrating or dominating style is not successful.
e. Temporary solution to a complex problem is needed.

In general, integrating and, to some extent compromising, styles are appropriate for dealing with strategic issues. The remaining styles can be used to deal with tactical or day-to-day problems. The above discussion on the styles of handling conflict and the situations where they are appropriate or inappropriate is a normative approach to managing conflict.

## 5 Diagnosis

The management of organizational conflict involves the diagnosis of and intervention in conflict. A diagnosis of conflict in a system is important because the underlying sources and nature of conflicts may not be what they appear on the surface. If an intervention is made without a proper diagnosis, there is the probability that a change agent may try to solve a wrong problem. This may lead to what Mitroff and Featheringham (1974) call the error of the third kind. This error has been defined by them, "as the probability of having solved the wrong problem when one should have solved the right problem" (p. 383). The management of organizational conflict involves a systematic diagnosis of the problems in order to minimize the error of the third kind. A comprehensive diagnosis involves the **measurement** as follows:

a. The amount of conflict at the individual, group, and intergroup levels.
b. The styles of handling conflict of the organizational members with superior(s), subordinates, and peers.
c. The sources of (a) and (b).
d. Individual, group, and organizational effectiveness.

The **analysis** of diagnostic data should include:

a. The amount of conflict and conflict styles classified by departments, units, divisions, etc., and whether they are different from their corresponding norms.
b. The relationships of conflict and conflict styles to their sources.
c. The relationships of conflict and conflict styles to effectiveness.

The results of diagnosis should indicate whether there is any need for intervention and the type of intervention necessary for managing conflict. The results of diagnosis should be discussed preferably by a representative group of managers, who are concerned with the management of conflict, with the help of an outside expert who specializes in conflict research and training. A discussion of the results should enable the managers to identify the problems of conflict, if any, that must be dealt with.

The above discussion presented an approach that may be used to conduct a comprehensive diagnosis of conflict. This should not be taken to mean that every organization requires such a diagnosis. A management practitioner or consultant should decide when and to what extent a diagnosis is needed for a proper understanding of a conflict problem.

Recently two instruments were designed by Rahim (1983c,d) for measuring the amount of conflict at individual, group, and intergroup levels, and the five styles of handling interpersonal conflict. The Rahim Organizational Conflict Inventory-I (ROCISI) was designed to measure the self-report of intrapersonal, and the perception of intragroup, and intergroup conflicts. The

Rahim Organizational Conflict Inventory-II (ROCISII) contains three instruments for measuring the self-report of the styles of handling conflict of an organizational member with his or her superior(s) (Form A), subordinates (Form B), and peers (Form C). These instruments use a 5-point Likert scale to measure the amount of conflict at the three levels and the five styles of handling interpersonal conflict. A higher score represents perceptions of greater amount of one type of conflict or more use of a style of handling interpersonal conflict. The test-retest and internal consistency reliabilities and construct and empirical validities of the scales in these inventories were found to be quite adequate (Lee, 1990; Rahim, 1983a,b,e; Ting-Toomey, Gao, Trubisky, Yang, Kim, Lin, & Nishida, 1991). The ROCISI and ROCISII were used to collect data from two random national samples of 1,188 and 1,219 executives, respectively. The national percentile and reference group norms of the three types of conflict and five styles of handling interpersonal conflict have been reported (Rahim, 1983e).

Data collected through questionnaires should not be the sole basis of a diagnosis. In-depth interviews with the conflicting parties are needed to gain a better understanding of the nature of conflict and the type of intervention needed.

## 6 Intervention

An intervention may be needed if there is too little or too much conflict and/or the organizational members are not handling their conflict effectively. The national norms of conflict, discussed before, provide some rough guidelines to decide whether an organization has too little or too much of a particular type of conflict. In addition to the national norms, data from interviews should be used to determine the effectiveness of the styles of handling interpersonal conflict of the organizational members.

There are two basic approaches to intervention in conflict: process and structural (Rahim, 1992). The process approach attempts to improve organizational effectiveness by changing members' attitudes and behavior regarding conflict. The process approach is mainly designed to manage conflict by enabling the organizational participants to learn the five styles of handling interpersonal conflict and the situations where they are appropriate or inappropriate. The technique of role analysis may be used to enable organizational members deal with their intrapersonal conflict functionally. Other behavioral science techniques, such as transactional analysis, team building, and intergroup problem-solving may be used to enable the organizational members to deal with interpersonal, intragroup, and intergroup conflicts, respectively.

The structural approach attempts to improve organizational effectiveness by changing the organization's structural design characteristics--differentiation and integration mechanisms, system of communication, reward structure, etc. This approach mainly attempts to manage conflict by altering the amount of conflict experienced by the organizational members at various levels. The structural interventions, such as job design, provision for ombudsman, analysis of group tasks, and analysis of task interdependence of two or more groups may be used to reduce or generate conflict at intrapersonal, interpersonal, intragroup, and intergroup conflicts, respectively.

## 7 Discussion

Organizational conflict must not necessarily be reduced, suppressed, or eliminated, but managed to enhance individual, group, and organizational effectiveness. The management of conflict at the individual, interpersonal, group, and intergroup levels involves the maintenance of a moderate amount of conflict at each level and helping the organizational participants to learn the five styles of handling interpersonal conflict for dealing with different conflict situations effectively.

An effective management of organizational conflict involves diagnosis and intervention. A comprehensive diagnosis should include the measures of the amount of conflict, styles of handling interpersonal conflict, sources of conflict, and effectiveness. The analysis of diagnostic data should indicate the relationships of conflict and conflict styles to their sources and effectiveness. Intervention may be needed when there is too little or too much of intrapersonal, intragroup, and intergroup conflicts and/or the organizational members are not effectively using the five behavioral styles to deal with different situations effectively. The process intervention is mainly designed to manage conflict by enabling organizational participants to learn the various styles of handling conflict to deal with different situations effectively. The structural approach is mainly designed to manage conflict by changing the organization's structural design characteristics. A structural intervention aims mainly at maintaining a moderate amount of conflict by altering the structural sources of conflict.

## 8 References

Blake, R. R., & Mouton, J. S. (1964). **The managerial grid.** Houston, TX: Gulf Publishing.

Boulding, K. E. (1962). **Conflict and defense: A general theory.** New York: Harper & Row.

Burke, R. J. (1970). Methods of resolving superior-subordinate conflict: The constructive use of subordinate differences and disagreements. **Organizational Behavior and Human Performance,** 5, 393-411.

Follett, M. P. (1940). Constructive conflict. In H. C. Metcalf & L. Urwick (Eds.), **Dynamic administration: The collected papers of Mary Parker Follett** (pp. 30-49). New York: Harper & Row. [originally published 1926].

Hart, L. B. (1991). **Learning from conflict: A handbook for trainers and group leaders** (2nd ed.). Amherst, MA: Human Resource Development.

Lee, C.-W. (1990). Relative status of employees and styles of handling interpersonal conflict: An experimental study with Korean managers. **International Journal of Conflict Management**, 1, 327-340.

Mitroff, I. I., & Featheringham, T. R. (1974). On systemic problem solving and the error of the third kind. **Behavioral Science**, 19, 383-393.

Prein, H. C. M. (1976). Stijlen van conflicthantering [Styles of handling conflict]. **Nederlands Tijdschrift voor de Psychologie,** 31, 321-346.

Rahim, M. A. (1983a). Measurement of organizational conflict. **Journal of General Psychology**, 109, 189-199.

Rahim, M. A. (1983b). A measure of styles of handling interpersonal conflict. **Academy of Management Journal,** 26, 368-376.

Rahim, M. A. (1983c). **Rahim Organizational Conflict Inventory-I**. Palo Alto, CA: Consulting Psychologists Press.

Rahim, M. A. (1983d). **Rahim Organizational Conflict Inventory-II, Forms A,B, & C.** Palo Alto, CA: Consulting Psychologists Press.

Rahim, M. A. (1983e). **Rahim organizational conflict inventories: Professional manual.** Palo Alto, CA: Consulting Psychologists Press.

Rahim, M. A. (1992). **Managing conflict in organizations**. New York: Praeger.

Thomas. K. W. (1976). Conflict and conflict management. In M. D. Dunnette (Ed.), **Handbook of industrial and organizational psychology** (pp. 889-935). Chicago: Rand McNally.

Thomas, K. W. (1977). Toward multi-dimensional values in teaching: The example of conflict behaviors. **Academy of Management Review**, 2, 484-490.

Ting-Toomey, S., Gao, G., Trubisky, P., Yang, Z., Kim, H. S., Lin, S.-L., & Nishida, T. (1991). Culture, face maintenance, and conflict styles of handling interpersonal conflict: A study in five cultures. **International Journal of Conflict Management,** 2, 275-296.

Van de Vliert, E., & Kabanoff, B. (1990). Toward theory-based measures of conflict management. **Academy of Management Journal**, 33, 199-209.

Walton, R. E., & McKersie, R. B. (1965). **A behavioral theory of labor negotiations: An analysis of a social interaction system**. New York: McGraw-Hill.

# PLANNING FOR DISPUTES - EDUCATING CONSTRUCTION MANAGEMENT

DAVID F. L. BISHOP
David Bishop Associates, Doncaster, England

Abstract
Contracts described as farragos of obscurity, an array of procurement methods, multi disciplinary teams with operations from unskilled manual to high technology, claims mentality, a volatile supply and demand market - the Construction Industry is synonomous with disputes.

Historically, construction attitudes have tended toward a client on one side of the table, contractor on the other each with their own professional teams. The spirit of `them and us' perpetrating suspicion and mistrust.

Perhaps the recipe for disputes is already laid down in cultural roots with the educative process doing little to look critically at the quality of management techniques both in academic and workplace environments.

Does our approach towards problem solving create disputes rather than settle them? This paper focuses on the role of the management disciplines in contributing to conflict. Personal experience has revealed that the construction industry tends to tackle problems on a retrospective basis. Potential solutions become live problems before being recognised and dealt with. Conflict becomes inevitable, working relationships soured; the ingredients necessary for effective and efficient construction destroyed.

Construction is a complex operation. I believe not enough emphasis is placed on problem solving/avoidance involving specific management disciplines and promoting the positive benefits of teamwork.

This paper seeks to identify areas where management attitudes are inadequate in the construction process. In conclusion, I would suggest how my own professional discipline, Chartered Quantity Surveying could benefit from reviewing aspects of a Quantity Surveyor's academic and professional training in relation to the views proposed for the principles behind construction process conflict.
Keywords: Conflict, Disputes, Education, Management Skills, Psychology.

## 1 Introduction

`Men keep their agreements when it is an advantage to both parties not to break them' observed Solon over two and a half thousand years ago. Not much has changed - well, certainly not in our construction industry anyway!

Building activity is a complex process involving many disciplines with differing skills. Wherever a number of diverse elements are drawn together, grey areas open to alternative interpretation arise. The evolution and translation of a client's requirements into an on-site physical presence carries with it a risk of the unpredictable uncertainty leading to a difference of opinion. We freely describe the result as conflict or dispute. How we use these words is a good indication as to whether our behaviour is aggressive or conciliatory. The term `dispute' can have a hard or soft meaning. A softer use is to interpret a dispute as a contention in argument or to debate indicating a positive approach to reaching an agreeable conclusion. However, the word is often used to project strife, a conflict describing a clash of views, a negative battle of position. This is a subtle but important point with roots in psychology which I will return to later.

Quantity surveyors are a recognised profession within the British construction industry but their origins to a certain point in time remain the subject of conjecture. Tradesmen such as masons, carpenters and joiners can be traced back to the middle ages. The traditional system was for the future building owner (the client) to pay for materials delivered to the site direct and the various classes of artificer according to their skill. This relates to the modern equivalent of cost plus (but without the plus!) and incorporated a degree of design development during the building programme. It will be noted that the client had no certainty over the final cost.

Inigo Jones was one of the pioneering architects who, in the early seventeenth century developed complete building designs from which a new way of paying for buildings emerged. It was possible for master craftsmen to be paid based on known quantities of work initially valued after completion, hence the term `measure' and `value'; an intermediate step in controlling the value of the work done. This later developed into piecework rates agreed beforehand on quantities based on drawings of the completed building, finalised before construction work commenced, the responsibility of the measuring often being left to the craftsmen themselves. This is the basis of current procurement practice today although it should be remembered that the basis of execution was still as a series of sub contracts. It was not until the nineteenth century when the modern equivalent of single cell trade contracts with quantity surveyors representing both paymasters and builders came to the fore.

Although a quantity surveyor (the later day version of a measurer) is sometimes referred to as an independant professional, the natural tendancy is to look after one's client's interests. It is recorded that an emminent 17th century measurer had a dispute about `different modes of measurement' with a certain Mr. Leonard Sowersby on the subject of chimneys. It was also noted at the time, the custom was to pitch one measurer (acting for the master craftsmen) against another (acting for the paymaster). A `them' and `us' culture has existed for at least 300 years.

In the period from building work inception to completion, no document can fully cover every eventuality and interpretation of production. The resolution of problems can be a very personal or confidential affair and not surprisingly, there is a dearth of statistical information relating to the nature, frequency and scale of building disputes. Most differences can be resolved within the framework of a pricing control document and legal document. However, experience reveals that many seek to exploit vagueness and weaknesses in the system in the pursuit of self interest, ignoring the common purpose of the parties. I hear the words of Solon ringing in my head!

## 2 The Problem

It is not the purpose of this paper to explore in detail the reasons behind the approach of apparent self interest (by parties from either side of the `table'), that being a topic in its own right. Only those that run the full term to a court judgement are reported in detail. One can only summise at the scale of those that have been grudgingly settled along the way with the thoughts of `I'll never work with them again!' From experience, settlements are often protracted unecessarily causing additional costs and loss of goodwill which could invariably have been avoided with a different approach.

I believe that therein lies the heart of the matter. Problems arise because traditional views often perceive a different interpretation between the goals of self interest and the mutual interests of the working partnership. The hard edge meaning of disputes is adopted.

Negative psychology has pervaded our cultural base leaving us with real difficulties in the way we implement our construction process with conflict inevitable. Mistrust and suspicion are prevalent thoughts. In my view, the key to a more productive future is in the word ATTITUDE.

Successful activities demand a conscious understanding of management disciplines and techniques. I would categorise some of the main issues relevant to this paper thus:

1. Commercial management
2. Operational management
3. Personnel management
4. Problem solving

**2.1 Problem Solving**

Taking my point on attitude, let me illustrate this by contrasting Eastern and Western approaches taking point Number 4, problem solving as an exampled application.

For instance, Japanese philosophy in the case of differences of opinion is on the lines of `what a wonderful and perfect idea - it cannot be improved - now let us explore that view'.

Contrast this with arguments you may have had where, if honesty prevails, there has been little incentive for you to pick out the best of an opposing viewpoint and any coming together is regarded as a reluctant compromise or a stand down. Our historic approach to arguments can be broken down into a series of syndromes:

(a) trial of strength - it is the strength of a point of view (eg. you hold the money and can afford to fight) rather than the excellence of that view.
(b) entrenched position - each side becomes more rigid with a reluctance to develop ideas that are different from those that clash. This moves towards a standstill in progress which continues over time generating the human reaction of not wanting to be the first to `give in' or to be seen as weak by others.
(c) defeatist or negative comparing - creativity and ingenuity in thought is not used constructively in improving solutions but to destroy an opposing idea. The subtle result is that the idea that wins may be construed as stronger but is not necessarily the better one. Examples of this syndrome may be in finding an inaccuracy in one small part of another's case and exterpolating to cast doubt over the total case. Alternatively, concerns could be expressed on a more personal level bringing into question a person's capabilities. Another method is to work a hard/soft routine by opening discussions in cold dialogue often with threats, using intimidating body language or staging walkouts. This is reinforced by logistical assessment of future consequences say on the lines of `we will see you in court', or `you will never work in this town again'.

With disputes, it is not uncommon to set out to prove someone wrong, to link any weakness with casting doubt on the validity of other points of view (a magnitude effect). Tactics may become personal with character assassination or driven by the need to impress others, to set an emotional tone, or to doubt particular interpretation or confusing the issue by identifying possible consequences of projected future or deeper developments. Very simply, do we in the U.K. construction industry cause the problems by adopting an attitude of viewpoint ownership which we must then vigorously protect to maintain a personal status?

Compare this policy to Japanese tradition where:

(a) existing ideas are not attached and may be returned to discussions without discredit. Ownership problems are avoided so that opinions are not set against each other but are evaluated collectively.
(b) partner (or partners) are encouraged to adopt a lateral and creative exploration of the subject from the outset.
(c) there is no need to show fault or error as it is possible to introduce an alternative idea which has merit.
(d) all the time is spent in a positive manner being jointly designed and evaluated.

Is Edward De Bono right in suggesting `conflict is an expected and revered idiom in our civilisation?' Yet in reality `everyone is right and no one is ever right'. I would suggest that attitudes displayed within the construction industry create the right conditions to fulfill the first statement and what is needed is a greater recognition of the truth in the second view.

Conflict emerges from two corridors; one route where the management process has taken an early decision to create a situation in the certain knowledge of a fight. In difficult trading conditions we can see by tenders returned at below cost price that the likelihood of argument and exploitation is significantly increased in order to improve returns to the contractor. Client paymasters can also suffer the same affliction as worsened cashflows concentrate the minds on frustating payments.

Those that walk the other corridor, often do so in ignorance with breakdown in management practices leading to situations that require recovery retrospectively. Human error or inadequacies account for many of the problems. Referring to the building process and the way information is assembled, distributed, interpreted and acted upon, it is easy to envisage many potential weak links in the chain. I maintain that improved management skills overseen by Total Quality management systems with Quality Assurance as a major component significantly reduces the opportunity for reflective crises.

I have previously referred to four management areas crucial to an improved way of delivering building work and have already discussed a proposal for a changed attitude to problem solving. A recognition of other issues should include an understanding on:

**2.2 Commercial management/** corporately and individually the difference between efficiency and effectiveness should be clear. Both are linked but the balance between them varies on the type of product or service provision. Even within the construction industry, this balance will change between say the construction of a retail supermarket and the remodelling of an estate incorporating community involvement.

In commercial terms, efficiency is frequently measured by the bottom line, the profit level, whilst effectiveness is a measure of performance in achieving efficiency (eg. how much turnover has been secured or how a project was run and is everyone content). It is useful to note here the importance of this relationship in the context of the wider mutual interests - a thread woven into my opening remarks where the climate for mutual self interest is promoted.

**2.3 Operational management/** as an explanation of poor performance, the cry for better people is heard. Peter F. Druker highlights the fallacy of this belief with comment to the effect that `if we cannot organise the task so that it will be done adequately by people who only try hard, then it cannot be done at all'. More often than not, the fault is in the system not individuals. Having said this, the individual needs support in a broader awareness of the management skills which would compliment technical knowledge. This means communications should be meaningful, constructive and clear. One of the worst contributions to conflict arise from not setting out what is really meant. Intangible or unquantifiable objectives open to inferrment or loose interpretation are a recipe for disaster.

**2.4 Personnel management/** the way people are treated primarily affects their responses and attitudes. Personnel management governed by a monetory reward system only, to the exclusion of human reactions other than threats does not create a productive environment. This has clearly been established in the work of Elton Mayo in the mid 1920s with the Hawthorne experiments. What this research demonstrated was that there is `something far more important than hours, wages or physical conditions at work - something which increases output no matter what was done about physical conditions'.

In essence, the cause was attributed to a change in staff attitude both towards work and colleagues. By seeking help and co operation from staff, the senior management instilled a sense of importance in the workers. The attitude changed from individual actions to a cohesive group contributing to a corporate goal. Senses of stability, purpose, ownership, belonging, and recognition created a productive rather than destructive attitude.

## 3 Education and Training

Looking at the construction industry and considering various activities, whether it is a brickwork subcontract, gang or site construction management team, the architects or quantity surveyors' office teams, each plays an important role in the overall process. How much of the conflict situation is aggravated by poor personnel management I wonder? A recent case in point was where in the role of Employer's Agent under a design and build contract, I had cause to be critical of the standards of workmanship of finishing trades. Each trade was isolated and risked not functioning as part of a team. Without this sense of purpose, time, money and frustration was spent putting right work which, for one reason or another should have been right first time. WRONG ATTITUDES COST!

Education and training initiatives in the construction industry are continuing to improve standards, particularly in management skills. For supervisory or professional disciplines, the education system is often via a technical teaching establishment. This becomes less so with trades related skills which tend to have a different training regime concentrating on practical skills. It should be noted that continuous learning does not replace specific training. Workers require a need, to be motivated through responsibility in undertaking a job which should be capable of being carried out. Training provides the technical understanding to enable this sequence to happen and is workplace based with the employer leading the way. Continuous learning is about improving self performance arising from the cycle of responsibility in the way work is produced with feedback information providing a learning curve reapplied at the level of responsibility.

The Industry is evolving a series of benchmarks based on the National Vocational Qualification `building block' system. The professions particularly are examining the module content required for reaching approved standards. The construction professionals are also exploring aspects of commonality amongst teaching and training subjects in pursuit of a cohesive approach to excellence. By necessity individual elements are taught in isolation. It is crucial that business and educational establishments work closely with each other to ensure that these seperate elements are brought together generating practical experience in real life applications. It will enable students to appreciate the importance of thinking laterally and responding to the shift in attitude required in a teamwork environment.

So how does this sit with the education and training of Chartered Quantity Surveyors? The usual education route is through a polytechnic or university, either part time or on a sandwich basis (approximately one year in a working situation). I have been interested in the management content of a syllabus, taking the technical side as read. An introduction to management theory starts early in a student's course: in the second year of study, specific modules are introduced with ninety hours allocated for management theory and a further ninety hours concentrating on the practical application of these skills in construction management, how to implement and run a project. Further topics extend a candidate's knowledge in areas of problem solving including new techniques of Alternative Dispute Resolution. This work is complemented by interactive sessions combining the two areas referred to earlier of training and continuous learning.

With this background where do the problems occur? My thoughts on this are:

(a) Transferring this knowledge into a workplace environment that does not respond to or encourage its use.
(b) Dealing with other construction industry colleagues who do not share the same understanding.
(c) Quality of understanding, interpretation and application by an individual.
(d) Cultural attitudes.

## 4 Conclusion

By way of illustration I have observed from my own role as a practitioner in partnership arrangements, management consultancy and quantity surveying that most common areas of difficulty arise, notwithstanding experience levels, through lack of forward planning or foresight, breakdown in communications internally and with clients and co professionals, vague objective or task description, poor decision making approach and a failure to differentiate between self and mutual interests, particularly in the field of problem solving. To create an improved positive climate that will reduce the potential for disputes, the total education including training of Chartered Quantity Surveyors requires particular attention to the following:

(a) **Effective decision making** - western culture tends towards answering the questions. Eastern culture moves towards defining the question with answers flowing as a natural consequence. How many times do we find that we have answered the wrong question? Dissent and discussion is healthy, the freedom to express an opinion should be encouraged. Facts are important but

should be relevant to the discussion framework.
Having come to a decision:
Who has to know?
What action has to be taken?
Who has to take it?
What has to happen so that those who have to take the action can do it?

(b) **Communication** - there is a distinction between communication and information. It is the recipient who is the communicator because it is they who perceive (interpret) what is being conveyed. The way that the communication is made needs to be clear and concise; the way in which it is put across can affect the recipient's response. Studies suggest that the more emotional the demand through communication, the less is retained by the receiver. Information is definitive but communication is a perception; the two are interdependant.

(c) **Continual learning** - ask the following questions of yourself:
(ci) Have you learned to be more productive?
(cii) Have you learned how to perform the task better?
(ciii)Have you improved the method of achieving the task?
(civ)Do you need more knowledge to improve any aspect of (ci) to (ciii)?
(cv) Do you need additional tools to improve any aspect of (ci) to (ciii)?
(cvi)In preparation for additional learning do you need to adopt new methods of working, levels of understanding or performance capabilities?

The polytechnics and universities can play a pivotal role in drawing together the aspects of training and continual professional development of quantity surveyors working closely with employers and individuals in a structured programme geared to the workplace, eventually extending the target groups to clients and other professionals.

However, the most important aspect is how to use the skills creatively not destructively. From the early days of the measurer self interest has evolved as the platform for disputes, perpetuated to the present day. Indeed some may argue that the more we try to cater for every eventuality in our documentation, the greater the certainty of disagreement. Alternative Dispute Resolution is the first step to willing parties seeking solutions to problems via a skilled intermediary. We should use this initiative to modify our approach at the outset of establishing a working relationship. ATTITUDES MUST CHANGE. No amount of carefully constructed documentation will eradicate the opportunity for a dispute should one party have a mind to go down such a route.

Planning for disputes through improved attitudes and management skills starts with the early education process. The proliferation of partnership style projects demands a proactive response in management style and breaking down the barriers between self and mutual interest of all parties. This paper draws attention to areas that require special attention in the learning and work experience of a chartered quantity surveyor with the most controversial being the need to change attitudes.

I return to the words of Solon, clearly a man of vision `Men keep their agreements when it is an advantage to both parties not to break them'. Quantity Surveyors should take the lead in ensuring that no such advantage exists.

**References**


Argyle, Michael (1965) **The Psychology of Interpersonal Behaviour**, Pelican, London.

Brown, J.A.C. (1970) **The Social Psychology of Industry** Pelican, London

Bono, Edward De, **Conflicts, A Better Way to Resolve Them**, Harrap, London.

Drucker, F. Peter, **Management Tasks, Responsibilities Practices**, Heinemann, London.

Francis, Dave (1990), **Effective Problem Solving**, Routledge, London.

Plutarch, **Lives: Solon**

Thompson, FML (1968), **Chartered Surveyors, The Growth of a Profession**, Routledge and Keegan Paul, London.

Tillett, A., Kemper T. and Wills G. (1970), **Management Thinkers**, Pelican, London.



Acknowledgement to Barbara Colledge, module co ordinator, Leeds School of the Built Environment for background information of syllabus content of Quantity Surveying courses recognised and approved by the Royal Institution of Chartered Surveyors.

# CONFLICT IN THE CONTEXT OF EDUCATION IN BUILDING ETHICS

MICHAEL POWELL
Anglia Polytechnic, Essex and Cambridge, England

Abstract
This paper argues that, as conflict, whether promoted, resolved or prevented, is part of the building process, it is properly subjected to ethical scrutiny within professional education. The nature of building itself, ethics and conflict are considered and questions posed arising from the assumptions made. The ethical aspects of the relation between conflict and trust and between conflict and co-operation are explored. Examples are given of how the ethical discussion of conflict can be related to the stories of building projects and to illuminating scenarios from general English literature. Questions for discussion are included in an Appendix.
Keywords: Education, Building, Ethics, Conflict.

## 1 Introduction

This paper investigates how the subject of conflict can be addressed in the context of broad-based education in building ethics. The rationale for this is that the incidence, management, promotion, resolution and prevention of conflict as part of the building process cannot be excluded from ethical scrutiny.

In the UK this scrutiny takes place in the context of a pluralist society. Values, attitudes and views are not handed to us ready-made from any particular tradition or school of thought. Each person has to be reflective and reason them out for himself/herself. This paper suggests how that reflective and reasoning process may be developed.

We need to be clear what we mean by education. The Fontana Dictionary of Modern Thought (eds Bullock et al 1988) offers three definitions:

> A passing on of a cultural heritage.
> The initiation of the young into worthwhile ways of thinking and doing.

The fostering of the individual's growth.

While in building ethics we have a cultural heritage to pass on, such as that of many old-established family building firms, and while we must initiate the young, it is the fostering of individual growth and development in understanding and reasoning that matters most. It is individuals who will determine our diverse and pluralist future, in Building Ethics as much as in everything else.

## 2 The nature of building, ethics and conflict

If we are to consider the matter of conflict in relation to building ethics, we must first say what we mean by, or consider to be important about, building, ethics and conflict.

### 2.1 The nature of building

For the purpose of this paper, three things are particularly important about building. First, building involves everyone in the population. From the smallest child to the most senior citizen, we all see buildings, go into them, love or hate them and make 'spiritual' responses to 'material' things. We all have an interest at stake, either directly or indirectly, in good and ethical building.

Second, building is a part of life upon which many diverse sciences, arts and philosophies converge, for example the science of structures, the science of economics, the art of architecture, the art of management and the philosophy of ecology. Underlying this diversity, is a network of interacting values. I have discussed elsewhere (Powell 1991) the nature of these values. The essence is embodied in Fig.1. Building, the Figure suggests, is subject to ten categories of value, each of which is related to all the others.

Third, building is often too emotionally painful an activity to be engaged in the raw. We turn it into a game. For people and life, we substitute rules and roles. We dehumanise, arguably in the interests of humanity, one of the most human of activities.

Our ethical consideration of conflict, therefore, presupposes that building involves everyone, is value-laden and is often 'played as a game'.

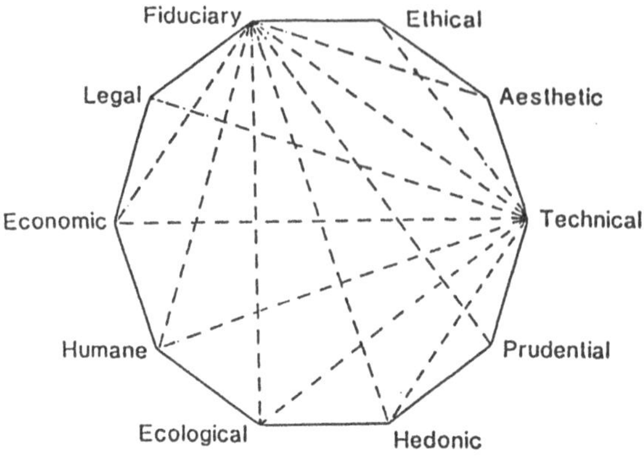

Fig.1. Values relating to building.

## 2.2 The nature of ethics

The Fontana Dictionary of Modern Thought (eds Bullock et al 1988) defines ethics as follows:

> Ethics is the branch of philosophy that investigates morality and, in particular, the varieties of thinking by which human conduct is guided and may be appraised. Its special concern is with rightness and wrongness of actions, the virtue or vice of the motives which prompt them, the praiseworthiness or blameworthiness of the agents which perform them and the goodness or badness of the consequences to which they give rise.

Building Ethics is this statement applied to the product and the process of building. Working out what it means, is both an intellectual and a practical process.

## 2.3 The nature of conflict

Robbins (1974) suggests that there are three approaches to conflict:

| | |
|---|---|
| TRADITIONAL | All conflicts must be avoided. |
| HUMAN RELATIONS | Conflict is natural and inevitable. |
| INTERACTIONIST | A positive force, essential for the best performance. |

The implication of this analysis is that the building world, or each microcosm of it, has to judge whether it wishes to avoid conflict, accept it, or benefit positively from encouraging it.

de Bono (1985) offers an alternative analysis. He criticises the Western tradition of elevating conflict on the basis that excellent syntheses between interests rarely occur. What normally happens is that one party wins and the other loses. He argues strongly for the collaborative and creative design of new possibilities as the way to resolve impasses. He recommends the PM1 method of mapping, in which pluses, minuses and interesting (neutral) possibilities in any situation are explored. He gives a low rating to negotiation as a means of resolving conflict. Our consideration of the ethics of conflict must take account of this diversity of view.

## 3 Questions arising

The assumption made in Section 2 about the nature of building, ethics and conflict give rise to ethical questions which are properly considered in the educational context.

### 3.1 The morality of conflict

We start by asking about the morality of conflict . Is the avoidance, acceptance or practice of conflict good or bad? If we avoid conflict, will that lead to a good outcome? It will if it maintains needed harmony. It will not if it sweeps under the carpet what should be revealed. If we place a high value on creative conflict, what will happen? Either a creative breakthrough may occur or a crisis which it would have been better not to have risked.

What are the motives that people bring to the decision on whether to avoid or stimulate conflict and how ethical are they? People may argue that an attitude to conflict that benefits a project is good. An attitude to conflict that advances one's own benefit may be seen as good. On the other hand, the avoidance or promotion of conflict, the only effect of which is to do harm to another, is, prima facie, unethical.

### 3.2 The universal interest in building

We have said that 'everyone' has an involvement and an interest in the fabric of our environment. What are these interests and what conflicts may pertain to them? At the most basic level, we are all interested in safe structures that are pleasant to look at and in builders who do not engage in antisocial behaviour such as putting mud all over the road. Often the public interest will go much deeper. Should a particular building be built at all? What importance is given, for example, to ecological considerations? What weight do we give to conflict (or the avoidance of it) that advances the broad, public interest? In what kinds of circumstance should conflict that is in the public interest override avoidance that is in the private interest (or vice versa)?

### 3.3 Conflict and values

Can we, ever properly understand conflict without first understanding the values and attitudes of those engaging in it? de Bono (1984) uses the illustration of three men each of whom is holding a piece of wood and lets go of it. In one case, the wood falls; in the second it rises; and in the third it stays where it is. de Bono explains that each man is in a different 'universe'. One is in a normal gravity-controlled universe, the second is underwater and the third is in space. We must ask whether we can make any judgements at all about conflict, unless we understand first the 'universes' of value and meaning inhabited by the people who involved in the conflict. Does not conflict arise, ultimately, at least on occasions, out of our perception of what matters? If, for example, we were to believe with Sieff (1990) that 'people matter most' followed by quality, that would colour all our thinking.

### 3.4 Conflict and building 'games'

For many reasons, people have chosen to develop 'games' for building. The various rules and procedures relating to building contracts may be regarded as games. Where games are played, people can become confused with roles, and ethics confused with mere rules. If conflict arises out of the roles and the rules, we must ask about the ethics of the assumptions underlying them. Is immorality ever justifiable in a set of contractual rules, or in the application of them in particular situations? Is the ultimate issue what the rule says, or what is right? When does right override rule? Should we not judge conflict against right, as well as against rule?

### 3.5 Creative design to resolve conflict

de Bono, we have seen, argues that the creative design of new possibilities is the best response to conflict. What does it mean to be creative? What does it mean to design? It means to see, compare and choose between the known possibilities that everyone, or at least some other people, can see and other possibilities that other people cannot see. When we are thinking about designing buildings, we are inclined to see conceptual design as a one person activity but, in relation to extricating ourselves from conflict, de Bono warns against this. To be effective design must be collaborative. How big a challenge is that for people in building coming from different 'universes' of value?

## 4 Widening the discussion

So far, this discussion has been limited to questions

about conflict in direct relation to building and ethics. That is very artificial. To be more realistic, we must spread the discussion more widely.

In a module or study unit on building ethics, conflict will be one of a number of themes. We need to relate conflict to other themes. Two such relationships are now considered.

**4.1 Conflict and trust**

I have summarised elsewhere (Powell, 1990) the broad, initial findings of some research on Trust in Building. Trust pertains to relationships, the future, confidence, morality and mutuality. When I say that I trust my builder, I am saying that I have confidence that tomorrow, whatever circumstances arise, he will behave well towards me, recognising that we both have a stake in our relationship. I need to have my house built properly at a fair price and he needs to maintain his business and enhance his reputation. Generally, trust only comes about gradually and painstakingly. It may be perceived as a kind of climate.

Now, can we have both trust and conflict at the same time? Suppose that both my builder, who is building me a house, and I want to come out of our present relationship, knowing that our trust in one another has been vindicated and has grown. In such a case, any risk or incidence of conflict must be managed in the interest of trust. The worst thing that can happen is that an issue is swept under the carpet. Trust requires that the issue be faced. Two main possibilities present themselves. The first is to let the conflict have its head, up to the point at which it starts to erode trust. This is a high risk strategy. Working through the conflict may increase trust or it may, like a forest fire, get out of hand and destroy everything in its path, including trust. The second possibility is to say that the relationship of trust is worth more than conflict could produce and, further, that its very nature makes possible the collaborative and creative design so strongly recommended by de Bono.

**4.2 Conflict and cooperation**

A formative thinker on cooperation as a mode of human behaviour is Axelrod (1984). He believes that people can be taught to behave cooperatively, not because of some dream or even belief, but because of practicality. His researches, using the computer game Prisoner's Dilemma, have shown that a presumption in favour of cooperative behaviour is beneficial to both players of a game. The recommended tit-for-tat strategy depends on each player starting to be cooperative and not being the first to renage.

Lloyd (1990) is concerned about the persistent view that companies are 'nasty, brutal and thoroughly untrustworthy creatures'. His remedy for this undesirable situation is not to pretend that companies can become moral but that, following Axelrod, they should learn the selfish benefits of cooperative modes of behaviour. In their own selfish interests they should cultivate a personality that is perceived to be honest, fair, responsible and generally well mannered.

How does this relate to conflict? If, in my relationship with my builder, I abandon cooperation and cut and run to establish an advantage over him, he has two choices of action. The first is to get out of his relationship with me as quickly as he can, and that might be prudent, or he has to stand and run the risk of fighting, and that, to a greater or less extent, is to engage in conflict. In other words, he has a choice between being prudent and seeking justice for himself. If he goes away, he leaves other builders, and the butcher and baker, prey to my undesirable behaviour. If he seeks justice for himself, he is, arguably, working for a greater good than just his own. That argument has to be weighed against de Bono's thesis that, by and large, engagement in conflict does not produce good and creative results.

## 5 Relating the discussion to concrete situations

The discussion so far in this paper has been conducted in relatively abstract or conceptual terms. I make no apology for that, believing it to be an important part of the education of the honours graduate, and, perhaps all professionals, to engage in such debate. However, unless such discussion is focussed on specific, concrete situations, it is liable to remain inconclusive and speculative. How then can we make the discussion concrete?

### 5.1 Personal experience

Students undertaking sandwich courses, part-time or post experience courses, are likely to have experience of their own which raises questions about conflict and ethics. It is essential that connections be made with such experience. If they are not, there is a risk that the student will keep the ethical discussion of conflict in a separate mental compartment from practical experience. That is disaster. Experience has to be used as part of the educational process.

However, raw experience may be unusable for educational purposes. It may be too crude and too imbued with emotions, such as satisfaction or anger. Subjective perceptions may be too dominant and dominating. To be usable educationally, experience has to be reflected upon. The student has to step outside the situation and view it with more objectivity. This can be done but it is a difficult process.

### 5.2 Building stories

Stories of particular building situations, with which the student is not connected, may provide better material with which to work. Positive stories about the absence or resolution of conflict are available in, for example, the reports of the annual Building Manager of the Year Awards or special reports such as those of the Broadgate and Canary Wharf projects.

In the case of Canary Wharf, it may be argued that the most fundamental ethical question concerns the relationship of the development and the developers with the indigenous population of the area. Should there have been conflict about the effects of the development on the community? Was there? If it was managed or resolved, how, and to whose benefit? Pro-development reports may not discuss such questions but pro-resident reports may. Both need to be considered.

Longer and more detailed building stories such as 'Skyscraper' (Summagh, 1989) provide wider scope for study. Half a dozen major conflicts or risks of conflict can be identified. Students can be asked to comment upon them. A useful exercise may be to use the text as a source, from which to construct a mini-play relating to the handling of the conflict-prone situation.

Live, local stories may yield good material. As I am writing this paper, a developer is speaking to a colleague of mine offering to do a one hour talk on how provision for conflict avoidance or resolution is being made in the contract terms for a major East Anglian project. This would be an interesting example of how building game rules are being adapted to preclude or minimise conflict. From an ethical point of view, the needed response is not necessarily a strong 'hoorah' but questions about how power is being used. It may be that the developer is using his power justly, as regards himself and his builder. On the other hand, he may not. What price would then be being paid for the avoidance of conflict?

### 5.3 General literature

I am strongly committed to the view that the discussion of ethical issues must begin at the beginning of a student's career. Occasionally a student will take readily to conceptual discussion but most students are unlikely to do so. They need concrete anchors. Practical experience is either non-existent or minimal and even where it exists, it is subject to the limitations indicated above. While some of the simpler building stories may be usable in discussion, only a limited amount of the story is usually available, and perhaps from only one point of view. Given this dilemma, I have started to turn to English literature as a resource and am finding it effective.

Richard Adams' 'Watership Down'(1972) has proved eminently usable as a first year under-graduate management text and students have needed little prompting to see the implicit ethical discussion. They can identify issues such as the conflict between the various rabbit communities and between the individual members of the group escaping to Watership Down as, through times of danger, risk and uncertainty over technical issues, they weld themselves into a team and a community. It is somewhat poignant that the need for the exodus in the first place is the arrival near their home of a housing developer's nameboard. As with Canary Wharf, the most basic conflict is between the interests of the indigenous community and those of the developer and the clients to whom he will sell the houses.

Reactions to Primo Levi's 'The Wrench' (1987) have been more mixed. It is the story, written autobiographically, of a rigger on major projects such as bridges. It gives excellent insights into the actual workman's experience in construction. In a legal and contractual sense, he has little power or dignity but as a person of experience, courage and intelligence, he has both. This is an example of conflict between role and person. Some of the most telling parts of the book are about the conflicts of perception between rigger and engineering designers. The book does not take us much into the realm of formal conflict and dispute but, as good literature should, it concerns itself with people as people. This opens up the question of whether role and rule create formalised conflict or whether they modify and give some order to the stresses that may or may not arise between real, raw people.

A variety of insights is available from The Oxford Book of Essays (ed John Gross 1991). A 1979 essay entitled 'To Err is Human' argues that while computers may get everything 'right' within their limits, humans are created to argue, make mistakes and be fallible. This brings us to the heart of the conflict issue. Do we expect ourselves and other people to be computer-like in a kind of limited 'rightness', or do we enter the human domain of argument, wrongness, fallibility, exploration and progress? Is there something here undergirding Robbins' view of conflict as creativity, which we have noted earlier? If there is, how do we handle it in relation to building?

A century earlier, Anthony Trollope wrote his essay 'The Plumber'. He invites us to think of the plumber who, in the interests of feeding his wife and children, ensures by the way he chooses to carry out repairs, that his services will go on being needed by his wealthy householder customer, although it is within his capacity to make himself superfluous for five years. While the plumber faces a conflict of loyalties between family and customer, the householder faces the conflict between his own self-interest and his social responsibility. The irony is that 'on the next morning, being by profession a respectable solicitor, he [the householder] is hard at work at Lincoln's Inn, paving the way for fresh litigation'. Perhaps literature is an educational medium, not only for the young, but also for those of experience who have the courage to face the questions it raises!

## 6 Questions for discussion

Appendix A to this paper contains a set of discussion questions based on sections 3, 4 and 5. There are no answers in the back of the book! The learning and personal growth will be in the discussion and the thinking that they stimulate.

## 7 Conclusion

This paper has shown that the issue of conflict can be considered within the framework of education in building ethics. Such a consideration, carried out reflectively and non-dogmatically, will lead to personal growth and development.

## 8 References


Adams, Richard (1972) **Watership Down**, Penguin Books.

Axelrod, Robert (1984) **The Evolution of Cooperation**, Penguin Books.

Bullock, Alan et al eds (1988) **The Fontana Dictionary of Modern Thought**, Fontana Press.

de Bono, Edward (1986) **Conflicts: A Better Way to Resolve Them**, Penguin Books.

Gross, John ed (1991) **The Oxford Book of Essays**, Oxford University Press.

Levi, Primo (1987) **The Wrench**, Sphere Books.

Lloyd, Tom (1990) **The Nice Company**, Bloomsbury Press.

Powell, Michael (1990) Establishing Relations of Trust in Building Projects, **ARCOM Conference Proceedings.**
Powell, Michael (1991) Axiology (The Science of Values) Related to Building, **ARCOM Conference Proceedings.**
Robbins, S.F. (1974) **Managing Organisational Conflict: A Non-Traditional Approach**, Prentice Hall NY.
Sieff, Marcus (1990) **On Management**, Weidenfold and Nicholson.
Summagh, Karl (1989) **Skyscraper**, Macmillan/Channel 4.
Broadgate and Liverpool Street Station, **New Builder**, Special Supplement, December 1991.
Canary Wharf, **Building**, Special Supplement, October 1991.

## 9 Appendix A. Questions for discussion

1. 'Building involves diverse sciences, arts, philosophies and traditions'. Does this diversity by its very nature create conflict? What is the cost of suppressing such conflict?
2. 'A building may be conceived in conflict between a developer who wants it and a local community that does not.' Discuss.
3. Conditions of building contract are rules for games to be played by rival teams, one of whom must win and the other lose. Evaluate, relating your answer to ethical aspects of conflict.
4. When is conflict moral?
5. Robert Axelrod has written '...an important way to promote cooperation is to arrange that the same two individuals will meet each other again, be able to recognise each other from the past and recall how the other has behaved until now'. To what extent is this an effective and ethical way of handling the risk of conflict?
6. Does trust create or prevent conflict?
7. Edward de Bono has written 'Cost should be the major determinant of the feasibility of a conflict. In practice, it rarely comes in at all, because of this notion that money and rights are two separate universes'. Are they, and which universe is more important ethically?
8. In Watership Down, the relationship between the crow, Kehaar, and the rabbits was one of cooperation. What made it so? What might have happened if it had turned into conflict?
9. In Watership Down, what is the relationship between fear and conflict?

10 'I have always thought that bridges are the most beautiful work there is, because you're sure they'll never do anybody harm; in fact they do good, because people pass over bridges.....bridges are sort of the opposite of boundaries, and boundaries are where wars start.' (Primo Levi in 'The Wrench'). Consider how positive and ethical approaches to construction work may affect attitudes to any conflicts that may arise.

## Educating Construction Professionals to Improve the Built Environment

MICHAEL HANCOCK
Construction Study Unit, School of Architecture & Building Engineering, University of Bath, United Kingdom.

Abstract
It is argued that many of the problems and conflicts within the construction industry are a result of misunderstanding and a lack of perception founded in our education of construction industry professionals. Currently the education of these people is of a form which encourages blind adherence to sets of rules and procedures that are both inflexible and often irrelevant.

Our education system is aiding and abetting a social system that is causing the psychological breakdown of society. This is being done through our teaching of art, science and technology in a way that is destroying the faculty of thought, breaks down individual will, impairs feeling and insists on the adoption of a view of the world from which man is alienated.

As courses in education increasingly become seen as the provision of a "meal-ticket" there is a concomitant denigration of the learning experience to one that is concerned only with the search for facts and the joy of discovery is reduced to a search for the utilization of what is discovered. Thinking is dispelled and replaced by mindlessness and conciousness is totally ignored. This is the work of intellect and it penetrates every aspect of our education system from beginning to end.

As a vehicle for this paper, consideration is given to our present education system and to the philosophy that underpins it. The effects of specialization and technology are underlined in supporting an argument for a form of education that does not rely on external solutions, but which provides construction professionals with the ability to act morally and independently thereby reducing the incidence of conflict within the industry.
Keywords: Conflict, Education, Technology, Specialization

### 1 Introduction

Construction personnel fulfilling "professional" roles are both restrained and guided by sets of rules and established procedures. Some of these rules and procedures are founded in the law, some in the regulations and codes of conduct published by professional institutions, some are required by individual employing organizations, others by accepted ritual or custom. Rules and procedures might reasonably be seen as ensuring "fair play" and constituting safeguards for all parties affected by the behaviour of these people. Nevertheless, despite all of these rules and procedures the construction industry is continually dogged with conflict. Why? Surely if "professionals" are all educated human beings then their personal inner moral code will prevent the greed or alternative interpretation of legally binding rules and established procedures that leads to conflict?

## 2 Education causing Conflict

### 2.1 Rationalism & Philosophy

Since the seventeenth century humans have become progressively dominated by Intellect. This is the mental faculty which demands that we think about and rationalize problems with which we are faced at the expense of everything that is *human* within us. From this period of western philosophical history onwards philosophers have asserted that the only basis for a valid knowledge of reality is rationalism. As a result, revelation and intuition are rejected as a source of genuine knowledge. Our feelings are not allowed to corrupt the taking of an objective view and this leads to a dehumanizing of society. This denial of the intrinsic human being and the exclusive cultivation of intellect has led to an external approach to life. This in turn has destroyed our ability to solve problems that should be dependant upon feeling and intuition, and our ability to act morally and independantly. Note for instance the recent rise in popularity of quality assurance. Quality at one time assured by the morality and inner feeling of those providing a service is now assured through a set of formalised external procedures and structures that control the work and organizations involved in a wholly rationalized and unthinking way. Flexibility is not permitted in such a system and conflict resolution is reduced to an allocation of blame based on a review of rationalized procedures and legal systems. The law that presides over dispute resolution is not concerned to address the cause of any dispute, only to treat the symptoms. There is no human concept of what law is, only a technical one.

This paper does not suggest that there is anything wrong with a pragmatic handling of dispute. What it does argue is that there is a self-defeating and mental abnormality in the current single-minded insistence on accepting only external solutions. That is not enough to correct human problems such as conflict.

What then can we do to improve the situation? First, we need to understand what has brought us to our present condition. How did the role of intellect, consciousness, feeling, will and drives change over time and bring us to our present social existence and modern life? In a short paper such as this it is impossible to address every aspect of this question, but we might usefully consider the roles of education and technology in the affair.

In order to achieve a degree of commonality; essential for the coherence and stability required for the successful completion of a construction contract, construction professionals are trained to make decisions that conform to certain standards. These are standards that have arisen in concert with the development of science and technology; which are themselves based on a philosophy of western logic and rationalization. The problem is that if disputes arise, as they surely do, then something has gone awry.

This century has been one of enormous scientific and technological advance. This would not have been possible without an education system that devoted itself predominantly to these areas of study. Science and technology are an integral part of our society, as is material well-being, but at what cost? As far back as 1933, the slogan of the World Fair was, "Science Finds, Industry Applies, Man Conforms. The implications are quite clear: the rational logic which underpins science and technology have separated humankind from the natural world. Modern education places value on these areas of study to the detriment of human feeling and will, it is unconcerned with the development of independent thought, self-knowledge and appreciation of human obligations. Searching for the facts prevails and the joy of discovery is overcome by concern for the utility of what is found. Thinking and consciousness are either devalued or ignored in an education system where the prime motivation is the securing of employment.

### 2.2 Willpower

The devaluation referred to above is taken a step further in the case of willpower. This is a much misunderstood area of the mind, frequently confused with instinct, impulse and drive. It is actually a basic mental faculty that responds to exercise in a manner comparable to physical muscles. The effort required to develop this "muscle" must come from the individual for development to take place. Artificial (external) stimulation can only leave the will in a flaccid state.

Experience has shown that both educators and the institutions to which they belong ignore the importance of will, preferring instead to resort to behaviourism to develop the student. They continually repress or force, rather than provide the guidance or stimulus for the exercise of will. Consequently, will (a vital mental ability) is at best weakened, at worst, destroyed under a system of regimentation and control. For a student to become successful within our education system requires him/her to follow directions and listen unquestioningly to the tutor. In other words, to surrender his/her will. Self-determination, perseverance and individual initiative have been effectively replaced by external guidance systems.

This state of affairs is neither accidental nor coincidental. The products of our current education system fit very well into an economic system founded on submission to bureaucratic organization and competitive economic self-interest.

**2.3 Competition & Specialization**

In the construction industry, it is through competition that firms traditionally get work. Once a project is underway, the relationship between the representatives of client and contractor is essentially of a competitive nature (frequently disguised as co-operation). When a conflict arises the disguises are shed to reveal open competition often requiring the services of a referee in the form of an arbitrator. So why do we continue with such a system? Some will argue that competition satisfies public accountability. This must surely be a suspect notion if the result of competition is a cost increasing conflict. Others will allege that competition stimulates performance. It might however be argued that what it actually does is replace honesty, morality and self-initiated accomplishment with a conditioning that compels "response" performance; the very opposite of what is claimed.

Individuals can only compete for possessions, power and prestige. They cannot compete for their own being, quality or scope of achievement. Competition then is a form of passivity in which the individual never changes and is only recognised as a reflection of what s/he is required to do on behalf of his/her employers. To be successful we are required to compete harder. The fiercer the competition, the less room there is for real human development. This surely denotes an immature society and one that results from an education system that has been intentionally and not accidentally introduced.

In order to achieve what we have during this century it has been necessary to insist on increasing specialization in all aspects of industry and education. No longer is art the holistic study of grammar, logic, rhetoric, arithmetic, geometry, music and astronomy. In today's education system these areas have been divided up into the separate disciplines of language, mathematics, philosophy, music and astronomy. This separation of art and science and the subdivisions of subjects within those categories has been necessary for the division of labour required by industrial capitalism in order to make the technological advances that we have seen. The problem is that we now face a crisis. Do we allow technology to continue to set the agenda for our lives, or can we do something about it before we become automatons?

**2.4 Education, Training & Technology**

It would be counter-productive and ridiculous to suggest that we now have enough technology and that what is required is a return to the "good old days" (which probably weren't so good anyway). This line would constitute "throwing out the baby with the bath water". What is needed is an improved balance between the technological and human requirements of society. A return to a less specialized form of education and a clear understanding of the difference between education and training.

Training is a particular kind of activity or process which is designed to equip people with the necessary skills for a job. This presupposes an extrinsic end. Education requires no specific activity and its aims are the development of individual potentials, intellect and/or character. These constitute intrinsic ends. Both definitions imply that something worthwhile takes place; but those things are different. Training refers to the imparting of proficiency in a particular skill or way of thinking. Education refers to an association with and understanding of a wider system of beliefs. There are no grounds for suggesting that vocational training should not also be educational, but a difference remains. In essence this difference is concerned with an understanding of the reason "why" of things. To be able to answer examination questions about the law of contract, economics, structures, building technology, management or any of the other subjects studied by building professionals does not necessarily affect the way in which an individual looks at buildings, institutions or social relationships. The ability to pass the examinations set by the so called educators does not render an individual educated any more than learning how to operate a lathe or welding equipment. These are forms of training and their product is knowledge not an educated mind.

And what of the demands of technology? The very great danger here lies in the mechanistic and depersonalised nature of technolgy; in the idea that technology is somehow neutral. We have already seen that for its development there has been a necessary specialization in education and training. Humans left devoid of will are prepared to sacrifice their very individuality and are satisfied by the material and monetary products of technology. As far as technology is concerned, humanness with its variability and originality is a threat.

Our current system of education is increasingly based on the needs of technology. We provide the required predictable human components by ignoring consciousness in the processing of information. This ensures complete control, reliability and productivity of people. As a result, construction professionals are simply and automatically responding to sets of rationalized techniques and procedures designed to resolve conflict when it arises. Nothing in our present system is aimed at the mental activity, emotional and critical investigation, ethics or morality that might help prevent conflict from arising in the first place.

Of course, conflict is a human problem and not one of technology. Maybe, in time, if man continues to learn how to respond in an unthinking and automatic way there will be no more conflict....there certainly wont be anything human left in man.

## 3 Selected Bibliography


Illich Ivan (1973) Deschooling Society, Penguin, Harmondsworth

Lewis John (1975) Max Weber & Value-Free Sociology, Lawrence & Wishart, London

Lupton Tom (1971) Management & the Social Sciences, Penguin, Harmondsworth

McHugh P. (1971) On the Failure of Positivism, in Understanding Everyday Life: Toward the Reconstruction of Sociological Knowledge, (ed J.D. Douglas), Routledge & Kegan Paul, London

Peters R.S. (1966) Ethics & Education, Unwin, London

Russell Bertrand (1961) History of Western Philosophy (2nd Ed), George Allen & Unwin, London

Salaman G & Thompson K. (1980) Control & Ideology in Organizations, Open University Press, Milton Keynes

# CONSTRUCTION CONFLICT MANAGEMENT - THE ROLE OF EDUCATION AND TRAINING

JAMES FRANKS
South Bank University, London, England

Abstract
The report of the Minister of Works' initiated Banwell Committee (1964) stated that; '<u>the relationship between those responsible for design and those who actually build must be improved through common education</u>'. Earlier and subsequent reports have made similar recommendations.

This paper is concerned with the role of education and training in creating harmonious attitudes in those to be employed in the construction industry with the aim of reducing the incidence of conflict on building projects for the benefit of the industry's clients.

The hypothesis is that confrontational attitudes among the 'professionals' on building projects can be significantly reduced by education and training which engenders collaboration between the professionals and that the industry's clients are the beneficiaries of such harmony.

The paper analyses the findings of a survey of the views of heads of departments/schools of architecture and building in universities and polytechnics in Britain, made in February 1992, on the subject of common education for students of architecture and building. The findings suggest that some schools have moved towards common education for architects and builders but that almost thirty years after Banwell there is still far to go. Meanwhile, conflict, delays and dissatisfied clients abound.

The paper refers to 'interdisciplinary study' experienced at Birmingham Polytechnic and it outlines projects developed at South Bank University, commencing in 1971, (which it is not suggested are unique), aimed at developing collaborative attitudes between final year students on first degree architectural and building courses. It suggests that these have been successful in their aims but that collaboration should start earlier in courses.

The paper recognises that it would take at least a generation to change attitudes but it suggests that existing, established confrontational attitudes may be leavened by the promotion of multi-disciplinary post-qualification courses such as the MSc Construction Management course at South Bank where emphasis is placed on the advantages of co-operation and the cultivation of common interests rather than contractual rights.

The paper concludes with an appeal for recognition by those responsible for educating and training tomorrow's architects and builders of the need to create an educational environment conducive to encouraging harmony rather than conflict.
Keywords: Construction Conflict; Conflict Avoidance, Role of Education

## 1 Introduction

This conference is concerned with managing construction conflict. Conflict should be regarded as a disease which like any disease can be prevented by removing the causes. Education is the principal if not the only means of prevention. Education is the only way in which conflict can be effectively managed by prevention. The other concern of this conference, conflict resolution is treatment of a disease which has reached epidemic proportion. The proposition is that divisive 'professional' attitudes are the principal cause of conflict on building projects and that common education and training for architects and builders are the only medium and long term remedy.

It follows that the industry's clients frequently suffer as a result of the conflict because it leads to projects costing more than they should, to them not being completed on time and to them failing to meeting quality standards.

The 'professionals', (which term is taken to embrace architects, builders, engineers, surveyors, etc), devote much time and energy to considering secondary causes of conflict such as contractual arrangements and means of resolving disputes arising from the conflict whilst ignoring the root causes. Clearly, improved contractual arrangements, (which are devised by the professionals and should therefore be within their power to perfect), are important as, indeed, are methods for resolution of disputes but they are treatments for ailments rather than measures to prevent root causes of the conflict.

## 2 Professional isolation - a root cause of conflict

What are the root causes?

The roots go deep.

As the professions were formalised towards the end of the eighteenth century responsibility for the design and the construction functions were separated and attitudes hardened as architects and builders identified their separate roles.

By the end of the first quarter of the 19th century the distinction between the roles of architect and builder was established in the minds of those involved in the industry even if not in the minds of the general public.

An account of a brush between an architect and a counsel during a court case in 1817 illustrates what was clearly the architects perception of his role and standing;

'You are a builder, I believe?'
'No Sir, I am not a builder; I am an architect'
'Ah well, builder or architect, architect or builder - they are pretty much the same, I suppose?'
'I beg your pardon; they are totally different'
'Oh indeed: Perhaps you will state wherein this difference consists'
'An architect, Sir, conceives the design, prepares the plan, draws out the specification - in short supplies the mind. The builder is merely the machine; the architect the power that puts the machine together and sets it going'
'Oh, very well, Mr Architect, that will do. A very ingenious distinction without a difference ....' Colvin (1954).

To what extent have such perceptions persisted?

It is unlikely that many architects in later years would regard the builder as 'merely a machine' but the majority of architects and builders found little in common as Sir Harold Emmerson was to discover in 1962 when he reported to the Minister of Works, Emmerson (1962), that he found a 'lack of cohesion between the architect and his professional colleagues and the builder and that, 'there is all too often a lack of confidence between architect and builder amounting at its worst to distrust and mutual recrimination. Even at their best, relations are affected by an aloofness which cannot make for efficiency, and the building owner suffers. In no other important industry is the responsibility for design so far removed from the responsibility for production'.

Emmerson's view was supported by the report of the Banwell Committee published in 1964, Banwell (1964).

The client suffers.

In 1989, a quarter of a century after Banwell, clients' perceptions of the building industry's weaknesses were essentially the same as those of the Banwell Committee.

A survey of fifty clients to ascertain their needs and expectations of the building team found that their second most numerous comments were concerned with, poor communications. As one client put it; 'poor communication exists .... between all members of the team at all stages and in all forms of communication, written, oral etc' and another; 'there is a lack of liaison between members of the team and lack of feedback'. One client used the words of the Emmerson Report and referred to the divorce of design from production being the cause of so many of the clients' dissatisfaction with the industry's performance, Franks (1990).

## 3 Conflict culture indoctrination

Architect Max Hutchinson, a past president of the RIBA, said in an article in Building, November 1991, that 'the conflict culture starts in the lecture room, at the seminar or in the studio. The bug is spread at the very inception of the construction industry's skill base'.

Indoctrination with hostile attitudes commences early in the education and training of the 'professionals! At the time when they are most impressionable. 'Give me a child for the first seven years ....;'say the jesuits or more recently, as a character in the musical 'South Pacific' rhymes; 'you've got to be taught how to hate .... before you are six or seven or eight'. The indoctrination of building professionals does not start at seven but it is nevertheless effective.

But even if hostile attitudes are not actually inculcated, separate education and training makes communication and mutuality difficult if not impossible.

## 4 Communication

Common - shared - experience, which includes cultural, social, educational and environmental experience, is identified by Drucker, (1977), to mention but one management guru, as a fundamental requirement of good communication.

Architect Hellman's cartoon, Fig.1., illustrates with brilliant perception and clarity the cultural, social and educational voids between architect and builder which make communication - the transfer of meaning - so difficult. The cartoon raises a question. Clearly the man on the left is the architect, this would be apparent even if one were unfamiliar with Hellman's cartoons. But who is the builder? Is he the man in the centre, the interpreter, the enabler, or the man on the right with the trowel? Common education provides an opportunity to share experience, to breach the communications barriers.

**Fig.1. Louis Hellman, Architects' Journal, 19 April 1972.**

## 5 Conflict humour

Distrust and divisive attitudes are part of the industry's culture. Even the humour is potentially divisive.

'I'd been in building twenty years before I discovered 'bloody architect' was not one word'; says the builder in the rather sad joke which has been resurrected from time to time during the more than 40 years I have 'been in building!

The story becomes more alliterative but no less sad if the roles in the joke are reversed and the architect refers to 'bloody builder!

Common education is a potent prophylactic to distrust and devision.

## 6 Aspects of teamwork

The procurement of buildings requires teamwork of a high order. Effective multi-disciplinary small groups are the key to success.

Conflict and poor communications, the building industry's chronic ailments, are not conducive to effective small group operation.

The client suffers.

Research into group dynamics and psychosocial systems confirms that sustained conflict has, typically, been thought of as bad for organisations although it is recognised that it is unlikely that any group can attain a completely conflict-free situation.

Research identifies four stages of group development as the people concerned progress from being a 'bunch' to become a group. It suggests that during development of the group personal relations 'move from apprehension, tentative interactions with dependence on leaders or institutions through confusion and conflict (either overt or covert) to cohesiveness and ultimate interdependence', Kast and Rosenzweig (1986).

It suggests that it is necessary to recognise dysfunctional aspects of small group formation and to guard against them whilst recognising and encouraging the functional aspects. Clearly the conflict inherent in the building industry is a dysfunctional aspect to be guarded against.

Blake and Mouton's managerial grid for the measurement of optimal integration of 'concern for people' with 'concern for production' identifies 'team management' as the most effective organisational form. In this form, work is 'accomplished by committed people with interdependence through a common stake in organisation purpose which leads to relationships of trust and respect', Kast and Rosenzweig (1986).

The building industry's professional attitudes do not fit comfortably into that team management model and the client suffers.

Emmerson's and Banwell's findings were essentially the same with regard to the need for an improvement in the relationship between architect and builder and the action to be taken to improve the relationship, namely, common education.

## 7 The need for common education

Emmerson (1962) noted that; 'In the building industry builders and architects are trained quite separately' despite recommendations of the 1950 Building Working Party that those entering the building industry as architects or managers should take a common course of study for an initial period.

He added that the 1950 recommendations had not been acted upon owing to lack of agreement although a special conference held in 1956 had resolved that; 'the industry could improve its standards and raise productivity by inter-relating the training of its constituents administrating branches'.

The Banwell Committee, making a study of the problems identified by Emmerson, noted that; 'those who continue to regard design and construction as separate fields of endeavour are mistaken' and added 'there is, (the need for), such a close and growing inter-relationship between (those engaged in design and those engaged in building) that each could with advantage possess a working knowledge of many of the processes hitherto known only to the other'.

The Committee's report included a recommendation that; 'the relationship between those responsible for design and those who actually build must be improved through common education'.

There is, then, ample recognition from the past of the need for common education and the report 'Investing in Building 2001', Centre for Strategic Studies in Construction (1989), has as part of its 'New education strategy' the establishment of Centres for the Built Environment at universities to provide common first degrees, accepted equally for entry to architecture, building, engineering and surveying; and masters courses for all the specialist building disciplines.

What action has been taken? What action should be taken?

## 8 The present condition

A survey was carried out in February 1992, Franks (1992), with the aim of exploring the 'present condition' of common education. The response comprised a 57% sample of heads of departments/schools of architecture and building. The thirty four respondents were architects, builders (surveyors and 'others' (typically from engineering disciplines) almost equally represented.

The questionnaire stated that its' aim was to ascertain the extent to which common education for architects and builders had been or is being introduced. It was concerned with first degree, with masters courses and with the respondents' view on the subject of the survey.

The survey revealed that 58% of the schools which offer courses in both architecture and building provide some common teaching. In most, the extent is not great. Several which do not at present offer common education for architects and builders have plans to do so in the future.

Almost all built environment 'design team' disciplines are taught in common for at least part of their courses, certainly more than the two disciplines with which this paper is concerned.

Only two courses covered in the survey sample teach architectural and building students together at masters level. The extent of collaboration does not exceed 50% of the time.

94% of respondents were aware of recommendations that common education and training for architects and builders is desirable.

The Reading University 2001 reports, the CNAA/Birmingham Study, Central Government, CIC and joint meetings of professional bodies, particularly in the Midlands, were mentioned as sources of recommendations. There was reference to 'pressure group lobbying' and to the need for such common education being 'common knowledge'. 56% of the respondents 'personally' consider common teaching to be desirable while 9% regard it as 'acceptable'. Building and 'other' respondents were more favourably disposed towards common teaching than architect respondents. Respondents gave a number of reasons for supporting common education but some expressed reservations. A small minority were strongly opposed.

The problems of administering common education for architects and builders is one reason given for not introducing it but it is a reason which does not appear fully valid because common teaching occurs on other courses which have similar knowledge needs.

An analysis of responses to the questionnaire was sent to all respondents with a request that they complete a second 'short' questionnaire which took as it's hypothesis Max Hutchinson's statement, referred to under heading '3' above, that the conflict culture starts in the lecture room etc.

The questionnaire received a 50% response, architecture, building and 'other' disciplines being quite evenly represented.

In answer to a question regarding their agreement with Hutchinson's statement the respondents were almost equally divided with comments from those who did not agree with him that there is no firm evidence supporting him, that the fault lies in the industry, that conflict does not exist and that it starts in the womb.

61% considered common education for architectural and building students to be a significant factor in reducing the conflict culture. Supporters suggested that common education led to a better appreciation of the other's views and that learning together should lead to working together but one respondent asked;

'would you favour common education of pianists and piano tuners?'.

Of the 83% who favoured common education, 73% considered, it is best commenced as early as possible but the remainder favoured commencement at post-graduate level.

Respondents urged the need to take care and to integrate any common teaching whilst another suggested that it should not be instituted for merely political reasons.

The survey demonstrates that a substantial majority of the educators consider common education for architectural and building students to be a significant factor in reducing the conflict culture.

What has been the experience of those concerned with common education?

Professor Collier, (1992), drawing on experience obtained at Birmingham Polytechnic identifies the pressures and difficulties which beset 'interdisciplinary study' but suggests that the opportunities are immense for students, teachers, course leaders, practitioners/employers and professional bodies in both the short and the long term.

He concludes that there are pressures on educational establishments for innovative change, that they have the capability to innovate and change, that there are many opportunities for innovation and change to take place, that there are few threats and that the main obstacles can be identified. He identifies the key steps which can be taken to establish a common framework for the validation of all built environment courses within which 'real diversity and specialisms' can flourish.

Following publication of the Banwell Report in 1964 there was considerable debate on implementation of the report's recommendation.

In 1971 the, then, departments of architecture and of building at Brixton School of Building (now part of South Bank University), recognised the need to bring together the disciplines and designed a project which was called 'Design, cost and build' in which final year architectural and building students worked in small teams to do what the title suggests.

An important element of the project was that it was a largely extra-mural study which required the disparate students to work together in unfamiliar settings such as building sites, students' digs and (unfamiliar to building students) studios. Students formed their own teams, selected leaders and allocated tasks. The traditional architectural critiques were extended to include presentations by the builders on time, cost and construction methodology.

Their tasks were, of course, similar to those which design and build firms now do as a matter of course but in 1971 the concept was little known. Twenty years on the 'Collaborative Exercise' is an established and CNAA commended part of the curricula of the courses. It is rewarding to watch how the different disciplines approach the project with mutual suspicion but usually end with respect for their opposite number's expertise. It is recognised that the 'final year' is really too late to start bringing disciplines together but it is better than never and as the current trend towards 'unit-based' courses blossoms, students of different disciplines can more readily be taught together earlier in their courses. After all, construction is construction and acoustics are acoustics whether one aims to be an architect, a builder, an engineer or a surveyor.

Writing in 1976 , about learning through projects at South Bank which involve 'integrative teaching', A S Morgan (1976) notes; 'Perhaps because the participants are still students, they do not appear to have the same difficulties that many experienced professionals have in working in a design team. The teams readily accept collective responsibilities for their designs and defend them with conviction... Given that one really does want to develop integrative project work, staff confidence seems to be the key factor'.

The South Bank MSc Construction Management course attempts to reconcile differences which might have developed by recruiting, as a matter of policy, all disciplines onto the course. When team projects are involved, students are encouraged to work in interdisciplinary teams. The dispute management unit recognises that many disputes can be avoided by adoption of good contract procedures and by a willingness to cultivate common interests rather than press for contractual rights.

By the time students reach a masters course attitudes have almost certainly hardened, the conflict culture is established, but studying together and sharing experiences can only assist in alleviating the 'them and us' attitudes so prevalent in the construction industry.

## 9 The way ahead

The proposition is that divisive professional attitudes are the principal cause of conflict and that the conflict culture starts in the lecture room - if it is not already established.

There is a strong (but not unanimous) view, a 'common knowledge' albeit without proven evidence, that common education and training of architects and builders is a way ahead. That 'learning together should lead to working together', to quote a respondent to a questionnaire.

The present approach to the education of construction industry 'professionals' is based on the concept that the different disciplines have different educational needs. Is this a valid concept?

Karl Popper (1963) holds that, 'disciplines are distinguished partly for historical reasons and reasons of administrative convenience and partly because the theories which we construct to solve our problems have a tendency to grow into unified systems. But all this classification and distinction is a comparatively unimportant and superficial affair, we are not students of some subject matter but students of problems. And problems may cut right across the borders of any subject matter or discipline'.

Certainly this is the nature of construction industry problems.

Clearly, Popper spurns discrete disciplines. If one accepts his view it follows that separate education is an anachronism.

Popper refers to 'administrative convenience' as a reason for having disciplines and it would be foolish to ignore this reason. But administrative convenience must, surely, be subordinate to academic and professional needs.

The way ahead requires a determined combination of the industry led, perhaps, by the Construction Industry Council and by a body representing the educators which has a genuine commitment to common education and the power and influence to implement action.

Unless some such positive action is taken it is difficult to see an end to the conflict culture which has bedeviled the construction industry for far too long.

## 10 References


Banwell H, (1964), **The planning and management of contracts for building and civil engineering work**, HMSO.

Centre for Strategic Studies in Construction, (1989), **Investing in building 2001**, Reading University.

Collier T, (1992), **Enhancing interdisciplinary study**, paper presented at conference, whom do we serve, Construction Industry Council, 11 March 1992.

Colvin H, (1954), **Biographical Dictionary of English Architects**, 1660 - 1840, John Murphy.

Drucker P, (1977), **Management, Tasks, Responsibilities, Practices**, Heinemann.

Emmerson H, (1962), **Survey of problems before the construction industries**, HMSO.

Franks J, (1990), **Building Procurement Systems**, Chartered Institute of Building.

Franks J, (1992), **Common education for architects and builders analysis of responses to a questionnaire circulated to 'educators'**, South Bank Polytechnic.

Kast F E and Rosenzweig J E, (1986), **Organisation and Management**, McGraw-Hill.

Morgan A S, (1976), Learning through projects, **Studies in Higher Education**, Vol 1 No 1.

Popper K, (1963), **Conjectures and refutations**, Oxford University Press.

# THE CONSTRUCTION INDUSTRY'S MALE CULTURE MUST FEMINIZE IF CONFLICT IS TO BE REDUCED: THE ROLE OF EDUCATION AS GATEKEEPER TO A MALE CONSTRUCTION INDUSTRY.

A.W. GALE
Department of Building Engineering, UMIST, Manchester, England

Abstract
This paper draws on insights gained from current research on the relationship between the image of the British construction industry and the entry of women into occupations as professional construction managers.

It is argued that departments of construction in the education system and their courses act as gatekeepers to the culture of the construction industry. They maintain and promote values, images and practices axiomatic of construction culture. It is suggested that the construction industry has a particularly male culture. The argument is developed that male gender values are inherently more conflictual than female gender values.

The role of education is identified as key in supporting the management of change necessary to bring about a fundamental shift in the construction industry's culture.

It is argued that educational philosophy and curriculum need to be changed. Recommendations for further research are suggested.
Keywords: Conflict, Education, Culture, Image, Women.

## 1 Introduction

This paper puts forward the radical position that the construction industry is conflictual because it has a male culture. Further, it argues that the vested interests of those in the industry and the educational departments associated with the industry may prevent change from occurring to this culture. It is argued that the educational departments are gatekeepers to the construction industry's culture and it is here that changes to the philosophy and curriculum could bring about the feminization of the construction culture.

In the first part of the paper concepts and empirical work are discussed. Then the importance of the concept and understanding of gender is debated. A distinction between male and female culture is discussed, drawing on the work of Judi Marshall (1984), who defines the notions of agency and communion.

The paper goes on to talk about how the construction culture is maintained, drawing on the findings of Clara Greed (1991). It is argued that the culture of the industry is characterised by crisis, conflict and masculinity.

Conclusions are drawn relating to the role of educational gatekeepers to the construction culture and an argument is made for research into the measurement of the industry's image and culture.

## 2 Theoretical concepts and empirical work

The interpretation of empirical work on women in construction is grounded in theoretical concepts used in the definition and understanding of: gender, motivation, culture, image and determinants of occupational choice.

This paper adopts Handy's definition (1985) of conflict. He says that argument, competition and conflict are manifestations of differences between people and groups in organizations. Argument and competition, although seen by some as symptoms to, or precursors of, conflict are generally positive. However, they can degenerate into conflict which some would argue to be inevitable. Assuming conflict to be negative, its symptoms are: poor communication, inter-group hostility and jealousy, inter-personal friction, low morale, proliferation of rules and an escalation in arbitration; the referral of a dispute to a higher level in the organizational hierarchy. Causes of conflict stem from differences in objectives and ideologies and/or territory or role. The former are related to world views and beliefs, goals, standards and rules. The latter are metaphors for psychological, rather than physical, aspects of work, status and influence.

Many of the ideas perspectives and constructs in this paper arise from insights gained from the my ongoing doctoral research begun in 1987 on women in the construction industry (Gale 1990a and b). The aim of the research is to establish why there are only a few women occupied as building professionals in the British construction industry. The objectives are firstly, to investigate whether there is a relationship between perceived image of the construction industry and to identify any differences in that perception according to sex. Secondly, to investigate whether there is a relationship (and, if so what it is) between the perceived image of the construction industry and recruitment levels of men and women to positions as building professionals. The working population is defined in two dimensions; one industrial, containing a near comparator industry (engineering) and far comparator (banking and finance). The other, the education-work continuum; going from school, through higher education to employment. Data have been gathered using depth and structured interview instruments. A questionnaire survey yielded 286 responses.

Action research carried out in parallel and interacting with survey research yielded qualitative and quantitative data on the image and culture of the construction industry. Insight courses for careers advisors, careers teachers and school students were evaluated. Useful insights were gained into gender differences in problem solving in the construction context as well as more general findings. Insight courses are defined here as short, two or three day, courses during which participants are exposed to female role models from the construction industry and undertake experiential learning exercises simulating the construction process. Structured site visits were a feature of the courses, followed by discussions and workshops to deal with questions arising. The Engineering Industry Training Board developed the concept of "Insight" courses (Peacock and Shinkins, 1983, Viscardi, 1987).

## 3 The importance of gender

It is important to distinguish between gender and sex. Essentially people can be biologically sex typed male or female but a gender role is then learnt or socially determined (Garrett, 1987). The question of how gender is related to views, values, motivations, attitudes and approaches to work is complex.

The evaluation of one Insight course yielded interesting findings on differences of approach between males and females to group working and problem solving in construction project management exercises. The course was run in 1991 for twenty (10 male, 10 female) Advanced Level students, aged 17-18 years, who were considering reading for a degree in a construction discipline. Comments made by participants as well as direct observations and data from questionnaires yielded insights (Gale, 1991b and c).

The participants were formed into single sex groups. The boys discussed their roles in the groups in relation to each other whilst the girls related their roles to the task. The boys appeared more decisive but did not always make good decisions. Girls were concerned with achieving absolute agreement consistently making better decisions, winning more points than boys for group tasks.

Girls seemed to be more self critical than their male counterparts. It was interesting to observe that the girls complimented each other on their skills, a behaviour not observed in boy's groups.

One exercise required the design and costing of facilities in an airport for people with disabilities modifications to an airport. The girls concentrated on the needs of disabled people whereas the boys concentrated on financial aspects. This competitive exercise was won by a female team. The boys appeared to focus on one leader and had very little discussion. Girls seemed to be more democratic. The boys chose their "best" presenter to talk about the project at presentation stage. Girls tended to all take turns to talk about their projects. Interestingly boys saw the girls as being better organized than them and working together well.

These findings are supported internationally, having been the focus of much discussion at successive Gender and Science and Technology (GASAT) conferences.

According to Marshall (1984) two separate strategies with which individuals respond to threats in the environment are "agency" and "communion". She explains that Bakan (1966) describes these as two fundamental principles of psychological functioning. Agency is associated with the ego and there are four processes based on externalising difficulties that one can not cope with. They are; separation, mastery, denial and beholding.

Separation is concerned with distinguishing between likes and dislikes. The individual projects these onto groups, people or objects leading to mastery of the environment supported by denial of aspects of the environment and feelings which threaten mastery. The last stage is paradoxical, involving the beholding or encountering what was originally separated; reuniting what was split apart. Marshall argues that:

> "Men "agentically" project threatening qualities onto women, and master these threats by suppressing and devaluing women."
>
> (pp 65-6)

Communion is said to occur all at once. Marshall explains:

> "Grounded in its orientation of union, communion's perception is naturalistic, reflecting the nature and patterns of the environment, and is only minimally guided by prior analytic classifications."
>
> (pp 66-7)

Communion can be broadly regarded as a female and agency a male principle. Neither sex is exclusively associated with these principles but in general they are associated thus contributing to the difference between male and female cultures.

Marshall (1984) also describes women's characteristics and culture. She proposes a model of the five dimensions of womanhood. These are; emotional grounding, perception of the world, creativity, continuity and nurturing. Examples of how these tend to contribute to a less conflictual culture include the outward aspect of emotional grounding. The likelihood that women are often good facilitators to others' emotional expression and development. They are "emotional shock absorbers".

The problem is that because men interpret women's culture using the frame of reference and values of male culture their is no language to describe what they see. Therefore, the lack of open conflict observed in women's culture is interpreted as avoidance behaviour. I would argue that construction culture is male and inherently conflictual. It would benefit the industry for a more female culture to be developed . This would reduce conflict and attract a more varied intake of people to positions in the industry. This would lead to a change in working practices and management style. As the industry feminized it would be demonstrably less male dominated in the physical and psychological sense; gender inclusive.

It seems that if construction culture needs to be feminized the curriculum must address the question of interpretation and gender differences so as to free up the thinking of construction students.

## 4 Maintaining the culture

According to Handy (1985) cultures can not be defined precisely. They can, however, be differentiated and a good fit between an organization's prevailing culture and an individuals cultural preference leads to a satisfying social contract. Citing Hofstede (1983) he explains that masculinity, one of four cultural dimensions, of a culture is connected with ambition, the desire to achieve and to earn more, whereas its opposite, femininity, is more concerned with inter-personal relationships, the environment and a sense of service. Men tend to be concerned with quantitative and women qualitative aspects of life. These are stereotypic and whilst possibly demonstrable are challenged as necessarily being linked with gender. The differences may be explained by the difference between the material experiences of men and women (cf Dex,1988, Cockburn,1983 and 1985). However, different cultural types can be described and compared. Cultural differences can be defined at different levels: national, industrial, organizational and individual.

Greed (1991) gives many graphic examples of the male surveying culture, a subculture of construction, and how women cope. There are identifiable differences between various subcultures. These may in some part be due to the relative levels of professionalization, historical determinants and prevalent commercial organizational types.

Greed describes negotiation:

> "When men meet (based on what men have told me) they are likely to spend a while discussing the weather, cricket, women, cars, etc., and then almost as an afterthought say, "my goodness look at the time, let's see what I can do for you". There then follows a prolonged period of competitive discussion in which both sides want to save face and protect their egos. Men tell me that men always like to haggle and there are unwritten rules about offering high unrealistic figures first to protect the pride of each side - Men have always got to have the last word and win, or choose to concede." (p151)

She goes on to describe the approaches taken by women. They either consciously try to emulate men or are more likely to be more direct in putting their final offer. Also a characteristic of women's negotiation seems to be to put her final offer forward at the beginning of the negotiation. Greed finds that those who have negotiated with women find the experience much more straightforward, with far less posturing and no male egos to protect.

Greed explains that surveying culture exists in the surveying departments of educational establishments and writes at length about this phenomenon and how women cope with it.

Roger Pauli, Group Managing Director of Stuart Crystal, speaking at to The Women's Education Conference (Pauli,1989) made some brave and revealing comments about male culture in British industry. These could certainly apply to the construction industry.

The following is compiled from Roger Pauli's speech to a predominantly female audience:

> "I am not sure that I believe that men don't know what they are doing when they stereotype women. Our organizational structures have been devised within a male dominated society. They are primarily command structures. I believe they are to do with how to gain and retain power: how to get people to do what you want them to do. It is not necessarily how to organize the carrying out of tasks. I think that is one of the reasons women can feel so uncomfortable in industry.
>
> We borrowed these organizational structures from the army - the only organization available when industry started. We are talking mostly about people's thinking directing the activities of others.
>
> In a healthy environment, people thought about what they felt and did. In a command structure thinking was done at the top and people lower down in the organization were not expected to think about what they did. How, then could they develop any decent feeling or conscience about what they were doing?
>
> We need to help society to recover from this terrible malaise we have got. Hold fast to your courage, stay female, and help us men to find more of the feminine within us."
>
> (p.23)

The construction industry is demonstrably male (Gale,1991a). In terms of the horizontal sex segregation of the labour market women represented 6.7% (EOC, 1988) of the full-time work force in 1981. No reliable vertical segregation data are gathered. However an analysis of 1981 General Census data shows that only 8.4% (Rainbird,1989) of those women in the construction industry occupy managerial positions; 82.2% of them were employed in secretarial or clerical jobs. These percentages are reflected in higher education, where in 1989, 8% (Beacock et all, 1989) of building students were female (Gale,1991a). Also, the professions demonstrate the male characteristics of the industry with 40 women out of 8452 corporate members of the Chartered Institute of Building (CIOB) in 1992 and 594 women out of a total of 32569 at all grades; representing less than 2% women (CIOB,1992). It should be said that the CIOB is attempting to address this situation; a point demonstrated by the fact that these data are a function of the Institute's self critical review. A Women in Building Consultative Committee meets annually to discuss this question.

Not only is the workforce male, the prevailing culture and ethos of the industry appears to be extremely male, characterised by comments like the ones below. These were taken from the responses of males to a questionnaire survey of young construction industry trainee professional managers studying on a part-time basis for professional examinations (Gale, 1987):

"The natural male instinct for attraction to women with implied sexual "innuendo" helps women considerably in their careers."

"The women fundamentally are cheap, docile, unionised clerical labour."

"I don't really understand why women want to work in a traditionally male industry - compare with nursing. (I don't have anything against women)." (sic)

These are not just important because they demonstrate how male, almost how misogynist, the construction industry is but because they are young people who, it could be argued, might be expected to hold more enlightened views.

After years of working in the construction industry for British and American companies in the UK and overseas, plus interviewing both male and female members of the industry it seems to me that the following key words would sum up the industry's culture:

(a) crisis
(b) conflict
(c) masculine

Although much research continues to be done in the area of risk management the underlying tendency in construction management seems to me to be that of a willingness to engage in crisis management if at all possible. In fact, reduction of risk and uncertainty is an anathema in construction culture. People do not join the industry to have an "easy" life, they thrive on crisis management.

Conflict is a part of every day life in the industry. The management of operative labour involves high expressed emotion and the image of the aggressive "barking" foreman is a generally held one both in and outside of the industry. The handling of conflict is seen as an important management skill at all levels of management. A great deal of space is taken up in the construction press with contractual matters. Conflict is almost assumed at the inception and completion of a contract.

There are a lot of men in construction; so it is obviously male. However, masculine is meant here in a gender value and behavioral sense. The customs and working practices of construction managers and operatives alike seem to be very male. The image is held both in and outside the industry of the hard drinking, sexist, hard playing male. To get on, women have to fit in to the culture. The mobility of the work force is particularly high in the construction industry. This contributes to some extent to the exclusivity of the male "clubiness".

All of the above characteristics are to a certain extent carried over into the subculture of the further and higher education departments relating through their strong vocational orientation to the construction industry. It is easier for those outside of this subculture but in the educational environment to observe and comment on this tendency. Further, I would suggest that the strength of this male construction culture is probably much stronger in the further education colleges. This is because their courses are strictly vocational, quite often with a strong craft and technician tradition. Also notions of academic freedom and autonomy in curriculum development, to all intents and purposes, do not exist. University departments are a lot less likely to promote actively male construction culture due to their different, more independent, ethos. However, the students in all educational establishments share a common orientation towards a career in the industry. Also, because a high proportion of them are, or have been, employed in the industry, they bring the industry's values into their courses and departments. In this way the "acceptable" codes of behaviour in construction departments are linked to those of the construction industry. After all, the construction industry is where these students are eventually to "belong" if they are to "succeed". It could be argued then that construction departments actively and/or passively promote and/or maintain the construction culture. If it is accepted that this culture is male and conflictual then it seems that the conflict will continue in future. If moves are made to change the cultures of the educational departments it could be argued that over time this may have an impact on the construction industry's culture. This could become particularly import as the proportion of the workforce that needs to be educated to a higher level increases, thus putting more importance on the role of construction departments.

It is worth mentioning here that as the proportion of women in construction increases it does not follow that the culture will automatically change. The concept of critical mass; the proportion of a minority that will cause change to occur in the culture, customs, working practices and behaviour of a previously male dominated situation is often argued to be around 35% (Kock,1990). She develops her argument based on the theory expounded by Dahlerup (1988). The problem is that there is no empirical evidence in construction that this would be the case. The theory states when that a 35% proportionality is reached stereotyping will diminish, there will be new role models for girls and women, the open resistance towards engineer-women in the labour market will disappear, women's professional decisions will be trusted except be elderly people and female values will be accepted as appropriate and natural. It has to be said that this theory is controversial and unproven. Just because women form an increasing proportion of the workforce it does not follow that female values will be promoted by them. One way in which women have learnt to cope in male dominated organizations is to emulate male approaches. The same can be said for men who would naturally prefer a different culture to the one they find themselves in. Because gender values can be described as a continuum ranging from male to female, it may be that men and women holding similar values are attracted to similar occupations. This would explain why not only women but men too may find the construction culture an unacceptable one in which to work. It follows from this that the image of the construction industry may be an important factor in the career selection process of young men and women. The image is based on the reality. If the reality is that the construction industry has a masculine culture then those that seek to be a part of that reality will join courses leading to careers in the industry. Further, there is then a vested interest in those who have chosen the culture to promote and maintain that culture and resist change. If male values include the propensity for conflict in human interaction, then conflict becomes locked into construction culture.

It seems that because of the likelihood of the male culture being perpetuated that construction culture may not be that much affected by an increase in the proportion of women over the next few years. Further, even a higher proportion of female construction graduates does not mean that they will end up in the most acutely male groupings in the industry because men and women alike seek their comfort zones within which to work.

If increasing the proportion of women in the construction industry can not necessarily be relied upon to change the culture to one that is more female and thus less conflictual the question must be; how can the culture be changed?

The process of changing, in this case feminizing, the construction culture is a long term prospect. The only real potential for meaningful change must surely be in the construction departments of the education system. One problem here is that the very nature of the vocational orientation of these departments presents a barrier to change. Courses vocationally orientated towards construction may well be seen in the long run to be counterproductive because by serving the industry's needs now future needs may remain unaddressed. This could arise because change would not be allowed to occur due of the vested interests of current construction culture. This means that the vocational orientation of courses, or at least departments, should be questioned.

The curriculum and ethos of courses leading to careers in the construction industry should be critically reviewed. It should be broader. The courses would therefore need to be longer. The style of teaching should also be scrutinised, particularly in the further education colleges. Anti-intellectual tendencies among some students and lectures should be challenged.

## 5 Conclusions and recommendations

I have argued that conflict is the degeneration of competition and argument. The context of the construction industry being one of a male and therefore necessarily conflictual culture. From this it can be suggested that conflict is the inevitable outcome to argument and competition occurring as part of the construction process. Further, it could be said that those entering the industry have a taste for conflict because they are male and because that is what they seek. The educational departments of construction, particularly those in the Further Education sector, are prone to very male cultures and are acting as gatekeepers to the construction culture.

This culture must feminize if a real change is to occur with respect to the problem of conflict in the industry. However, it seems that it is in the interests of those who have chosen to work in the industry to maintain the maleness of the culture, thus keeping conflict and crisis as preferred aspects of every day working life.

Attempts to limit the conflict endemic in the industry through alternative dispute procedures, however well meaning, are bound to have only a cosmetic effect because they are inherently superstructural. This means that such approaches and procedures are concerned with the periphery or symptoms and not the causes of conflict. An analogy is that of trying to improve communications when it is well known theoretically that poor communications are a symptom of an unhealthy organization.

If things are to change the construction industry must feminize in order to become psychologically and physically less male.

The construction departments should undertake critical reviews of their role in keeping the gate for the construction culture. The curriculum must be critically reviewed. It is probably over concerned with quantitative approaches and failing to deeply address the real understanding of human relations in the construction process.

It seems logical to challenge the concept of vocationally orientated courses. They may only serve the short term needs of an industry that is attempting to maintain, not change, its ways. However, the vested interests of those in the industry and educational departments associated with the industry may be too powerful. It is probably those practitioners in the industry with a deep and philosophical understanding for the need to change that may offer the best arguments for the educational gatekeepers to review their position.

My own ongoing research into the image of the construction industry as a determinant of women's entry to the industry gives insight to the possible value of research into the dynamics of the image and its relationship with the industry's culture. It would be worth validating an industry image scale and investigating further the subject of the industry's culture. This theoretical work could provide the key to change.

## 6 References

Beacock, P.M. Pearson, J.S.D. and Massey, H.P. (1989) **Characteristics of higher education for the construction professions, Development Service Project Report 24.** CNAA, London, 96-7.

CIOB (1992) Appendix A in **document AU(WIBCC)823**, Chartered Institute of Building, Ascot, England.

Cockburn, C. (1983) **Brothers: Male Dominance and Technological Change.** Pluto Press, London.

Cockburn, C. (1985) **Machinery of Dominance: Women, Men and Technical Know-how.** Pluto Press, London.

Dahlerup, D. (1988) **Vi har vent lange nok**. Kopenhavn

Dex, S. (1988) **Women's Attitudes towards Work.** The Macmillan Press Ltd., London.

Equal Opportunities Commission, (1988) **Women and Men in Britain A research profile.** EOC. Manchester, England, 37.

Gale, A.W. (1987) The Socio-economic Significance of an Increase in the Proportion of Women in the Construction Industry. **Coombe Lodge Report, Managing Construction Education.**, FESC, Bristol, England. Vol 20 No. 2, 113-124.

Gale, A.W. and Skitmore, M. (1990) Women in the Construction Industry: A discussion on findings of the two recent studies of the construction industry sector. in **Proceedings of The CIB W55/W65 International Symposium,** Sydney, Australia, 209-221.

Gale, A.W. (1990b) Women in Construction - Comparative survey of women and men in the construction, engineering and banking and finance industries, in **Proceedings of The 1990 GASAT Conference,** Jonkoping, Sweden, 63-72.

Gale, A.W. (1991a) What is good for women is good for men: Theoretical foundations for action research aimed at increasing the proportion of women in construction management, in **Practice Management: New perspectives for the construction professional** (eds P. Barrett and R. Males), Chapman and Hall, London, 26-34.

Gale, A.W. (1991b) **Construction industry careers experience course, 8-10 April 1991 at the University of Reading, All Female Participants, First Interim Report.** Construction Careers Service, Construction Industry Training Board, Bircham Newton, England.

Gale, A.W. (1991c) **Construction industry career experience course, 16-18 July 1991 at the University of Salford, Male and female participants in single sex separate groups, Second Interim Report.** Construction Careers Service, Construction Industry Training Board, Bircham Newton, England.

Garrett, S. (1987) **Gender**. Tavistock Publications, London.

Greed, C. (1991) **Surveying Sisters: Women in a Traditional Male Profession**. Routledge. London.

Handy, C (1985) **Understanding Organizations.** Penguin Books Ltd,. Harmondsworth, England.

Kock, H. (990) Frame 2 Part 2 in **GASAT 1990 Report Book**, Jonkoping, Sweeden, 44-47.

Marshall, J. (1984) **Women Managers Travellers in a Male World.** J. Wiley and Sons,. London.

Pauli, R. (1989) Masculine Management in Industry, in **Harnessing The Female Resource, The Women's Education Conference of 1989, Report of the conference held at the Queen Elizabeth II Conference Centre,** July 11, 1989, Published by The Women in Management Association, London. 22-3.

Peacock, S. and Shinkins, S. (1983) **Insight: A Review of Insight programmes to encourage more girls to become professional engineers.** EITB. Watford, England.

Rainbird, H. (1989) **Personal communication.**

Viscardi, K. (1987) **Insight: Encouraging girls to become professional engineers**, EITB, Watford, England.

# Index of keywords

For Product Safety Concerns and Information please contact our EU representative GPSR@taylorandfrancis.com Taylor & Francis Verlag GmbH, Kaufingerstraße 24, 80331 München, Germany